Systematic and Taxonomic Approaches in Palaeobotany

Proceedings of an International Symposium held at Goldsmiths' College, London, April 1985

The Systematics Association
Special Volume No. 31

Systematic and Taxonomic Approaches in Palaeobotany

Edited by

R.A. Spicer
Life Sciences Department,
Goldsmiths' College, London

and

B.A. Thomas
Botany Department,
National Museum of Wales, Cardiff

Published for the SYSTEMATICS ASSOCIATION by
CLARENDON PRESS · OXFORD
1986

Oxford University Press, Walton Street, Oxford OX2 6DP
Oxford New York Toronto
Delhi Bombay Calcutta Madras Karachi
Petaling Jaya Singapore Hong Kong Tokyo
Nairobi Dar es Salaam Cape Town
Melbourne Auckland
and associated companies in
Beirut Berlin Ibadan Nicosia

Oxford is a trade mark of Oxford University Press

Published in the United States
by Oxford University Press, New York

British Library Cataloguing in Publication Data
Spicer, R.A.
Systematic and taxonomic approaches in palaeobotany.
1. Palaeobotany——Classification
I. Title II. Thomas, Barry A.
III. Systematics Association
561 QE903
ISBN 0-19-857704-4

Library of Congress Cataloging in Publication Data
Systematic and taxonomic approaches in palaeobotany.
(The Systematics Association special volume; no. 31)
"Proceedings of an international symposium held at Goldsmiths' College, London, April 1985"——Half t.p. verso.
Bibliography; p.
Includes index.
1. Palaeobotany——Classification——Congresses.
I. Spicer, Robert A., 1950- . II. Thomas, Barry A.
III. Systematics Association. IV. Series.
QE901.S98 1986 561 86-5237
ISBN 0-19-857704-4

Set by Colset Private Ltd, Singapore
Printed in Great Britain by St Edmundsbury Press, Bury St Edmunds

Preface

Plant fossils are used in many ways for a variety of different purposes. They can be the subjects for detailed morphological or taxonomical studies and provide data for evolutionary theory, palaeoecological, palaeoenvironmental or palaeogeographical reconstructions and interpretations, or used in biostratigraphy. For all of these applications the identification and naming of the fossils are of paramount importance so it may seem strange that difficulties arise in their subsequent use. The key to understanding palaeobotanical nomenclature is the acceptance that their study is often bedevilled by problems of taxonomy and classification that are never encountered in comparable studies of extant plants. Plant fossils are usually only fragments of once living organisms that may be badly preserved through decay or incomplete fossilization and it is the palaeobotanical tradition of naming all these dispersed organs independently that often causes so much problem to the non specialist.

This meeting was conceived as an attempt to bring together palaeobotanists with differing interests to discuss their interrelated problems of taxonomy and systematics. In the event, over 80 people from 13 countries attended the conference on the '*Systematic and Taxonomic Approaches in Palaeobotany*' held at Goldsmiths' College, London (30 March — 3 April 1985). A field excursion to the Isle of Sheppey to collect Tertiary plant fossils started the proceedings, being ably led by Dr Margaret Collinson and assisted by Dr Hazel Wilkinson and Dr Gerry Hooker. Five half-day sessions of papers followed which were under the general headings of, 'Taxonomic problems due to preservation', 'Levels of available information', 'Approaches above the generic level', 'Taxonomy and plant classification in a dynamic evolutionary context', and 'The cases for alternative taxonomic approaches'.

This volume contains the papers presented at the meeting together with an overview by Dr Ken Alvin. It is hoped that the Volume, containing the various approaches adopted by the authors, will stimulate further research and that it will benefit the application of palaeobotanical studies in the wider context of botany and geology.

The conference was sponsored and supported by the Systematics Association and the Linnean Society of London. Additional support was received from the Royal Society of London, the British Council, British Petroleum, Shell UK Exploration and Production, and Goldsmiths' College. We are most grateful for all of their support.

London R.A.S.
April 1985 B.A.T.

Contents

Contributors

DAVID J. BATTEN
Department of Geology and Mineralogy, Marischal College, Aberdeen University, Aberdeen AB9 1AS, UK.

ROBYN J. BURNHAM
Department of Botany, KB-15, University of Washington, Seattle, Washington 98195, USA.

WILLIAM G. CHALONER
Department of Botany, Royal Holloway and Bedford New College, Egham Hill, Egham, Surrey TW20 0EX, UK.

CHRISTOPHER J. CLEAL
Geological Conservation Review Unit, Nature Conservancy Council, Pearl House, Bartholomew Street, Newbury, Berkshire RG14 5LS, UK.

MARGARET E. COLLINSON
Department of Biology, King's College London, 68 Half Moon Lane, Herne Hill, London SE24 9JF.

PETER R. CRANE
Department of Geology, Field Museum of Natural History, Roosevelt Road at Lake Shore Drive, Chicago, Illinois 60605, USA.

MICHAEL J. DONOGHUE
*Department of Biology, San Diego State University, San Diego, California 92182, USA.**

JAMES A. DOYLE
Department of Botany, University of California, Davis, California 95616, USA.

DIANNE EDWARDS
Department of Plant Science, University College, Cardiff, CF1 1XL, UK.

DAVID S. EDWARDS
*Department of Life Sciences, Goldsmiths' College, London SE8 3BU, UK.***

JEAN GALTIER
Laboratoire de Paléobotanique, Université des Sciences et Techniques du Languedoc, Place Bataillon, 34060 Montpellier, France.

PAUL R. GRANT
Geology Department, Royal School of Mines, Imperial College of Science and Technology, London SW7 2BP, UK.

CHRIS R. HILL
Department of Palaeontology, British Museum (Natural History), Cromwell Road, London SW7 5BD, UK.

NORMAN F. HUGHES
Department of Earth Sciences, Downing Street, Cambridge CB2 3EQ, UK.

GILLIAN M. REX
Laboratoire de Paléobotanique, Université des Sciences et Techniques du Languedoc, Place E. Bataillon, 34060 Montpellier, France.

GAR W. ROTHWELL
Department of Botany, Ohio University, Athens, OH 45701, USA.

CEDRIC H. SHUTE
British Museum (Natural History), Cromwell Road, London SW7 5BD, UK.

ROBERT A. SPICER
Life Sciences Department, Goldsmiths' College, Rachel McMillan Building, Creek Road, London SE8 3BU, UK.

SARA P. STUBBLEFIELD
Department of Botany, The Ohio State University, Columbus, Ohio 43210, USA.

THOMAS N. TAYLOR
Department of Botany, The Ohio State University, Columbus, Ohio 43210, USA.

BARRY A. THOMAS
*Life Sciences Department, Goldsmiths' College, Rachel McMillan Building, Creek Road, London SE8 3BU.****

Present Addresses:

* *Department of Ecology and Evolutionary Biology, University of Arizona, Tucson, Arizona 85721, USA.*

** *c/o University of Brunei, Darusalam, Bandar Seri Bagawan, Brunei Darusalam, S.E. Asia.*

*** *Botany Department, National Museum of Wales, Cathays Park, Cardiff, UK.*

1. Taxonomic problems due to preservation: comparing compression and permineralized taxa

J. GALTIER

Abstract

Most plant remains are preserved as compression fossils which provide characteristics of the general morphology; more rarely fossil plants are permineralized, yielding perfect anatomical details.

Of course the information from the two categories of fossils is largely complementary and it is not surprising that some most important advances in palaeobotany resulted from the discovery of plants preserved in both states or from established correlations between compression and permineralized taxa.

Using examples taken in several groups of palaeozoic plants we discuss the established or possible correlations between compression and permineralizations. There are important differences with regard to the level of the taxa correlated. Also significant are the differences between taxa which cannot be explained by the mode of preservation but probably represent adaptations to different original ecology.

Introduction

A number of fundamental advances in Palaeobotany, for example the demonstration of the identity of *Archaeopteris* and *Callixylon* at the origin of the concept of Progymnosperms (Beck 1960), resulted from the discovery of exceptional specimens exhibiting both types of preservation, i.e. compressions partly petrified (pyritized) or permineralizations showing external morphological features. But very few studies have been devoted exclusively to the comparison of permineralized and compression taxa. One of the best and the most recent analysis on this topic is the work by Batenburg (1982) on compression and permineralized species of *Sphenophyllum*. In this study the comparison is restricted to the specific level which supposes an a priori or previously demonstrated generic identification for specimens of both preservation states. This is not always the case. In the present review we have extended the analysis to the comparison of taxa at the generic level. Any comparison is inevitably restricted by the quantity and the quality of the data available for each category of taxon and also dependent on the preservation, as summarized on Table 1.

Selected examples taken from various taxonomic groups show that the possibility of successfully comparing or even identifying specimens from

Table 1.1. Biological data revealed by fossil plants using appropriate techniques.

Information provided by	Compression	Permineralization
GENERAL MORPHOLOGY including branching, distribution of fertile parts.	★ ★ ★	○ ★ *reconstructed from serial sections (r.s.s.)*
INDIVIDUAL ORGAN MORPHOLOGY (Leaf, sporangia, seed, pollen organs)	★ ★ ★	★ *r.s.s.* + *maceration*
TRIDIMENSIONAL ORGANIZATION	★ *(degagement)*	★ *r.s.s.* + *maceration*
EXTERNAL 'ANATOMY'		
— Epidermis-hairs-spines-glands-stomata	★ ★ ★ *(transfer, S.E.M.)*	★ *r.s.s.* + *maceration*
— Leaf bases-cushions	★ ★ ★	★
INTERNAL ANATOMY (stem-root-leaf)		
— Outer middle cortex	○ ★	★ ★ ★
— Inner cortex	○	★ ★ ★
— Phloem	○	★ ★ ★
— Secondary xylem	○ ★ *(S.E.M.)*	★ ★ ★
— Primary xylem	○ ★ *(S.E.M.* + *maceration)*	★ ★ ★
— Pith	○ ★	★ ★ ★
— Lamina (Mesophyll-vascular strand)	★	★ ★
ANATOMY of SPORANGIA ovules, seeds, pollen organs	★ *(S.E.M.)*	★ ★ ★
SPORES — POLLEN MORPHOLOGY	★ ★ ★ *(S.E.M.* + *maceration)*	★ ★ ★

★ ★ ★ Generally present
★ Rarely present or revealed by appropriate techniques
○ Absent

different preservation states is dependent on the taxonomic value of the shared characters.

Different techniques (some indicated on Table 1.) are used to improve the amount and quality of the information available. They may be summarized as follows:

1. With permineralizations: serial peel sections and to a lesser degree maceration, excavating and SEM allow detailed reconstruction of organ morphology but rarely general morphology and habit. Thick paradermal sections (of leaves, etc.) inform more precisely on epidermis, stomata, hairs, appendages, etc.
2. With compressions the *dégagement* technique (Leclercq, 1960) is the most useful way of elucidating the three-dimensional organisation of ultimate organs; sectioning of fertile organs (Halle 1933, Gastaldo 1981) and leaves (Rex 1983) helps in reconstructing their morphology and comparison with permineralizations; macerations, transfers, SEM and latex replicas, (Chaloner and Gay 1972) provide invaluable details on epidermis, stomata, sporangial and pinnule morphology, tracheidal pitting and spore morphology.

When dealing with compression taxa, whatever the technique used, there is little or no chance of revealing the xylem morphology (stele shape), the inner cortex, phloem or leaf mesophyll histology. It is only the first that is an important taxonomic character.

Comparing Non-identifiable Genera

The example of 'Calamites'

The generic name *Calamites* applies to stem compression and pith casts with the subgenera, sometimes raised to generic level, *Stylocalamites, Crucicalamites, Diplocalamites, Calamitina* corresponding to different modes of branching. This taxonomy of compressions does not apply to permineralized specimens as it has proven impossible to demonstrate precisely which petrified axis corresponds to a given *Calamites* compression. As a result, permineralized stems have been given different generic names, *Arthropitys, Calamodendron, Arthroxylon*. These stem genera (probably partly synonymous) are generally based upon principal ray organization as seen in tangential section (Boureau 1964). This anatomical criterion is one that we cannot expect to find in compressions. Here then is a most striking example of the impossibility of reconciling compression and permineralized genera.

Comparing Genera with a Good Chance of Future Identification

The following examples, taken amongst ferns, represent decreasing probabilities of identification.

1. *Comparing 'Senftenbergia' (compression) and 'Tedelea' (permineralization).*

For a long time the Carboniferous genus *Senftenbergia* has been regarded as the oldest member of the ancient fern family Schizaeaceae (Radforth 1939). In fact, the main feature used to identify the Schizaeaceous affinities of *Senftenbergia* was the sporangial structure and especially the presence of an apical annulus and longitudinal dehiscence. The sporangia are generally arranged in a double row, 6–12 per pecopterid pinnule.

The size of the sporangia is also variable from 0.5 mm long in *S. sturii* from the Lower Carboniferous to 0.98 mm long in the Upper Carboniferous *S. penneaformis* (Radforth 1939). Jennings and Eggert (1977) described a partly petrified compression pinna of *Senftenbergia* that revealed a rachis vascular trace intermediate between *Clepsydropsis* and *Ankyropteris*-type anatomy. This indicates a close affinity of this *Senftenbergia* with the filicalean fern *Tedelea* based on permineralization. *Tedelea* was created by Eggert and Taylor (1966) for fertile fronds of the well known American species *Ankyropteris glabra* Baxter. Fertile pinnules reconstructed from paradermal peel sections are morphologically similar to some forms of *Sphenopteris* or *Pecopteris*. Spore and sporangial morphology closely compare with *Senftenbergia* but the main difference is the sporangial grouping: in clusters in *Tedelea* whilst sporangia appear to be borne singly in *Senftenbergia*. The *Ankyropteris* type anatomy of *Tedelea*, now known to occur in at least one species of *Senftenbergia*, has been taken as an argument for the creation of the extinct family *Tedeleaceae*, thus favouring the anatomical characters above the sporangium morphological ones.

2. *Comparing 'Biscalitheca' with related genera 'Etapteris-Zygopteris', 'Nemejcopteris', 'Schizostachys'*

The genus *Biscalitheca* was founded on the permineralized species *B. musata* described by Mamay (1957) from a mass of annulate sporangia in an Upper Pennsylvanian coal ball of Illinois. The characters of the genus are mainly restricted to those of the highly complex sporangia and spore morphology. Furthermore Phillips and Andrews (1968) described more complete permineralized specimens exhibiting the last three orders of foliar members. In fact, quite similar permineralized sporangia had been described from the French Stephanian-Autunian by Renault (1876) who designated them under the name of *Zygopteris lacattei*, emphasizing the constant association of the sporangia with this petiole species. Although highly probable, the connection of the fertile organs with this *Zygopteris* petiole (= *Etapteris*) has never been established. Galtier and Grambast (1972) demonstrated that the distinction between

Biscalitheca and the French material resulted from the previous incomplete descriptions of the latter; Galtier (1978) later proposed to refer all of them to *Biscalitheca*.

The confusing history of *Biscalitheca* includes also its synonymy with the frond compression *Monoscalitheca* (Abbott 1961). The only significant difference left to separate *Monoscalitheca* was the supposed possession of a single dorsal annulus on the sporangium compared to the paired lateral annuli of *Biscalitheca*. Restudying the type specimen of *Monoscalitheca*, Cridland (1971) revealed an identical organization of the sporangia and concluded that this genus must be put in synonymy with *Biscalitheca*.

On the other hand, Barthel (1968) described under the name of *Nemejcopteris feminaeformis* some frond compressions from the Lower Permian of Germany with supposedly slightly different sporangia but this later proved not to be true; moreover, in a new reconstruction Barthel (*in* Haubold 1982) shows that in *Nemejcopteris* the pinnules and bifurcated aphlebiae are arranged similary to those described later by Millay and Rothwell (1983) in the permineralized *Biscalitheca*. However, the associated (? corresponding) vegetative fronds of *Nemejcopteris* are different with quadriseriate pinnae bearing *Pecopteris*-type pinnules (Barthel 1968). Other frond compressions bearing similar clusters of these very complex sporangia, have been described under different names (*Schizostachys, Zygopteris*) in France and Germany for over a century. Barthel (1968) suggests all of them should be referred to the generic name *Schizostachys* Grand'Eury (1877). The question of the identity of all the compression and permineralized species will be cleared up by work in international cooperation but all these plants surely belong to a single natural group (? genus).

The comparison of compression and permineralized taxa related to *Biscalitheca* has still not brought a satisfying answer to whole plant reconstruction. In permineralized *Biscalitheca* the largest frond member, called 'rachis' by Phillips and Andrews (1968), is elliptical in cross section (15 × 3mm). These dimensions correspond to those of most of the known compressions. Millay and Rothwell (1983) consider that this type of rachis was no doubt borne in alternating opposite pairs along a petiole of the *Etapteris* type. However, connection of the *Biscalitheca* rachis with the *Etapteris* petiole has yet to be demonstrated. A puzzling feature is that the large size of the pinna rachis (about 1cm in diameter in both permineralizations and compressions) suggests that the *Etapteris* petiole, on which the pairs are borne, must exceed 2cm in diameter; at present such large specimens are not known.

Compared and Identified Genera Corresponding to Similar Organs

Comparing 'Pothocites' and 'Protocalamostachys'

The cones attached to *Archaeocalamites'* leafy stems from compressed material have been generally referred to the genus *Pothocites* Paterson. Walton (1949) created the genus *Protocalamostachys* for permineralized strobili (*P. arranensis*) resembling *Pothocites* in many important respects. Chaphekar (1963) described a new *Protocalamostachys* species and noted that the cone axis in both species of *Protocalamostachys* is similar in general structure to slender axes of *Archaeocalamites* and that 'there is thus good reason to believe that cones of *Protocalamostachys* kind were borne on *Archaeocalamites* stems'. Later Chaphekar (1965) compared the compression species *Pothocites grantonii* Paterson to permineralized species of *Protocalamostachys*; she came to the conclusion that they cannot be regarded as specifically identical but that there is no morphological distinction to separate them generically. She proposed to classify all these cones under the name having priority i.e. *Pothocites*, regardless of mode of preservation. It may be noted that the usage of a single generic name (e.g. *Calamostachys*) for cones of both preservation types is common in the Calamitaceae.

Identification of genera is sometimes recognized but the resulting synonymy is not admitted in practice. For example the medullosan permineralized seeds *Pachytesta* are correlated but not identified with the compressions or casts *Trigonocarpus*.

Identifying Genera Corresponding To Different Parts (Organs) of the Plant

1. The example of 'Lyginopteris'

As is well known, the recognition of the Pteridospermae by Oliver and Scott (1904) resulted from the evidence of a similar unique anatomical character (capitate glands) on morphologically different permineralized organs, the cupule of the seed *Lagenostoma lomaxii* Will. and the stems and leaf rachis of *Lyginopteris oldhamia* Binney. The permineralized leaf rachis had been previously compared with the fern-like compression *Sphenopteris hoeninghausii* Brongn. This comparison was based on pinnule morphology as seen in compression and in paradermal sections from coal balls and on the similar 'dictyoxylon' cortex of stems and petioles seen in both states of preservation.

2. 'Heterangium' (permineralized stems) '-Sphenopteris' (compression leaves)

Permineralized stems of *Heterangium grievii* Will. from the Lower Carboniferous of Scotland are small with decurrent leaf bases but long portions of fronds are not known attached. However, the foliage is believed (Scott 1923) to correspond to the compression *Sphenopteris elegans* Brongn. This correlation is based on the occurrence of horizontal bands in the cortex of compression rachis and stems which are believed to correspond to the sclerotic plates occuring in the middle cortex of the permineralized organs. This is an exceptional occurrence of anatomical features of the internal cortex recognizable in compression material. Quite differently Shadle and Stidd (1975) were able to obtain sufficient paradermal sections of connected pinnules of an American permineralized *Heterangium* to reconstruct these pinnules in detail and correlate them with the compression species *Sphenopteris obtusiloba* Brongn.

Many other examples belong to the same category, such as the identification of the *Archaeopteris* 'leaves' with *Callixylon* stems (Beck 1960). In every case palaeobotanists are faced with the problem of naming the 'whole plant' and in practice they do not uniformly follow the rules of priority of the *International Code of Botanical Nomenclature*.

Genera to be Split

This is the reverse situation where a 'broad' genus has been traditionally used as a repository for both compression and permineralized species.

The example of 'Telangium'.

From its initial creation by Benson (1904) the genus *Telangium* has been applied to a variety of microsporangiate fructifications occuring in various states of preservation. Eggert and Taylor (1971) proposed the name *Telangiopsis* as a genus for the compression synangia with a variable number of elongate sporangia fused at the base and which occur in stalked clusters borne terminally on branching axes lacking evidence of foliar structures. With the type species, *T. arkansanum*, from the Upper Mississipian of Arkansas they listed eight other species including *T. affine* L. and H. and *T. bifidum* L. and H. originally placed by Benson in *Telangium*. The name *Telangium* is thus retained only for the permineralized specimens, the type species being *T. scottii*, first listed by Benson (1904) from the Upper Carboniferous coal balls of England. Eggert and Taylor (1971) justified the creation of *Telangiopsis* by the fact that it was impossible to demonstrate identity of the petrified *T. scottii* with any of the compressions. Later Jennings (1976) described, from the Upper Mississipian of Illinois, exceptional material allowing the examination of both external morphology and internal anatomy. The specimens included synangiate fructifications corresponding to either

Telangium or *Telangiopsis*. Jennings considered that the petrified synangia corresponded to *Telangium*. However in *Telangium scottii* the eight sporangia are disposed in two rows, a feature questioned by Jennings who suggested that the apparent bilateral arrangement may be just the result of diagenetic flattening.

Reinvestigation of Benson's original material and description of new fructifications from the Lower Carboniferous of Scotland by B. Meyer-Berthaud and the present author tend to support the original interpretation of Benson that *T. scottii* was naturally bilaterally arranged. Synangia described by Jennings thus have characters of *Telangiopsis* and do not resemble *T. scottii*. The splitting of *Telangium* would thus be justified but is not based on preservational differences.

Interspecific Comparison Within a Genus Having a Compression Type Species

In this very common case, two conditions enable the comparison of permineralized specimens with the compression type species—either anatomical and morphological characters are preserved on the outer surface of the permineralization (e.g. in *Lepidodendron*, *Sigillaria*, *Sphenophyllum*)—or morphology of the permineralized species is very precisely reconstructed from serial peel sections (e.g. in *Psilophyton*, *Corynepteris*, *Palaeostachya*).

1. Comparing 'Lepidodendron' species

The type species is *Lepidodendron aculeatum* Sternberg, originally known only from compression specimens. In *Lepidodendron*, as in other arborescent lycopod genera, the number of species known from permineralizations is considerably smaller than that from compressions. As indicated by Chaloner (1967) most petrified *Lepidodendron* are not correlated with compression species either because the leaf cushions (the characteristics of which are the basis for separating compression species) are only observed in section or are badly preserved or because the leaf cushion is absent as a result of decortication. In the latter case the assignment of a petrified specimen to *Lepidodendron* is rather unfounded; attribution of petrified specimens to *Lepidodendron* has thus traditionally depended on the presence of leaf cushions, the same being true for *Sigillaria* (see examples in Chaloner 1967).

One major problem with *Lepidodendron* is the extreme overextension (DiMichele, 1983) of the name. Indeed the genus includes a broad range of anatomical and morphological (leaf cushion) variation. DiMichele tried to establish a more coherent circumscription of *Lepidodendron* based on the anatomically preserved species of the Upper Carboniferous. At first he distinguished three major types of anatomically preserved species (DiMichele 1981); later he proposed *Lepidodendron hickii* Watson as the only anatomically preserved carboniferous lepidodendrid that has leaf cushions with infrafoliar parichnos and general form referable to the compression type-species *L. aculeatum* (DiMichele, 1983). He also suggested that the vertically elongated *Lepidodendron* leaf cushion shape represents a plesiomorphic-ancestral character shared with *Sublepidophloios*. So the simple leaf cushion is of dubious value in the delimitation of *Lepidodendron*. In the same review DiMichele (1983) tried to correlate anatomically preserved stems with compressions. He considered that *Lepidodendron aculeatum-L. hickii* was more confined to clastic swamp and mesic-lowland environments in contrast to '*Lepidodendron*' *vasculare* and *Lepidophloios* which were centered in coal swamps. The different types of preservation representing these two kinds of vegetation is a complicating factor for taxonomy. Prudently DiMichele concluded that *L. hickii* is 'probably a preservational state of a species such as *L. aculeatum* or a closely related form with infrafoliar parichnos. A more exact correlation is not yet possible because the taxonomy of compression-impression lepidodendrids still needs to be reassessed in light of findings from petrifactions . . .'.

It is interesting to consider that all the above-mentioned comparisons of permineralized and compressed taxa rest upon generic and specific characters focused on one part of a plant organ, the leaf cushion; but it is also significant that they include characters (leaf scar, ligule pit, infrafoliar bladders) that have been selected separately by Thomas and Meyen (1984) as the most important for generic recognition in compression form-genera.

2. *Comparing 'Corynepteris' species*

The genus *Corynepteris* Baily was universally used after Zeiller (1888) added important information on the sorus and sporangial morphology. But it was not until 1976 that permineralized pinnae were attributed to *Corynepteris* and described independently as two new species, *C. involucrata* (Baxter and Baxendale 1976) from the Pennsylvanian of Kansas and *C. scottii* (Galtier and Holmes 1976) from the Lower Westphalian of England. In both cases identification with the compression genus was made on the basis of similarities in sporangial grouping into sessile clusters and on the details of sporangial morphology with a V-shaped annulus. The permineralized species of *Corynepteris* exhibit several features hitherto unrecorded for the genus previously based on compression species. They concern: vascular and cortical anatomy of the ultimate rachis and pinnules; attachment of the sorus to a common vascularized stalk; the highly specialized sporangial wall with small sclerotic nests. In addition an involucre was described in *C. involucrata* and later also recognized in *Corynepteris* cf. *scottii* from Belgium (Galtier and Scott 1979) which is so far the oldest known fern to possess such a structure.

Though first described in permineralized specimens, several of the above-mentioned features should be recognisable on compression specimens if present—this is particularly the case for the sclerotic nests scattered on the abaxial wall or the well developed involucre. The absence of these features in all the known compression specimens cannot be explained simply by incomplete preservation—the occurrence of an involucre in permineralized species may represent an adaptive feature reflecting original habitat differences in a similar manner to those discussed below in *Sphenophyllum*.

Interspecific Comparison Within a Genus Having a Permineralized Type Species

The occurrence of histological characters exceptionally preserved in compression material allows comparison and even identification to a permineralized species. This is the case in *Biscalitheca*, discussed above, with very complex sporangia preserved in both kinds of fossils and which justified the broadening of the genus to include the compression species *B. kansana* (Cridland 1966).

The problem of 'Botryopteris burgkensis'.

The genus *Botryopteris* Renault constitutes a large number of permineralized species that range from the Lower to Uppermost Carboniferous. A single compression species, *B. burgkensis* from the Permian of Germany (Barthel, 1975) has been attributed to the genus on the basis of the grouping and organization of the sporangia that resemble *B. forensis* Ren.—*B. globosa* Dar. Indeed *B. burgkensis* sporangia are spherical with a lateral band-like annulus resembling *B. forensis*; however this 'generalized' simple sporangial morphology, also found in osmundaceous ferns and several palaeozoic taxa (Galtier and Scott 1985), cannot be taken as a sufficient proof for identifying a compression specimen with the permineralized genus.

Can we Really Identify Compression and Permineralized Species

Rare examples have been documented of single species pairs synonymyzing both compressions and permineralizations; one may cite:

—The identity of *Lepidodendron hickii* and *L. aculeatum* discussed above and that of *Sigillaria brardii* Brongn and *S. spinulosa* Ren. (Chaloner 1967).

—*Biscalitheca musata* put in synonymy with the compression species *Monoscalitheca fasciculata* (Cridland 1971).

—An increasing number of pteridosperm foliage species first known as compressions have now been described anatomically. Leisman (1960) by splitting, peeling, and macerating coal balls was able to identify petrified foliage of *Callipteridium sullivantii* Lcsq. Using similar methods Oestry-Stidd (1979) described the anatomy of the well known compression species *Neuropteris rarinervis* Bunb., and Rothwell (1980) excavated small fragments of *Callistophyton boyssettii* Ren. from coal balls.

All these examples concern only very small parts of the plant species.

Limits of the comparison. Some recent work, particularly by Thomas (1970,1977) on different lycopod genera, has shown how epidermal characters (epidermal cell size and shape, stomatal size, number and distribution) are of immense value in identifying compression species. Unfortunately these characters are rarely available in permineralized specimens and until now they have not been used when comparing species of different preservation. On the other hand DiMichele (1979,1981,1983) is trying to fight the belief that anatomy is not a diagnostic taxonomic tool and does not correlate with distinctive leaf cushion characters. Actually, anatomy is demonstrated as a useful additional character at the generic level but not as yet for segregating lycopod species.

Specific Differences not Related to Preservation

Comparing 'Sphenophyllum' species.

Vegetative stems attributed to the broad genus *Sphenophyllum* Koenig are known by compression-impression and, more rarely, petrified material. Batenburg (1982) concluded that attempts to correlate compressions and permineralizations of *Sphenophyllum* at the specific level have been relatively few and none of the correlations based on external morphology alone have met with general acceptance. Batenburg presents a critical comparison of the vegetative anatomy and morphology of species in both types of preservation. The compression species have been established on the basis of various morphological criteria mainly regarding leaves. The number of currently recognized compression species is much too high due to underestimation of individual variation and poor knowledge of the leaf polymorphism. Many fewer species have been distinguished in permineralized material. Renault (1870,1873) first identified the internal structure of silicified stems from the Stephanian-Autunian of France. Batenburg (1982) reviewed all the characters that may be observed in both compressions and petrifactions. In stems and leaves they concern: inner tissue (xylem thickenings, outer cortex), epidermal emergences, external morphology, occurrence of decortication, branching, hydathodes, climber hooks, and leaf polymorphism. The available data show clearly that there are many similarities between compression and petrifaction species. Batenburg considers that 'although attempts to correlate compressions and petrifactions at the specific level have been relatively few and rather unsuccessful, there cannot be any doubt as to the affinities of both groups. However there are some marked differences which cannot be accounted for by the different modes of preservation'. The main differences are: extensive periderm formation with casting off of stem epiderm and primary cortex and shedding of the leaves in the petrifaction species while all these features are absent in compression species. Also, adventious roots are rather common in petrifactions and very rare in compressions. On the contrary foliar climber hooks, spines or multicellular hairs are widespread in compression species but absent with one exception in petrifactions. These differences may be explained if a majority of the compression species were scramblers whilst the large permineralized naked stems with their adventitious roots suggest that these plants had at least prostrate rooting main axes from which an aerial branch system arose. So Batenburg (1982) suggests a correlation of differences in habit and habitat, agreeing with Storch (1966), and he considers it probable that the compression species grew at the periphery of the coal swamp whereas the coal ball species were really coal swamp plants.

In the present state of our knowledge this palaeoecological interpretation is quite attractive and I wish to add an argument in its favor. The silicified *Sphenophyllum stephanense* (Renault 1873) from Grand-Croix, France, is the exception amongst permineralized species in possessing hairs or spines. This is not surprising as these plants do not conform to a classic coal ball type of preservation; in fact this silicified vegetation from the Stephanian represented a mixed assemblage partly distinct from the contemporaneous coal swamp and one in which there has been considerable transport of plant material (Galtier and Phillips, *in press*).

We mentioned above in *Corynepteris* some specific differences (presence/absence of an involucre, of sclerotic nests) unrelated to preservation and which may also be interpreted in the same way.

Conclusion

The restricted number of examples discussed in this paper are sufficient to demonstrate that significant differences exist with regard to the taxonomic level of specimens which can be correlated. They also illustrate the extreme importance of using a wide variety of techniques to improve the available information, i.e. the source of taxonomic criteria. The probability of identifying taxa of different preservation increases if the considered criteria include a group of characters or concern complex organs. In some circumstances the comparison leads to lumping or to splitting genera or to making major taxonomic changes. The latter more often result from the discovery of exceptional fossils showing both types of preservation (e.g. compressions partly pyritized) which are better documented and not discussed in detail in this paper. Reconstructing whole plants and correctly interpreting their taxonomy are two major aims of Palaeobotany. It is significant that many plants successfully reconstructed and sometimes taxonomically re-interpreted (e.g. *Psilophyton, Pseudosporochnus, Rhacophyton, Archaeopteris, Lyginopteris*) resulted from specimens exhibiting both types of preservation or a correlation between the two. In all these examples the vascular anatomy was a decisive criterion for taxonomy at or above the generic level, whilst it is rarely useful at the specific level. At last it is becoming highly desirable to search for new cases of specific differences unrelated to preservation states but which may be interpreted in terms of different original palaeoecology.

Acknowledgements

I thank B. Meyer-Berthaud, G. Rex, and J. Holmes for kindly commenting on the manuscript and for helpful discussions. I am indebted to J. Courbet for technical assistance.

References

Abbott, M.L.(1961) A coenopterid fern fructification from the Upper Freeport, No.7, coal in southeastern Ohio. *J.Paleont.* **35**, 981–5.

Barthel, M. (1968) 'Pecopteris' feminaeformis (Schlotheim) Sterzel und 'Araucarites' spiciformis Andrae in Germar—Coenopteriden des Stephan und Unteren Perms. *Paläontol. Abh. B. Paläobot.* **2**, 727–42.

—— (1975) Neue Erkenntnisse und Fragen über Farne des Oberkarbons und des Unteren Perms. *Wissenschaftl. Zeitschrift der Humboldt-Univ. Berlin, Math. —Nat. R.* **24**, 461–74.

Batenburg, L.H. (1982) 'Compression species' and 'petrifaction species' of *Sphenophyllum* compared. *Rev. Palaeobot. Palynol* **36**, 335–59.

Baxter, R.W. and Baxendale, R.W. (1976) *Corynepteris involucrata*, sp. nov., a new fertile fern of possible zygopterid affinities from the Pennsylvanian of Kansas. *Univ. Kansas Paleontol. Contrib.*, Paper 85, 1–8.

Beck, C.B. (1960) The identity of *Archaeopteris* and *Callixylon*. *Brittonia* **12**, 351–68.

Benson, M. (1904) *Telangium scotti*, a new species of *Telangium* (*Calymmatotheca*) showing structure. *Ann. Bot.* **18**, 161–77.

Boureau, E. (1964) *Traité de Paléobotanique.* **Vol. 3**. *Sphenophyta, Noeggerathiophyta*. Masson, Paris.

Chaloner, W.G. (1967) Lycophyta. In *Traité de Paléobotanique.* **Vol. 2**. (ed. E. Boureau), pp. 435–834, Masson, Paris.

Chaloner, W.G. and Gay, M.M. (1972) Scanning electron microscopy of latex casts of fossil plant impressions. *Palaeontology* **16**, 654–59.

Chaphekar, M. (1963) Some calamitean plants from the Lower Carboniferous of Scotland. *Palaeontology* **6**, 408–29.

—— (1965) On the genus *Pothocites* Paterson. *Palaeontology* **8**, 107–12.

Cridland, A.A. (1966) *Biscalitheca kansana* sp.n. (Coenopteridales, Zygopteridaceae), a compression from the Lawrence Shale (Upper Pennsylvanian), Kansas, U.S.A. *Amer. J. Bot.* **53**, 987–94.

—— (1971) *Monoscalitheca*, a synonym of *Biscalitheca* (Coenopteridales). *Bull. Torrey Bot. Club* **98**, 258–64.

DiMichele, W.A. (1979) Arborescent lycopods of Pennsylvanian age coals: *Lepidophloios*. *Palaeontographica B* **171**, 57–77.

—— (1981) Arborescent lycopods of Pennsylvanian age coals: *Lepidodendron* with description of a new species. *Palaeontographica B* **175**, 85–125.

—— (1983) *Lepidodendron hickii* and generic delimitation in Carboniferous lepidodendrid lycopods. *Systematic Botany* **8**, 317–33.

Eggert, D.A. and Taylor, T.N. (1966) Studies of Paleozoic ferns: on the genus *Tedelea* gen. nov. *Palaeontographica B* **120**, 169–80.

—— and —— (1971) *Telangiopsis* sp. nov., an Upper Pennsylvanian pollen organ from Arkansas. *Bot. Gaz.* **132**, 30–7.

Galtier, J. (1978) Précisions sur '*Zygopteris lacattei*' et '*Botryopteris dubius*', fougères très rares de l'Autunien d'Autun. *Bull. Soc. Hist. Nat. Autun* **88**, 17–25.

—— and Grambast, L. (1972) Observations nouvelles sur les structures reproductrices attribuées à *Zygopteris lacattei* (Coenopteridales de l'Autuno-Stéphanien français). *Rev. Palaeobot. Palynol.* **14**, 101–11.

—— and Holmes, J. (1976) Un *Corynepteris* à structure conservée du Westphalien d'Angleterre. *C. R. Acad. Sci. Paris* **282**, 1265–68.

—— and Phillips, T.L. (1985) Swamp vegetation from Grand-Croix (Stephanian) and Autun (Autunian), France and comparisons with coal-ball peats of the Illinois basin. *C.R. 9th Intern. Congr. Carb.Strat. Geol.*, Urbana, 1979, **5**, 13–24.

—— and Scott, A.C. (1979) Studies of Paleozoic ferns: on the genus

Corynepteris. A redescription of the type and some other European species. *Palaeontographica B* **170**, 81–125.
—— and —— (1985) Diversification of early ferns. *Proc. Roy. Soc. Edinb.* **86** B, 289–301.
Gastaldo, R.A. (1981) *Palaeostachya dircei* n. sp., an authigenically cemented equisetalean strobilus from the Middle Pennsylvanian of southern Illinois. *Amer. J. Bot.* **68**, 1306–18.
Grand, Eury, F.C. (1877) Flore carbonifère du département de la Loire et du centre de la France. *Mém. Acad. Sci. Inst. France* **24**, 624 p.
Halle, T.G. (1933) The structure of certain fossil spore-bearing organs believed to belong to pteridosperms. *Kgl. Svenska Vetenskapsakad. Handlingar*, ser. 3, **12**, 1–103.
Haubold, H. (1982) *Die Lebewelt des Rotliegenden*. A. Ziemsen Verl., Wittenberg.
Jennings, J.R. (1976) The morphology and relationships of *Rhodea*, *Telangium*, *Telangiopsis* and *Heterangium*. *Amer. J. Bot.* **63**, 1119–33.
—— and Eggert, D.A. (1977) Preliminary report on permineralized *Senftenbergia* from the Chester Series of Illinois. *Rev. Palaeobot. Palynol.* **24**, 221–25.
Leclercq, S. (1960) Refendage d'une roche fossilifère et dégagement de ses fossiles sous binoculaire. *Senckenbergiana Lethaea* **41**, 483–7.
Leisman, G.A. (1960) The morphology and anatomy of *Callipteridium sullivanti*. *Amer. J. Bot.* **48**, 281–7.
Mamay, S.H. (1957) *Biscalitheca*, a new genus of Pennsylvanian coenopterids based on its fructification. *Amer. J. Bot.* **44**, 229–39.
Millay, M.A. and Rothwell, G.W. (1983) Fertile pinnae of *Biscalitheca* (Zygopteridales) from the Upper Pennsylvanian of the Appalachain basin. *Bot. Gaz.* **144**, 589–99.
Oestry-Stidd, L.L. (1979) Anatomically preserved *Neuropteris rarinervis* from American coal balls. *J. Paleontol.* **53**, 37–43.
Oliver, F.W. and Scott, D.H. (1904) On the structure of the Paleozoic seed *Lagenostoma lomaxi*, with a statement of the evidence upon which it is referred to Lyginodendron. *Phil. Trans. Roy. Soc. London B* **197**, 193–247.
Phillips, T.L. and Andrews, H.N. (1968) *Biscalitheca* (Coenopteridales) from the Upper Pennsylvanian of Illinois. *Palaeontology* **11**, 104–15.
Radforth, N.W. (1939) Further contributions to our knowledge of the fossil Schizaeaceae; genus *Senftenbergia*. *Trans. Roy. Soc. Edinb.* **59**, 745–61.
Renault, B. (1870) Mémoire sur l'organisation de rameaux silicifiés appartenant probablement à un *Sphenophyllum* *C. R. Acad. Sci. Paris* **70**, 1158–60.
—— (1873) Recherches sur l'organisation des *Sphenophyllum* et des *Annularia*. *Ann. Sci. Nat. sér. 5, Bot.* **18**, 5–22.
—— (1876) Recherches sur la fructification de quelques végétaux provenant des gisements silicifiés d'Autun et de Saint-Etienne. *Ann. Sci. Nat. sér. 6, Bot.* **3**, 5–29.
Rex, G. (1983) The compression state of preservation of Carboniferous lepidodendrid leaves. *Rev. Palaeobot. Palynol.* **39**, 65–85.
Rothwell, G.W. (1980) A technique for revealing the surface features of

permineralized 'coal ball' plants. *J. Paleontol.* **54**, 1131–33.

Scott, D.H. (1923) *Studies in Fossil Botany*. Vol.2. *Spermophyta*. Black, London.

Shadle, G.L. and Stidd, B.M. (1975) The frond of *Heterangium*. *Amer. J. Bot.* **62**, 67–75.

Storch, D. (1966) Die Arten der Gattung *Sphenophyllum* Brongniart im Zwickau-Lugau-Oelsnitzer Steinkohlenrevier: Ein Beitrag zur Revision der Gattung. *Paläontol. Abh. B* **2**, 195–326.

Thomas, B.A. (1970) Epidermal studies in the interpretation of *Lepidodendron* species. *Palaeontology* **13**, 145–73.

—— (1977) Epidermal studies in the interpretation of *Lepidophloios* species. *Palaeontology* **20**, 273–328.

—— and Meyen, S.V. (1984) A system of form-genera for the Upper Palaeozoic lepidophyte stems represented by compression-impression material. *Rev. Paleobot Palynol.* **41**, 273–81.

Walton, J. (1949) On some Lower Carboniferous Equisetineae from the Clyde area. *Trans. Roy. Soc. Edinb.* **61**, 729–36.

Zeiller, R. (1888) *Bassin houiller de Valenciennes. Description de la flore fossile. Etude des gites minéraux de la France*. Quantin, Paris.

2. Experimental modelling as an aid to interpreting the original three-dimensional structures of compressions.

G.M. REX

Abstract

Using experimental modelling techniques combined with a close examination of plant compression fossils, particularly in transverse section, some of the mechanisms that are important in this fossilization process have been elucidated. These are discussed with reference to three compression/impression fossils in particular. The lycopod leaf compression, *Cyperites bicarinatus* is shown to have a complex transverse section which can be related to the uncompressed form of the leaf. The redescription of this leaf has highlighted some of the problems of lycopod leaf taxonomy. The pteridosperm leaf compression *Alethopteris serli* demonstrates the behaviour of rigid tissues during compression and their subsequent effect on the form of the fossil. Finally the '*Scutum*'-type of ovulate glossopterid fructifications are discussed as a result of experimental modelling of the proposed structures of the uncompressed fructifications and the permineralised specimens. It is not easy to interpret all the species assigned to this group of fossils in terms of one common structure. An understanding of the processes involved when a plant organ is compressed in sediment enables correct interpretations to be made on the original form of the uncompressed plant organ, this therefore in-turn enables better taxonomic assignments of plant compression fossils.

Introduction

A number of different experimental techniques have been used to gain an understanding of the processes involved when plant organs are compressed in sediment (Rex 1983a, b, Rex 1985, Rex 1986, Rex and Chaloner 1983). During the course of this study on plant compression fossils it has become evident that the process of fossilization is complex and the final form of the fossil as seen on the rock surface results from the interaction of a number of processes. It is not possible to demonstrate all of these here but some of the important factors are discussed with reference to fossils that have been investigated using experimental techniques, sectioning, and detailed examination of specimens.

Cyperites bicarinatus Lindley and Hutton provides a very good test for the type of taxonomic problems that surround the interpretations of plant compression fossils. It is demonstrated that this apparently flat lycopod leaf compression has a surprising complexity and that its structure differs from that previously described. The uncompressed form of the leaf is related to the compression using modelling techniques.

The pteridosperm leaf compression, *Alethopteris serli* Brongniart is described to demonstrate the behaviour of rigid tissues within the original organ and how this influences the form of the compression. The fossilization of the leaf is discussed by relating the compression to the permineralised leaf. It is shown that some structures described on species of *Alethopteris* are simply structures formed as a result of compression and are not original features of the leaf.

Finally, the problematic group of *Scutum*-type compression/impression fossils are discussed. The original structure of these fossils has been highly controversial since they were first described by Plumstead (1952). Discovery of the permineralised fructifications associated with *Glossopteris* Brongniart leaves by Gould and Delevoryas (1977) showed a fossil with a radically different form that was unlike any of the proposed reconstructions. Examination of some specimens of these fossils together with experimental modelling of all the postulated models of these fructifications and of the permineralised specimens indicates that the structure of the original uncompressed organ is still not easily interpreted in terms of a single form. The range of structures within the fossils and between species seems to indicate that there were a number of different original uncompressed structures which formed a '*Scutum*'-type fossil.

(1) The structure and mechanics of fossilization of the lycopod leaf compression 'Cyperites bicarinatus' and its taxonomic implications

Cyperites is the very common grass-like lycopod leaf compression frequently found in shales of the Coal Measures. The specimens described here were collected from the shales of the Radstock coalfield (Westphalian D). *Cyperites* usually occurs as a mass of coaly leaves randomly distributed over the rock surface (Fig. 2.1A). The leaves vary in length but generally are 2–3mm wide. The leaves are parallel sided and leaf bases and apices are rarely found. *Cyperites* was first described by Lindley and

Hutton (1833) and the few subsequent diagnoses of the genus (Crookall 1966, Chaloner and Boureau 1967) refer to it as long linear leaf compression containing a central vein bounded by two lines either side. These lines have been attributed to stomatal grooves in the original leaf (Crookall 1966, Chaloner and Boureau 1967, Kosanke 1979).

A close examination of specimens of *Cyperites* demonstrated that there was a vast variation in the structure of the leaf surface. It was not possible to consistently recognise the 'midrib' and 'stomatal grooves' that had been previously described on the surface of the leaf compression. In fact the leaf surface was composed of a series of 'ridges' and 'furrows' which varied in number and extend along and between specimens.

Since the external structure of the leaf compression was difficult to characterise as a result of this variation in the 'ridges' and 'furrows' along the left, the leaves were sectioned while still intact in the matrix, i.e. before the leaves were revealed by a fracture plane. The transverse sections of these leaves showed that they were not flat coaly structures as would have been expected from the observations on the exposed leaf, as in transverse section the leaves had a surprising complexity. The transverse section of the leaf was 'butterfly' in appearance consisting of four coaly 'wings' each separated by matrix and all linked by a slightly thicker central coaly bar (Fig. 2.1B). There was considerable variation in the form of this 'butterfly' between leaves and often the coaly wings were branched, thickened or incurved at the tips (Fig. 2.1C).

The leaf compression in transverse section and the exposed leaf were related by reassembling the part and counterpart, and using a rock saw to section through the leaves. The fracture plane follows a pathway through the compression which leaves the 'butterfly' structure hidden in the matrix revealing the leaf as an apparently flat layer of coal (Fig. 2.1D–G). The fracture plane follows one of the coaly 'wings' of the 'butterfly' and the ridge/furrow structures on the exposed leaf surface result from the fracture plane picking up the junction between the 'wings' and the central horizontal plate of coal. The fracture plane passes out over the diagonally opposite wing. The variation in the extent and the position of the ridges and furrows on the exposed leaf surface are purely a function of the angle of the coaly wings to the horizontal coaly plate and the overall angle of the 'butterfly' in the matrix.

One of the problems encountered during the course of this work was relating the transverse section of the leaf compression to the original form of the uncompressed leaf. It has been postulated (Graham 1935, Crookall 1966) that *Sigillariopsis* is the petrified state of *Cyperites*. There is, though, considerable variation in the transverse sections of the described species of *Sigillariopsis*. All, however, possess stomatal grooves on the leaf margin but the over-all form of the leaf varies from rhombohedral to triangular. This proposed relationship between *Cyperites* and *Sigillariopsis* was investigated using the modelling

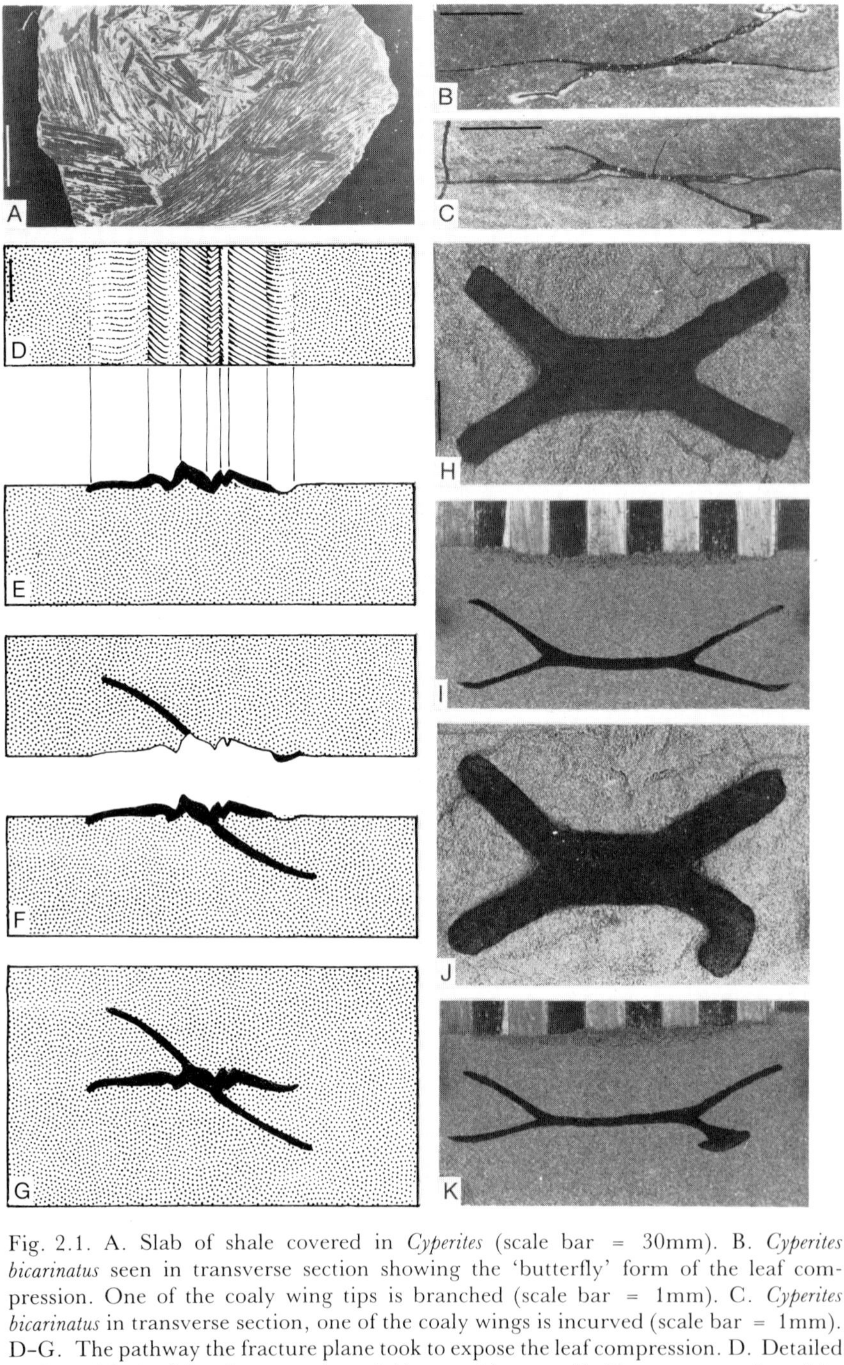

Fig. 2.1. A. Slab of shale covered in *Cyperites* (scale bar = 30mm). B. *Cyperites bicarinatus* seen in transverse section showing the 'butterfly' form of the leaf compression. One of the coaly wing tips is branched (scale bar = 1mm). C. *Cyperites bicarinatus* in transverse section, one of the coaly wings is incurved (scale bar = 1mm). D–G. The pathway the fracture plane took to expose the leaf compression. D. Detailed surface of the leaf showing a number of ridges and furrows. E. Transverse section of the

techniques described in Rex and Chaloner (1983). Models were constructed in foam rubber of all the described transverse sections of *Sigillariopsis* (Graham 1935, Reed 1941, Lemoigne 1961) and compressed in fine sawdust. None of the previously described sections of *Sigillariopsis* on compression produced the 'butterfly' form of the leaf compression *Cyperites*. Only the leaf described by Lemoigne (1961) produced a form slightly resembling the 'butterfly'. This was because in this leaf the original lobes were of equivalent width and when these were compressed gave a slightly four-winged form. This leaf form was modified and on compression produced the 'butterfly' form (Fig. 2.1H–I). The leaf compression had no obvious orientation in the matrix and the proposed model for the uncompressed leaf is symmetrical. Therefore, it could have landed either way up in the sediment and produced the 'butterfly' structure. Incurling or crumpling of the leaf margin would produce the hook-like terminations on the wing margins (Fig. 2.1J–K).

Discussion of *Cyperites*

It has been demonstrated that *Cyperites* has a more complicated transverse structure than has been realised. This work has therefore enabled the genus *Cyperites* and its type *Cyperites bicarinatus* to be redefined; a neotype has been erected elsewhere (Rex 1983a). The diagnosis of *Cyperites* now includes the 'butterfly' form in transverse section but can no longer include the presence of a midvein or stomatal grooves running along the length of the leaves. The structures pertaining to these on *Cyperites* bear no relation to these structures in the original leaf. They are simply the reflection of the pathway of the fracture plane through the 'butterfly' form. The number and extend of 'veins' on the leaf surface are dependent on the angle of the leaf compression in the sediment. The grooves lying either side of the leaf margins may well represent the stomatal grooves in the original leaf but latex replicas and subsequent SEM showed no evidence of stomata. The nature of the vascular strand i.e. whether single or double, cannot be recognised from the leaf compression.

Therefore, there are several points that can be made about lycopod leaf taxonomy as a result of this work.(i) There are, at least, two distinct genera of lycopod leaf compression, *Cyperites* and those leaves found

---

same leaf showing its surface topography. F. Transverse section of the part and counterpart showing the pathway of the fracture plane through the compression. G. Transverse section of the intact leaf compression (all at the same magnification, scale bar = 1mm). H–K. Experimental modelling of the uncompressed leaf. H. The leaf model (foam rubber) is embedded in sawdust. I. After compression, the characteristic butterfly form of the leaf compression is produced. J. The model is embedded in the matrix with one incurved lobe. K. After compression the butterfly shape is produced with a flange formed on one wing tip (all at the same magnification, scale bar = 30mm).

attached to lepidodendrid stems. Previous work (Rex 1983a) has demonstrated that the small, narrow tapered leaves found attached to lycopod stems e.g.*Lepidodendron wortheni* have a very simple transverse section consisting of a simple 'V' of coal. These leaves are probably the compressed form of the permineralised leaves *Lepidodendron latifolium* Graham and *L.minor* Graham; both being leaves without stomatal grooves. These leaves would probably be placed in *Lepidophylloides* by most authors if found unattached (Stewart 1983). (ii) At the present time the status of the genus *Lepidophylloides* is uncertain. It was erected by Snigerevskya (1958) for those fossils which were incorrectly assigned to *Lepidophyllum* Brongniart for this name had previously been given by Cassini to a living plant. Unfortunately in the typification she sited both a compression species (*Lepidophyllum lineare*) and a permineralised species (*L.aciculum*) in its synomony. It seems that it would be appropriate at some point to erect a new genus for compressed lycopod leaves since at the moment *Lepidophylloides* contains unrelated compression and permineralised species; the permineralised species may or may not contain stomatal grooves. (iii) The taxonomy is further complicated since permineralised isolated lycopod sporophyll lamina when observed in transverse section may have a very similar structure to the transverse section of a lycopod leaf without stomatal grooves. Therefore in some cases it may be difficult to distinguish between lycopod leaves and sporophyll lamina. (iv) *Sigillariopsis* as a genus seems clear and is reserved for permineralised leaves containing a double vascular strand; this name does not imply that the leaves were only borne by *Sigillaria* (Graham 1935,Crookall 1966). It is considered here that there is evidence that *Cyperites* was borne by both *Lepidodendron* (Kosanke 1979) and *Sigillaria* (Lindley and Hutton 1833). (v) Frequently petrifaction genera and species are poorly defined since they were frequently erected on the basis of single sections. The variation in the structure of the vascular strand along the leaf has never been fully investigated.

(2) The structure and fossilization of the pteridosperm leaf compression 'Alethopteris serli'

Several important aspects of the compression process can be illustrated by reference to the fern-like compression foliage borne by pteridosperms. Here one leaf compression is described in particular, *Alethopteris serli* collected from the shales of the Radstock coalfield (Westphalien D).

This leaf compression is usually in the form of one or more pinnae, usually composed of numerous pinnate lobes or pinnules. Each pinna has a central rachis from which midveins branch off and extend into the pinnules. Within the pinnules the midveins branch into numerous secondary veins which extend to the pinnule margin (Fig. 2.2A). Examination of the parts and counterparts of specimens of *A.serli* indicated that the 'compression' was always composed of positive pinnules that extended above the fracture surface within which the rachis and midveins were always depressions. The secondary veins on the pinnules were always in the form of positive coaly ridges (Fig. 2.2A). In the corresponding

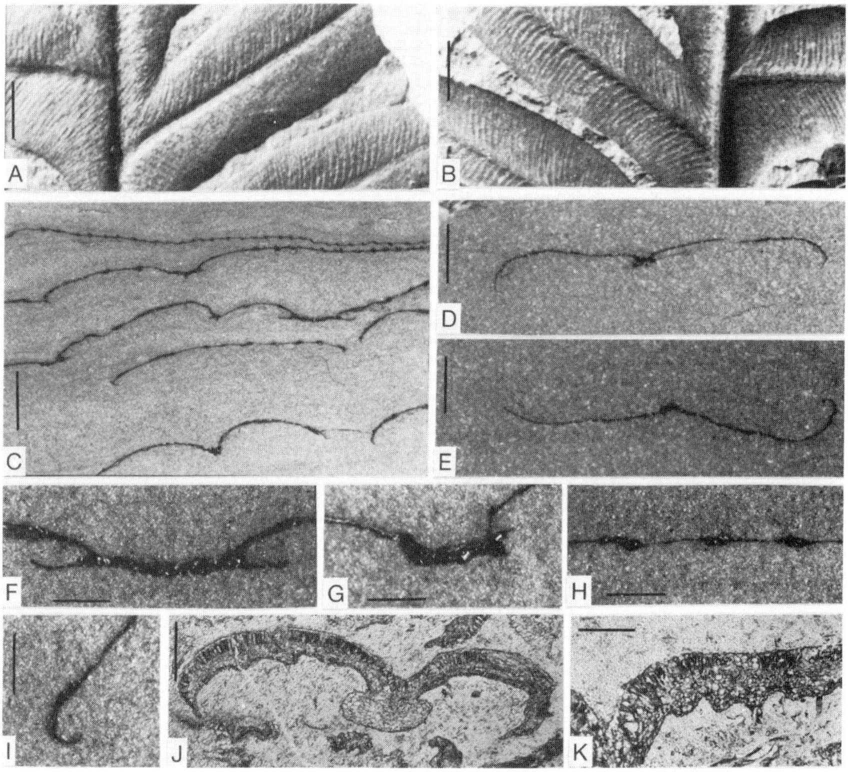

Fig. 2.2. A. *Alethopteris serli*, the part composed of positive pinnules containing negative rachis and midveins and positive secondary veins. B. *Alethopteris serli*, the counterpart containing negative pinnules, positive rachis and midveins and negative secondary veins (In A and B scale bar = 2.8mm). C. Transverse section and Longitudinal section of *Alethopteris serli* pinnules sectioned intact in the matrix (scale bar = 1.5mm). D. Transverse section of a compressed pinnule orientated the 'right way-up' in the matrix (scale bar = 1.5mm). E. Transverse section of a compressed pinnule orientated 'upside-down' in the matrix, this pinnule shows the same structure as that in D (scale bar = 1.5mm). F. Transverse section of the rachis of a pinna showing the coaly plate extending into the matrix (scale bar = 0.6mm). G. Transverse section of a pinnule midrib composed of a thick coaly plate (scale bar = 0.4mm). H. Transverse section of a pinnule showing the coaly knots of the secondary veins (scale bar = 0.4mm). I. The incurved hook of the pinnule margin (scale bar = 0.5mm). J. Transverse section of a permineralised *Alethopteris* pinnule from Ste Etienne, France (scale bar = 7mm). K. Transverse section of part of an *Alethopteris* pinnule showing the smooth upper epidermis and ribbed lower epidermis (scale bar = 0.5mm).

'impression' these structures were topographically opposite i.e. negative pinnules, positive rachis and midveins and negative secondary veins (Fig.2.2B).

A series of transverse and longitudinal sections were made through intact compressed pinnae. The transverse section of the pinna showed considerable relief. The pinnule consists of a fine coaly line which is convexed upwards (Fig. 2.2C). The midvein of the pinnule formed a U-shaped depression in the centre of the pinnule and this is underlain by a small plate of coal (Fig. 2.2G). The margins of the pinnule are usually strongly incurved forming small hooks directed towards the midvein (Fig. 2.2I). At higher magnifications of the longitudinal sections of the pinna it can be seen that the pinnule does not consists of a smooth line of coal but is composed of numerous coaly knots each measuring 0.1mm in diameter (Fig. 2.2H). These are separated from each other by a fine coaly line reaching up to 0.01mm thick. These knots are not arranged on either surface of the coaly line but generally cross it. Transverse sections of the compressed rachis shows that it has a distinct U-shaped structure underlain by a thick elongated plate of coal from which fine coaly lines often extended into the martrix (Fig. 2.2F). The distinct structure of the pinnule is maintained regardless of the original orientation of the leaf in the matrix (Fig. 2.2D,E). The fracture plane always takes the same pathway to expose the leaf leaving the coaly matter on the convexed surface of the pinnule and the hooks, coaly plates of the midvein and rachis, hidden in the matrix (Fig. 2.3D).

Alethopteris is more commonly known as a compression but it is known permineralised (Leisman 1960, Baxter and Willhite 1969, Mickle and Rothwell 1982). The typical form of an uncompressed *Alethopteris* pinnule is of a strongly revolute leaf with a sunken midrib (Fig. 2.2J). The upper epidermal surface of the pinnules is usually smooth and the lower may be smooth or slightly undulating (Fig. 2.2K). The pinnules are predominantly composed of mesophyll with the few xylem strands of the secondary veins lying close to the lower epidermis.

Hence, since the structure of the uncompressed pinnule is already known the fossilization of the compressed pinnule can be visualised. The leaf was buried in fine sediment which filled the midvein and rachis depressions (Fig. 2.3A). On compression, the sediment preserved these depressions as well as the overall convexed shape of the leaf (Fig. 2.3B–D). The midvein and rachis were compressed to elongated coaly plates underlying the sediment filled depressions.

It is the actual compression of the pinnule that is interesting. The secondary veins within it were rigid enough to withstand the compression process, compared to the remainder of the leaf mesophyll. The veins are now in the form of coaly knots separated by fine coaly lines that represent the compression of the bulk of the leaf (Fig. 2.3C). The pinnule

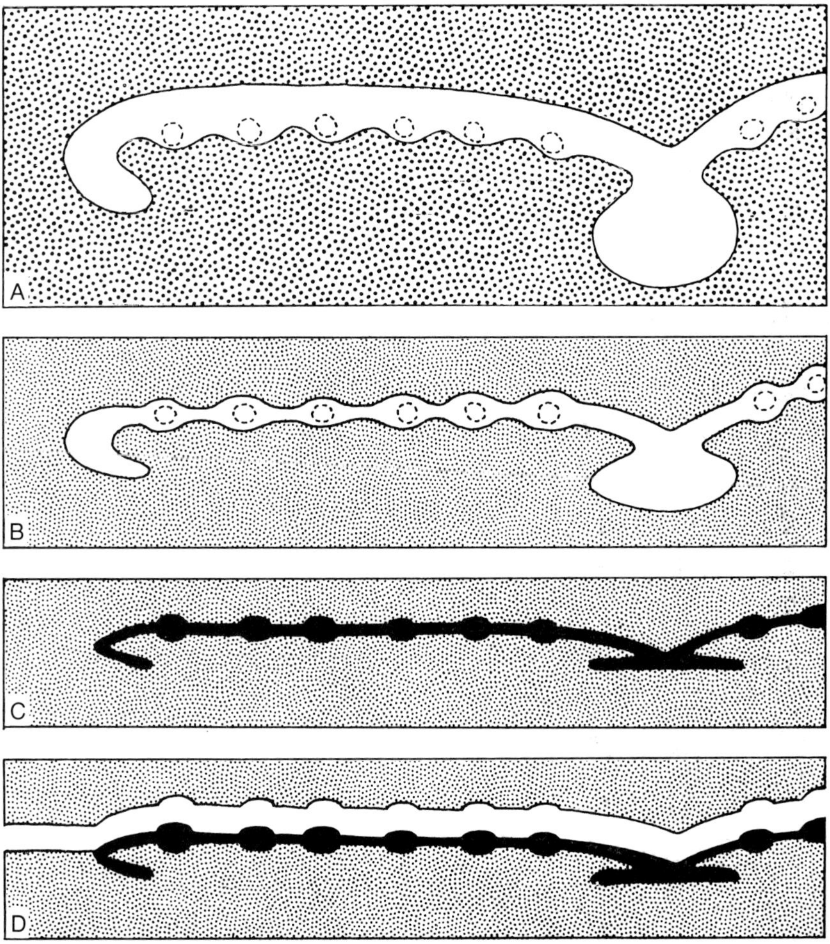

Fig. 2.3. The fossilization of an *Alethopteris* pinnule. A. Longitudinal section of an uncompressed pinnule. B. As the sediment compacts the leaf compresses around the secondary veins. C. The compressed pinnule in which the bulk of the leaf lamina has been reduced to a fine coaly line. The secondary veins have formed distinct coaly knots within the leaf pinnule. The rachis has been compressed to a coaly plate that underlies the sediment filled rachis depression and the pinnule margins have formed coaly hooks. D. The pathway the fracture plane takes to expose the leaf giving a compression with a positive topography, in which the hooks and compressed rachis remain hidden in the matrix, and an impression with a negative topography.

compression, therefore, in terms of its secondary venation has a more accentuated topography than the original leaf where the pinnule was almost smooth. In contrast the midvein and rachis have the same topographic relationship within the pinna as they had in the original leaf (Fig. 2.3D).

One of the few accounts that has considered the compression of *Alethopteris* was that by Walton (1936) although at this time no permineralised specimens had been described. He made the important point in his discussion of leaf compressions; that the resulting form of the compression depended on the way up the leaf landed in the sediment. If it landed the 'right way up' compression borders would be produced during the fossilisation of the leaf that would be recognised in the fossil. These would not be produced if the leaf landed the 'wrong' way up in the sediment. In fact, this does not occur during the fossilization of *Alethopteris* and the same leaf compression results from either orientation in the matrix. The compression borders described by Walton are not produced during the fossilization. Cridland (1968) described the compression *Alethopteris ambigua* and illustrated a blade-like extension to the midvein which he believed was an original feature of the leaf. This in fact, is simply the compressed midvein which is frequently not seen since it is hidden in the matrix and not exposed by the fracture plane.

The leaf compression *Alethopteris serli* has illustrated several important features concerning the compression process. The actual form of the original leaf is unchanged during the compression, the leaf retains a revolute structure and the midveins and rachis remain as depressions. This is because the sediment protects the leaf from the flattening process. It is only the relationship of the secondary veins to the pinnules that changes, these become considerably accentuated during the compression giving the pinnule a surface topography that it did not have in the uncompressed leaf. Some structures believed to be original characters of the leaf are simply compression phenomena.

(3) The uncompressed structure of the ovulate glossopterid fructifications

One of the important tests for conducting this type of investigation into the mechanics of the formation of plant compression fossils is to apply the results to a fossil or group of fossils that is only known in this state of preservation and to attempt to draw some conclusions on its original form. The highly controversial *'Scutum'*-type of ovulate fructifications preserved merely as impressions (i.e. devoid of coaly matter) were for a long time only known in this preservation state. They were first described by Plumstead (1952) attached to *Glossopteris* leaves and since that time many interpretations as to their original form have been postulated. The discovery of permineralised fructifications associated with *Glossopteris* leaves by Schopf (1970) and subsequently Gould and Delevoryas (1977) from the Permian of Queensland, Australia showed a form of an ovulate fructification that had not been previously suggested by palaeobotanists. Little attempt, though, has been made to accurately reconcile the fossils preserved as permineralizations and impressions.

The impression fossils are not easy to interpret because of several important problems, firstly, they are usually devoid of coal which significantly lowers the level of information obtainable from the fossil. Secondly, these fossils probably result from the compression of a complex original three-dimensional structure which has suffered extensive reduction in at least its vertical dimension. Thirdly, parts of the fructification, primarily the seeds, have been filled with fine sediment, this may have strongly influenced the form of the fossil. Finally, there are various pathways the fracture plane could have taken to expose the impression fossil. Experimental modelling has been conducted for all the proposed models of the '*Scutum*' fossils and for the described permineralised forms. This has been combined with an examination of specimens of these fossils housed in the British Museum, Natural History. The fossilisation of '*Scutum*' and its relationship to the permineralised specimens is discussed in the light of this work.

There are several genera that have been assigned to the ovulate glossopterid fructifications by Plumstead (1952, 56 and others (Surange and Chandra 1973, Chandra and Surange 1976, 1977, Rigby 1978). These include *Scutum, Hirsutum, Cistella* (now *Plumsteadia), Ottokaria, Lanceolatus* and *Dictyopteridium*. These fossils all have been described as having a similar, basic structural plan consisting of a 'head' region containing a number of oval sacs. This maybe surrounded by a flatter wing region of fused segments (*Scutum* (Fig. 2.4A), *Hirsutum* (Fig. 2.4B)) or free segments (*Ottokaria* (Fig. 2.4F)) or this maybe absent *Cistella* (Fig. 2.4C), *Lanceolatus* (Fig. 2.4E) and *Dictyopteridium* (Fig 4D).

There have been four models proposed for the uncompressed structure of these fossils:-

(a) Bilateral cupule model (Fig. 2.4G): structure of *Scutum* proposed by Plumstead (1952, 1956) when she first described these fructifications. Plumstead (1952) considered that *Scutum* was a 'bilaterally symmetrical cupule consisting of a fertile half containing a number of sacs and an empty veined half'. She based this interpretation on the fact that when the shale was split open to reveal the fossil, the two impressions contained different structures, this she concluded represented the two halves of a cupule. Plumstead's interpretation of *Scutum* and the other ovulate fructifications seems to indicate that these organs were flat in life and had suffered little during the fossilization process in terms of flattening. She does not discuss the details of the fossilization process.

(b) Cone/strobilus model (Fig. 2.4H): Published along with Plumstead's original description of the glossopterid fructifications was a series of written discussions by several palaeobotanists. Of these, Walton offered a radically different explanation of the uncompressed structure of *Scutum*. He offered a pre- and post- fossilization model and is one of the few to have done so. He proposed that *Scutum* was originally cone-like, radially

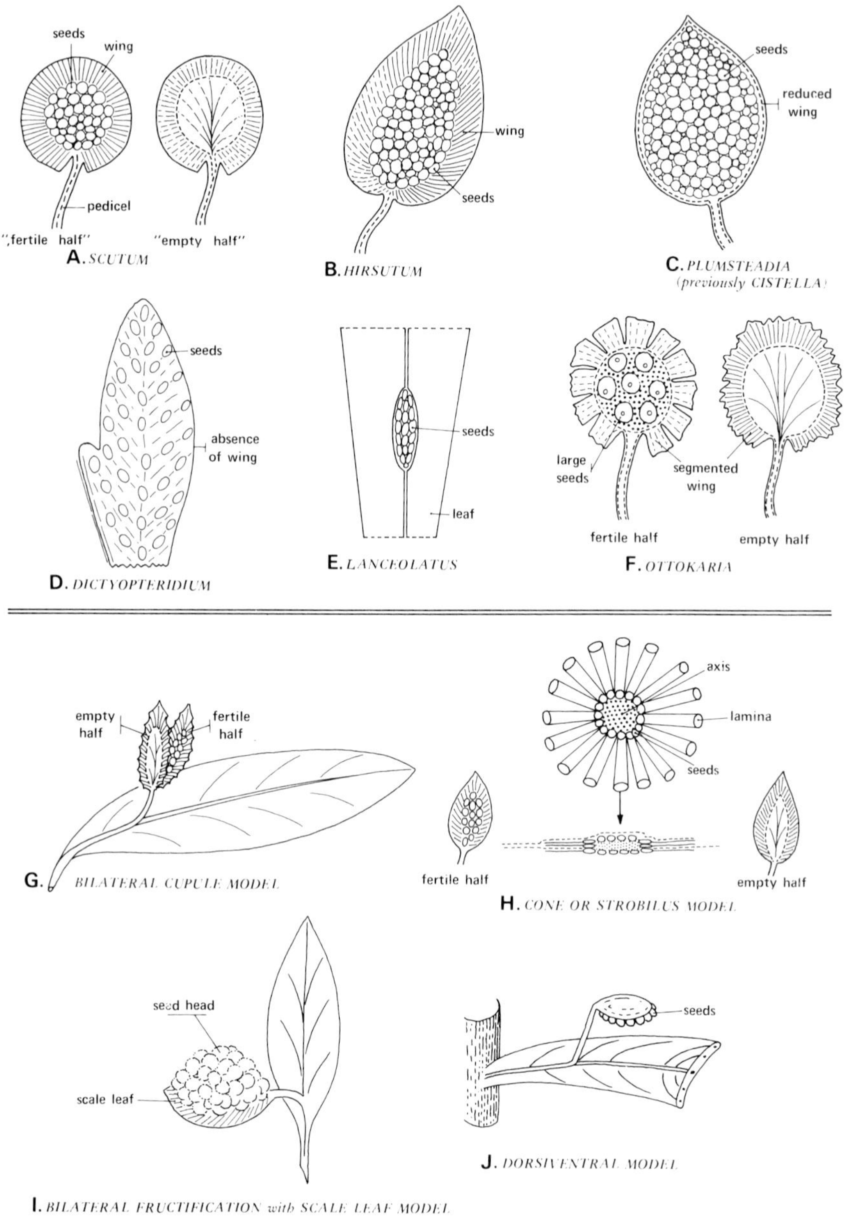

Fig. 2.4. A–F. Diagrammatic representations of the ovulate glossopterid fructifications. A. *Scutum* showing the distinct head and wing structure but with different structures within the two halves of the fossil. B. *Hirsutum* contains numerous but smaller seeds than in *Scutum*, the wing has a slightly hairy appearance. C. *Plumsteadia* contains many seeds but has a very reduced wing region. D. *Dictyopteridium* lacks a wing region and is composed of numerous spirally arranged seeds. E. *Lanceolatus* the fructification is borne

symmetrical in structure, with each seed attached to a 'projecting lamina which may have been tubular'. On compression these lamina would produce the 'wing' structure of *Scutum*. At different levels of fracturing through the compressed strobilus structures such as the 'veined' and 'empty' half of Plumstead's cupule would be produced.

Support for Walton's cone-hypothesis has come mainly from the Indian workers. Surange and Chandra (1973) in their discussion of *Dictyopteridium* reconstructed the fructification as a cone composed of seeds arranged in spirals around an axis. They concluded that other ovulate fructifications such as *Scutum* and *Cistella* had the same type of organisation but represented axis from which the seeds had been shed.

(c) Bilateral fructification with a scale leaf model (Fig. 2.4I): this model was proposed by Surange and Chandra (1974) as a result mainly of cuticle studies on the fructifications. Previously they had believed that the two halves of the *Scutum* fossils were identical. Further work though had demonstrated that this was not the case and they reconstructed *Scutum* as having a seed-bearing head and a scale leaf, both being borne on the same pedicel. During the fossilization they postulated that this scale leaf would have 'fallen over' the seed-head completely covering it. The wing structure in *Scutum* would have been produced as the result of the lateral compression of the seeds borne at the margins of the seed-head.

(d) The dorsiventral model (Fig. 2.4J): this model was proposed by Schopf (1976) and subsequently expanded by Rigby (1978). Schopf (1976) believed that all the ovulate fructifications of the glossopterids had a dorsiventeral structure in which the upper surface was striated or rugose, and the lower surface bore ovules. Rigby (1978) offered a similar explanation. He considered that the different structures present on the two halves of these fossils is a result of the fracture plane running through the centre of a dorsiventral structure. He postulated that during the fossilization of *Plumsteadia* the fructification was compressed onto the leaf bract so that they appeared fused. *Lanceolatus*, believed by Plumstead (1952) to have grown out of the leaf, he explains as having been 'squeezed' through the leaf during the compression process.

---

directly on the leaf. F. *Ottokaria* the wing region is composed of several free segments and the head region contains large but few seeds, the part and counterpart contain different structures. G–J. Proposed models for the uncompressed structure of *Scutum*. G. Bilateral cupule model (after Plumstead 1952), the two halves of the cupule are borne on a pedicel which develops from the midrib of the leaf. H. Cone or strobilus model (after Walton, in Plumstead 1952), the fructification consists of numerous lamina, each with a seed at its base, borne around an axis. On compression and fracturing the *Scutum* fossil is formed. I. Bilateral fructification with scale leaf model (after Surange and Chandra 1974), composed of a seed-bearing head borne on the same pedicel as a scale leaf. J. Dorsiventral model (after Schopf 1976) in which the fructification is borne on a pedicel, seeds are borne on one surface only.

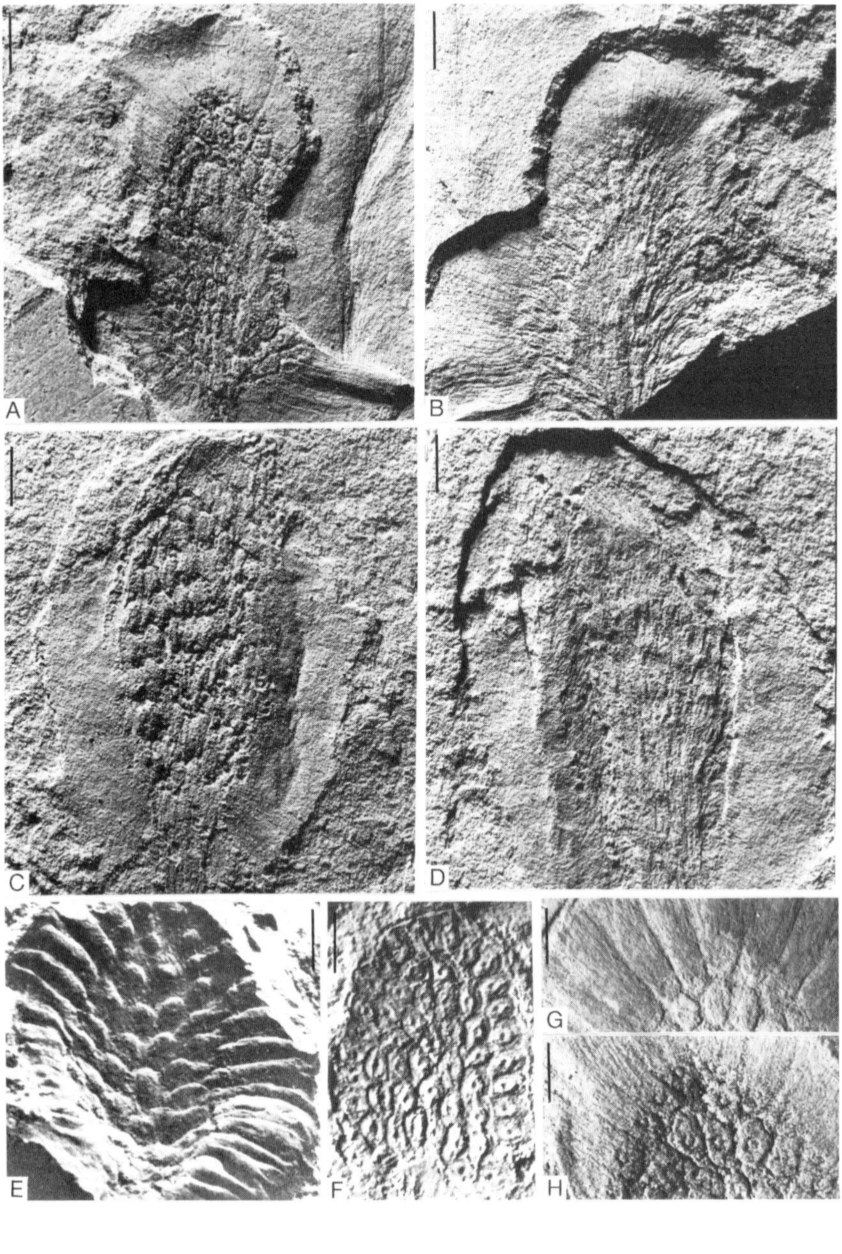
A
B
C
D
E
F
G
H

Alongside all these proposed models for the uncompressed *Scutum* fossils is the permineralized fructification described by Gould and Delevoryas (1977) found associated with *Glossopteris*. The seed bearing structure was completely different from any of the structures described earlier. It consisted of a *Glossopteris* leaf with infolded lateral margins which enclosed seeds. In some of the specimens the megasporophyll was opened out and this they argued could have been a possible dispersal stage.

Evidence from specimens

A series of impression fossils were examined of *Scutum* and *Lanceolatus* which had been deposited in the British Museum, Natural History by Plumstead. In a few cases both the part and the counterpart were available for examination. These clearly demonstrate that the part and counterpart contained different structures, the wing regions were opposite but conformable in contrast to the head regions (Fig. 2.5A,B). On one half the fossil contained positive seed-like structures whereas these were completely absent even as impressions in the counterpart (Fig. 2.5C,D). The evidence, though, from the literature is not consistent on the nature or differences in the two halves of these impression fossils, frequently because the two halves of the fossil have not been illustrated or because of poor photographic quality. Specimens illustrated by Plumstead (1952, 1956) seem to show, in some cases, that the two halves of the fossil are different. Many of the specimens illustrated from India do not show part and counterpart and those that are illustrated show similar structures (Chandra and Surange 1977, Plates 1 and 2). These Indian specimens do show some features very different to those from

---

Fig. 2.5. A. *Scutum* sp (v. 31161): one half of the fructification containing positive seed casts surrounded by a striated wing. B. *Scutum* sp (v. 31161): counterpart of A in which it can be seen that the head region contains a different structure, it has a veined appearance. The wing region is comformable with that in the part (in A and B the scale bar = 3.5mm). C. *Scutum stowanum* (v. 31210): one half of the fructification containing the seed casts bounded by a wing. D. *Scutum stowanum* (v. 31210): counterpart of C showing the striated structure of the head region, again the structures differ in the two halves of the fossil (in C & D the scale bar = 3mm). E. *Scutum indicum* (from Surange and Chandra 1974): the head region of this fossil seems to be composed of distorted seeds similar to those in the head region (scale bar = 5mm). F. *Plumsteadia* (from Plumstead 1958): the wing region seems to be composed of seeds showing the same structure as those in the head region (scale bar = 5mm). G. *Scutum rubridgeum* (v. 31205): The wing region has a segmented structure, each segment has a rounded base similar to the seeds in the head region (scale bar = 3mm). H. *Scutum* sp (v. 31161): the base of the wing segments contains seed casts similar in structure to those in the head region (scale bar = 1.5mm).

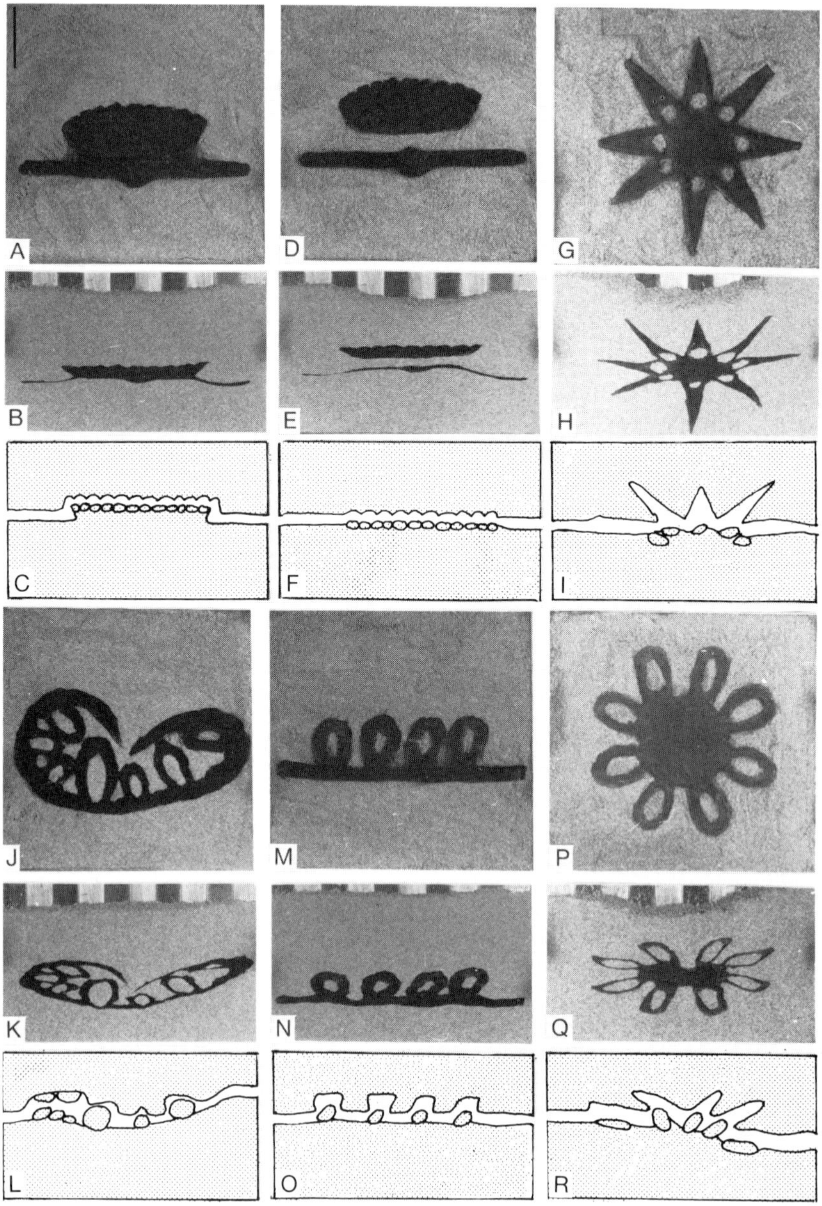

Fig. 2.6. Experimental modelling of the uncompressed structure of the ovulate glossopterid fructifications (all at same magnification, scale bar = 44mm). A–F. Experimental modelling of the dorsiventral model (after Rigby 1978). A. The fructification (foam rubber) is positioned over the leaf bract with no matrix between. B. After compression the two organs appear close together but the fructification is not 'squeezed' through the leaf bract. C. Loss of coaly matter and pathway the fracture plane has to take to reveal a fossil with seed casts on one surface; both halves of the fossil

South Africa. The fructifications have considerably more topography to them and the wing region has the distinct structure of elongated seeds (see Surange and Chandra 1974, Plate 3,4) giving support to the seed-head structure of *Scutum* proposed by Surange and Chandra (1974).

One of Plumstead's (1952) main arguments for the two-valved structure of *Scutum* was that the 'head' and the 'wing' region were distinct, separate structures. Examination of *Scutum* specimens (v.31205, v.31195) showed that the wing often had a segmented structure (Fig. 2.5G). Each segment having a rounded base which merged into the seeds in the 'head' region (Fig. 2.5G,H). Even in some of Plumstead's (1952) illustrations the wing can be clearly seen to contain elongated 'seed' fills the same as those present in the 'head' region (Fig. 2.5F). Frequently this seems to be the case with many of the Indian specimens (Fig. 2.5E).

If indeed the seeds were arranged in a strobilus structure as proposed by Walton (see discussion in Plumstead 1952) and Surange and Chandra (1973, 74) then there should be some evidence of distinct arrangement of the seeds in the impression fossils. Again the evidence is not clear on this. Some specimens that have been illustrated do show some evidence of spirality (Rigby 1978, Surange and Chandra 1973, 74) but this is often not the case and the seeds may have no distinct or consistent arrangement.

Evidence from experimental modelling:

(a) Plumstead's model: this is difficult to model since she never proposed an actual reconstruction of the fossil. The structure of the uncompressed fructification she proposed showed no difference from the fossil.

---

contain the same but opposite structures. D. The fructification is positioned above the leaf bract with sediment between. E. After compression the fructification remains separate from the leaf. F. Loss of coaly matter and pathway the fracture plane has to take to reveal a fossil with seed casts on one surface; again the two halves contain opposite but similar structures. G. Model of Walton's strobilus model, each lamina has a sediment filled seed at its base. H. After compression the marginal sporophylls form a wing region and those on the upper and lower surface foreshortern. I. Loss of coaly matter and subsequent fracturing produces a fossil with seed casts on one surface and a veined structure on the other. J. Model of the infolded megasporophyll described by Gould and Delevoryas (1977). The seeds and fructification are filled with sediment. K. After compression the seeds have distorted but no obvious wing structure has formed. L. Loss of coaly matter and one of the possible fracture planes through the impression, a *Scutum* fossil is not produced. M. Model of the unfolded megasporophyll with the seeds filled with sediment. N. After compression a wing structure is produced bearing positive seed casts. O. Subsequent loss of coaly matter and fracturing to give a fossil with positive seed casts produces a part and counterpart containing similar but opposite structures. P. Model of a strobilus structure bearing a number of naked seeds borne around an axis (from Surange and Chandra 1974). Q. On compression the marginal seeds distort to give a wing structure. R. Loss of coaly matter and subsequent fracturing reveals a fossil with similar structures to some of the Indian species of *Scutum*.

(b) Strobilus model: in modelling this form of the fructification the seeds, borne around the axis, were filled with the same sediment the model was incorporated in. This seems to be the case with the actual specimens since the seeds are always in the form of casts. In Walton's proposed model the seeds were attached to laminar appendages and these have been included (Fig. 2.6G). On compression the lamina have foreshortened on the upper and lower surfaces of the fructification (Fig. 2.6H). At the margins of the compression the lamina have been compressed to very thin structures. Subsequent loss of coaly matter and fracturing would give a part containing a wing structure with a central region containing positive seed fills (Fig. 2.6I). The counterpart would contain a conformable wing but also contain the impression of the compressed lamina. Therefore the part and the counterpart would contain different structures. This model therefore seems to produce an impression fossil that closely resembles *Scutum*.

A second model on the same theme as that above but following Surange and Chandra (1974) is of a fructification composed of a number of naked seeds, each filled with sediment, borne around an axis (Fig. 2.6P). Subsequent compression results in the formation of a fossil with a wing-like structure formed from the compression of the marginal seeds and a central region containing seed casts (Fig. 2.6Q). Loss of coaly matter, though, produces a fossil containing the same but topographically opposite structures (Fig. 2.6R). This, though, does seem to fit the evidence of some of the Indian and Australian specimens. The scale leaf has been ommitted from the model but it can be envisaged that if present and no sediment settled between the head and leaf then perhaps the resulting fossil would fit the evidence from the cuticle pulls. In many of the specimens illustrated it seems unnecessary to invoke a scale leaf structure.

(c) Dorsiventral model: the model of this hypothesis consisted simply of a dorsiventral structure lying under or above a leaf. The proposed model (Rigby 1978) seems to accept that no sediment would have settled between these two structures. This is hard to believe since sediment has filled the seeds and is highly likely to have, in some cases at least, settled between the two organs. The presence of incompressible sediment would have stopped the fusing of the two structures (Fig. 2.6C,D). Even if sediment was absent it is not possible to produce a fossil with different structures in the part and counterpart as the fracture plane would have to pass over the positive seed fills. This would result in a part and counterpart containing similar but opposite structures (Fig. 2.6C–F).

(d) Infolded megasporophyll: two models were made in this case; one of the infolded megasporophyll and one of it opened out. It is very difficult to produce a *Scutum* structure by compression of the unfolded megasporophyll, loss of coaly matter, and subsequent fracturing. There are many possible fracture pathways through the infolded form (Fig. 2.6J,K). A wing structure was not produced (Fig. 2.6L). The unrolled form of the permineralization produces a more *Scutum*-like structure on compression but here again loss of coaly matter and subsequent fracturing would result in an impression composed of two similar, but oppposite, halves (Fig. 2.6M,N,O). A wing region would have formed but this also would have the appearance of a *Glossopteris* leaf.

Discussion of the uncompressed structure of 'Scutum'

Examination of *Scutum* fossils from specimens and the literature has shown that this group of fructifications has no consistency in the structure of the impressions. The evidence is conflicting for in some examples the part and counterpart contain different structures, as illustrated (Fig. 2.5A–D), whereas in others they are remarkably similar as in the Indian material illustrated by Surange and Chandra (1977 *et seq*). The arrangement of the seeds is highly variable being either random or showing faint spirality. In all cases though, the seeds have been filled with the same fine sediment that the fructification was buried in giving positive seed casts on one half of the fossil. The wing region, when present is also highly variable in character and may be segmented, with seeds borne at the base of the segments, structureless, or composed of laterally compressed seeds.

Experimental modelling of the proposed forms of these fructifications has demonstrated that it is not simple, after fracturing and loss of organic material to produce an impression fossil which corresponds to the range of evidence shown by the fossils. Gould and Delevoryas (1977) indicated that the infolded megasporophyll envelope or its unfolded form corresponded to the described impression genera. Modelling has shown that this is not the case and it would seem on the basis of the present features that the permineralized and compressed genera represent different taxons.

Of all the models proposed perhaps that of Walton's seed-bearing strobilus most readily fits many of the impression fossils. Some material though, particularly from India, seems to have features compatible with the compression of a naked seed head. Modelling of the dorsi-ventral form did not produce the features seen in these fossils and seems the least likely uncompressed structure for these fructifications. In conclusion, it appears that all the genera and species assigned to this group of fossils cannot be explained in terms of a single uncompressed structure and that several forms need to be invoked to fit the range of *Scutum*-type impression fossils.

Conclusions

The formation of plant compression fossils has been investigated using experimental techniques. This has been combined with the examination of fossils in transverse section. This is particularly useful in that it provides a comparable surface to that observed in the experimental work and also exposes the entire structure of the compression. The results of this work have been discussed with reference to three compression/impression fossils. *Cyperites bicarinatus*, a lycopod leaf compression, is shown to have a complex 'butterfly' transverse section. Structures on the exposed leaf surface such as veins and grooves are shown to be simply phenomena produced by the pathway of the fracture plane through the compression and are unrelated to these features in the original leaf. The compression is related to the uncompressed leaf by experimental modelling. The pteridosperm leaf compression *Alethopteris serli* demonstrates several factors important in the fossilization process. Firstly, the effect of rigid tissues e.g. xylem tracheids on the form of the compression. Secondly, how the 'way up' of the original leaf in the sediment effects the form of the fossil and finally the recognition of compression phenomena in the fossil which are unrelated to the uncompressed leaf. The fossilization of the leaf is discussed. A test case for this type of study on the compression process of fossilization is provided by the ovulate glossopterid fructifications preserved as impressions. The original uncompressed form of these fossils has been a highly controversial issue since they were first described. Many uncompressed structures have been proposed and these are discussed after experimental modelling. Discovery of permineralised fructifications believed to have been borne by *Glossopteris* showed a fossil with a radically different form. It is demonstrated that it is not obvious how the impression and permineralised fossils are related or how to relate this group of fossils to a single uncompressed structure.

Acknowledgements

This work was undertaken in the Department of Botany, Bedford College, University of London during the tenure of a NERC studentship which is gratefully acknowledged. I should like to thank Professor W.G. Chaloner for all his help and encouragement during the course of the studentship. I should also like to thank Dr J. Galtier (Montpellier) for his help in the preparation of this manuscript and for reading an earlier draft. Jill Courbet and Jacques Martin provided technical assistance and I thank them both. I also thank the Keeper of Palaeontology, British Museum (Natural History) for allowing me to study material in the collections.

References

Baxter, R.W. and Willhite, M.R. (1969) The morphology and anatomy of *Alethopteris serli* Wagner. *Univ. Kans. Sci. Bull.* **48**, 769–83.

Chaloner, W.G. and Boureau, J.E. (1967) '*Lycophyta*'. In Boureau J.E. (Editor). Traité de Paléobotanique, 2. Masson, Paris.

Crookall, R. (1966) Fossil plants of the Carboniferous rocks of Great Britain. *Mem. Geol. Surv. G.B. Palaeontol.* **4**, 217–354.

Cridland, A.A. (1968) *Alethopteris ambigua* Lesquereux, a Pennyslvanian pteridosperm from Missouri, USA. *J. Linn. Soc. (Bot)* **61**, 107–11.

Chandra, S. and Surange K.R. (1976) Cuticular studies on the reproductive organs of *Glossopteris.* Part 1—*Dictyopteridium* Feistmantel sp.nov. attached on *Glossopteris tenuinervis. Palaeontographica, Abt B.* **156**, 87–102.

—— (1977) Cuticular studies of the reproductive organs of *Glossopteris*. Part 2—*Cistella* type fructification *Plumsteadiaostrobus ellipticus* Gen.et. sp.nov. attached on *Glossopteris taenoides* Feistmantel. *Palaeontographica*, Abt B. **23**, 161–76.

Gould, R.E. and Delevoryas, T. (1977) The biology of *Glossopteris*: evidence from petrified seed-bearing and pollen-bearing organs. *Alcheringa* **1**, 387–99.

Graham, R. (1935) An anatomical study of the leaves of the Carboniferous arborescent lycopods. *Ann. Bot.* **49**, 587–608.

Kosanke, R.M. (1979) A long-leaved specimen of *Lepidodendron. Geol. Soc. Am. Bull.* **90**, 431–35.

Leisman, G.A. (1960) The morphology and anatomy of *Callipteridium sullivanti. Am. J. Bot.* **47**, 281–87.

Lemoigne, Y. (1961) Etudes analytiques et comparées des structures internes des Sigillaries. *Ann. Sci. Nat. Bot.* **12**, 473–578.

Lindley, J. and Hutton, W. (1833) *The fossil flora of Great Britain.* Vol 2.

Mickle, J.E. and Rothwell, G.W. (1982) Permineralised *Alethopteris* from the Upper Pennyslvanian of Ohio and Illinois. *J. Paleontol.* **56**, 281–328.

Plumstead, E.P. (1952) Description of two new genera and six species of fructifications borne on *Glossopteris* leaves. *Trans. geol. Soc. S. Afr.* **55**, 281–328.

—— (1956) Bisexual fructifications borne on *Glossopteris* leaves from South Africa. *Palaeontographica*, Abt B, **100**, 1–25.

—— (1958) Further fructifications of the Glossopteridae and a provisional classification based upon them. *Trans. geol. Soc. S. Afr.* **61**, 51–76.

Reed, F. (1941) Coal flora studies: Lepidodendrales. *Bot. Gaz.* **102**, 663–83.

Rex, G.M. (1983a) The compression state of preservation of Carboniferous lepidodendrid leaves. *Rev. Palaeobot. Palynol.* **39**, 65–85.

—— (1983b) *The formation of plant compression fossils: experimental and sedimentological investigations.* Unpublised PhD thesis, University of London.

—— (1985) A laboratory flume tank investigation of the formation of fossil stem fills. *Sedimentology*, **32**, 245–55.

—— (1986) Further experimental investigations on the formation of plant compression fossils. *Lethaia*, **19**, 143–159.

—— and Chaloner, W.G. (1983) The experimental formation of plant compression fossils. *Palaeontology*, **26**, 231–52.

Rigby, J.F. (1978) Permian Glossopterid and other Cycadopsid fructifications from Queensland. *Geol, Surv. Qd. Palaeont. Pap.* **41**, 3–21.

Schopf, J.M. (1970) A petrified peat from a Permian Coal bed in Antartica. *Science*, **169**, 274–77.

—— (1976) Morphologic interpretations of fertile structures in Glossopterid gymnosperms. *Rev. Palaeobot. Palynol.* **21**, 25–64.

Snigirevskaya, N.S. (1958) Anatomical investigations of fossil leaves (phylloids) of certain lycopsida in coal balls of the Donetz Basin coalfields. *Bot. Zh. Akad. Nauk, U.S.S.R.* **43**, 106–12.

Stewart, W.N. (1983) *Palaeobotany and the evolution of plants.* Cambridge University Press, New York.

Surange, K.R. and Chandra, S. (1973) *Dictyopteridium sporiferum* Feistmantel- a fertile cone from the Lower Gondwana of India. *Palaeobotanist*, **20**, 127–36.

—— (1974) Fructifications of Glossopteridae from India. *Palaeobotanist*, **21**, 1–18.

Walton, J. (1936) On the factors that influence the external form of fossil plants: with descriptions of the foliage of some species of the Palaeozoic Equisitalean genus *Annularia* Sternberg. *Phil. Trans. R. Soc.* **226**, 219–37.

3. The biochemical analysis of fossil plants and its use in taxonomy and systematics

B.A. THOMAS

Abstract

Palaeobiochemistry offers a means of obtaining further characters for the taxonomic and systematic appraisal of plant fossils. Many types of organic residues can be recovered from plant fossils but some are of more taxonomic use than others. A review is given of the major groups of taxonomically effective compounds together with examples of their effectiveness.

Introduction

Much of the early work on palaeobiochemistry centred on the analyses of dispersed organic molecules rather than on fossil plants. This interest was primarily stimulated by oil exploration, although it has never really helped the industry in any major commercial way. A parallel line of research centred on the analysis of coal for identification and stratigraphy and for methods of helping convert it into alternative petroleum products (e.g. Bouska 1981). Through work of this nature a great number of organic compounds have been shown to be widely dispersed through sedimentary rocks of all ages throughout the world (Mackenzie *et al.* 1982). Pigments such as haemin and chlorophyll (Fox and Anderson 1941) and amino acids and sugars (Degens, Reuter, and Shaw 1964) have been used as off-shore indicators. More recently the distributions of aromatic aldehydes and polycyclic aromatics (Clayton and Swetland 1978; Laflamme and Hites 1978; Lam and Pedersen 1978) and terpenoids and steroids (Brassell, Eglinton, and Maxwell 1983) have been investigated from a wide range of sedimentary rocks.

Later interest in analysing plant fossils has followed rather uneasily in the footsteps of both organic geochemistry and biochemical systematics.

This interest in biochemical systematics was itself prompted by the development of rapid screening techniques and the desire to move ever closer to what can be regarded as a natural classification of plants. The amount of biochemical information that is available is determined by the type of fossilization process that resulted in the fossil remains and the potential is greatest in compressions which resulted from compaction in anaerobic environments where the amount of microbial destruction of the plant tissues was very limited. In such plant fossils the biochemical remains undergo certain complicated structural changes involving dehydration, hydrogenation, aromatization, isomerization at chiral centres, molecular rearrangement, and ring opening. By these processes the molecules are gradually changed towards their most thermodynamically stable saturated and aromatic hydrocarbon counterparts (Brassel, Eglinton, and Maxwell 1983).

Palaeobiochemical analysis of such plant fossils can yield valuable extra information for both taxonomic and systematic purposes. However, Harborne and Turner (1984) in their textbook on *Plant Chemosystematics* devote only eight out of their 487 pages of text to the subject of palaeobiochemistry and make no reference to plant fossils elsewhere. Unfortunately this seems to reflect both a lack of interest in fossils by most workers in plant chemosystematics and limited appreciation of the value of plant fossil chemosystematics by palaeobotanists.

Chemosystematics of fossil plants has to reply on a narrower range of organic residues than is possible for the study of living plants. Most volatile secondary metabolites, defence substances and coloration pigments are usually lost or degraded before or during fossilization. There are well known exceptions to this of course for chlorophyll derivatives have been recovered from Middle Eocene and Miocene leaves (Dilcher, 1967; Dilcher, Pavlick, and Mitchel 1970, Niklas and Brown 1981). Although these residues are interesting in that they demonstrate how complete preservation can sometimes be, they have been of no use for taxonomy. Similarly the study of isolated fossil pigments in postglacial lake sediments has been for ecological rather than taxonomical benefits (Sanger and Crowl 1979).

Long chain hydrocarbons also can persist almost unaltered in fossil plants. Palaeozoic plant fossils analysed by Swain, Bratt and Kirkwood (1967, 1968) showed that there were differences in galactose/glucose ratios between plants. They suggested that the changes in ratios followed our ideas of plant evolution as the more 'advanced' groups contained relatively more glucose.

Niklas, with co-workers, examined a range of Palaeozoic plant groups which indicated the presence of a very wide range of organic residues (Niklas and Gensel 1976, 1977, 1978 and references therein). Through the use of principle component analysis, plant fossils were grouped into what were called 'chemotaxonomic associations' with the clustering pattern of 27 palaeozoic plants indicating a similar grouping to that based on morphological characters. They then used a system of weighting the data by giving priority to aromatic acids and eliminating consideration of compounds thought to be mobile in rock strata (e.g. carbohydrates, amino acids, short chain hydrocarbons and fatty acids). These data again generally supported pre-existing ideas of classifications although there were some discrepancies (e.g. the cladoxylalean *Pseudosporochnus* and the lycophyte *Leclercqia* had similar ordinations as did the ginkgophyte *Solenites* and the progymnosperm *Archaeopteris*). Such discrepancies clearly highlight the dangers of using such ordination techniques to help classify any unknown plant fossils.

This sort of blanket analysis of every recognizable organic residue is therefore probably not the best way to proceed. It seems to be not very reliable at the broad generic level when results are from only one species and possibly only one sample. This approach is not being used in the chemotaxonomy of living plants and would be better abandoned in future work on fossils. To achieve results of any lasting value, the approach needs to be narrower and either confined to one group of compounds or concentrated on one group of plants. Repetition and repeatability are crucial as, clearly, is the identification of the plant fossil.

Methods

Harborne in his book on *Phytochemical Methods* (1973) states: 'Ideally, fresh plant tissues should be used for phytochemical analysis and the material should be plunged into boiling alcohol within minutes of its collection'. Luckily for the palaeobotanist this is not essential as other workers on living material have found that in most instances dried herbarium material can be used to great effect.

Accurate identification of the plants to be studied, whether living, herbarium or fossil, is of necessity crucial to biochemical analysis. Uncontaminated and undiseased materials are similarly vital if the results are to be beyond dispute. To this end fossils should preferably be newly collected, unweathered, untouched, and transported in a 'safe' wrapper such as kitchen foil. Paper and plastic should be avoided at all cost. In the laboratory samples should be similarly treated with care and stored in thoroughly clean, glass containers. Fossil plants can contain only very small amounts of recoverable organic compounds so that even the smallest amount of contaminant will affect the results.

Methods of extraction and separation of the organic residues of fossils are essentially those employed for living materials. In many instances however, it is possible only to use those techniques that can cope with the very small amounts of residues that can be recovered from fossils. When there are reasonable quantitaties of recoverable organics, paper chromatography or thin layer chromatography can be used. In the majority of instances it is preferable or even necessary to use gas liquid chromatography. Ultraviolet, visible and infra-red spectroscopy, mass spectroscopy and nuclear resonance spectroscopy are also methods of identification. The techniques for these methods can be gained from many published sources so I will not dwell on them here.

Results

Some of the major types of organic residues that are preserved in fossils or dispersed in sediments are of much more taxonomic value than others. This may be for one of several reasons. Some residues may be of great value in living plant studies while others are of too broad a distribution to be of use. Certain biochemicals are stable for long periods of time while others break down into constituent parts which are taxonomically of no significance.

Techniques are also important in determining which compounds are used for some are much easier to isolate and identify. I have therefore attempted to survey the major groups of biochemical compounds that have been successfully used in palaeobiochemical taxonomy. In doing this I have excluded a great deal of palaeobiochemical information that is not of direct relevance to taxonomy. Because of the interrelated nature of the different aspects of this work much of the other palaeobiochemical approaches can be reached through the references quoted here.

Flavonoids

Flavonoids are structurally diverse and stable metabolites that are widespread in higher plants, but absent from almost all fungi and algae. The biochemical complexity of flavonoids paralles the evolution of the land plants, although many of the more highly advanced angiosperms have simple flavonoids (Harborne 1978).

Lycophytes, sphenophytes and ferns have different types of flavonoids (Cooper-Driver 1980) but as yet there has been no attempt to survey the fossil pteridophytes for compatibility.

Angiosperms have been widely surveyed for flavonoids. They are very abundant in both the leaves and flowers of angiosperms (Harborne, Mabry and Mabry 1975, Harborne and Mabry 1982). Of these the colourless flavonoids are more likely to remain in fossils than the water soluble yellow flavonoids.

Flavonoids have been recovered from a number of angiosperm fossils. In a series of papers, flavonoids and steroids have been described from several genera of Miocene angiosperms (Niklas and Giannasi 1978, Giannasi and Niklas 1981, Niklas and Brown 1983, and included references therein). Giannasi (1978) has also looked at the distribution of flavonol aglycones, glycoflavones and leucoanthocyanidins of 80 species in 19 genera of the Ulmaceae. This survey showed that each living genus is characterized by the production of flavonoid. There is also a dichotomy in the results which is very compatible with the generally accepted division of the family into the sub-families Ulmoideae and Celtidoideae.

The Succor Creek flora from the Miocene of Oregon consists of some outstandingly well preserved leaf compressions. The leaves were buried in very fine sediments before any microbial activity had a chance to break down any of their tissues. This resulted in the preservation of internal anatomy and in many instances even colour. Three types of leaves referable on their morphologies to extant genera in the Ulmaceae have been shown to have similar flavonol contents to their living relatives. Fossil and living *Celtis* yielded only glycoflavones, while those referred to *Ulmus* and *Zelkova* gave only flavones. Other leaves assigned to *Acer* and *Quercus* have also been shown to have similar flavonoids to those of living species of these genera. Niklas and Giannasi (1978) have also suggested that the chemical evidence of three of their genera of Miocene fossil plants shows them to have east Asian affinities.

A great many morphological and biochemical intermediates must have existed and it is these which will help us to understand the relationships between genera. One such example comes from the Miocene Clarkia Lake deposits of northern Idaho: *Pseudofagus idahoensis* which was thought to belong to the Fagaceae on morphological features (Smiley and Huggins 1981). Analysis, based on stereochemistry, suggests that the plant does indeed belong to the Fagaceae, although the presence of an isorhamnetin (3-methyloxyquercetin), 3-Oglycoside, onocerane and 5-cholestane separates it from extant species of *Fagus* (Giannsi and Niklas 1981).

Flavonoids clearly seem to have immense value in biochemical taxonomy, but it remains to be seen if flavonoid analysis of fossils can be generally applied to their study. It is not really logical to assume that flavonoids must have remained unaltered while the plants were themselves evolving. Crawford (1978) has stressed that studies are needed of the genetics and enzymology of flavonoid compounds in different groups of living plants if flavonoids are to be used in a phylogenetic or evolutionary sense. Clearly a great many fossils need to be analysed before any firm deductions can be made about their taxonomic stability or any understanding can be achieved of flavonoid evolutionary change through time.

The best approach for this will probably be to follow Crawford and

Levy (1978). Instead of assuming that all flavonoids are equally complex and systematically unrelated, they looked for and evaluated the similarities within members of a family of biochemicals. This appears to be a more refined approach in that it allows a comparison of the relative complexities of both shared and disparate compounds in flavonoid profiles.

Fatty acids

Fatty acids occur mainly in esterified forms of fats or lipids. They are especially important in providing an important source of storage energy in seeds. The degree of unsaturation of the fatty acids determines the type of oil that can be recovered from the seed. For this reason most surveys have been carried out for economic purposes rather than for taxonomic ones. Nevertheless there is an expanding amount of information available for this purpose.

From this work it is apparent that in many cases a particular fatty acid pattern is characteristic of a family. Families can also be grouped on the basis of their fatty acids, but it is crucial to realise that this grouping is in no way related to their systematic positioning on other more conventional grounds. A few surveys of families have been very useful for taxonomic purposes, although the most successful have been those limited to the smaller plant groups. It seems it is at this level that future work would best be concentrated (Harborne and Turner 1984).

There are examples of fossil seeds still retaining identifiable fatty acids. Priestly, Galinat, and Leopold (1981) have reported on oleic and linolic acids in 1500 year old *Zea mays*, which agrees with the contents of modern seeds. Niklas, Tiffney, and Leopold (1982) have also recovered the same unsaturated fatty acids, several types of seeds, and fruits from the Eocene and Oligocene excellently preserved in the Geiselthal Flora of the DDR and the Brandon Lignite of Vermont, USA.

There is scope for further exploration work in analysing the fatty acid contents of fossil seeds and possibly even of other organs. Gemmrich (1977) has demonstrated that the lipid content of fern spores is constant within the genus and can be as high as 79 per cent in *Polypodium*. It would be very interesting to see if fossil fructifications or spore masses can give any worthwhile results.

There are a couple of problems however that urge caution in interpreting any results. In most cases where both leaves and seeds from the same plant have been analysed the fatty acid patterns have been found to be different. There is also the acute problem of variability within seeds of the same species. For this reason Stone *et al.* (1969) have suggested that up to 30 seeds need to be analysed to get any reliable average.

Terpenoids

A wide variety of terpenoids can be recovered from sediments with some surviving in their original form (Brassel, Eglinton, and Maxwell 1983). Their origins are from terrestrial higher plants, unicellular algae such as diatoms and dinoflagellates, and bacteria. The most useful in terms of their information content are resins.

Resins are one of a number of protective materials that have probably evolved from other protective compounds in plant epidermises as the plants lived longer and became larger. Resins are widespread and can apparently occur in almost any group of higher plants, although they are most common in conifers and angiosperms (Harborne 1970). The main components of resins are usually diterpenes and triterpenes which can change from free flowing emulsions and oils to tough impenetrable membranes through vinyl polymerisation that can survive fossilization.

There are several studies of fossil resins that can give relevant biochemical evidence in our search for extra taxonomic and systematic characters. Resins are potentially of great value in helping to relate Tertiary conifer woods and shoots to modern taxa. There can be great problems with these shoots for modern names may have no real relevance at all. As plant organs almost certainly evolve at different rates, comparisons between fossil and living sterile shoots are fraught with difficulties. Mesozoic workers have tackled this problem by erecting entirely artificial form genera rather than attempting to relate them to families (Harris 1969, 1979). Many Tertiary workers still however prefer to use modern taxa.

The composition of Baltic amber and its source are still matters for debate. One argument is that it resulted from acid catalysed transformation of terpenes usually found in pine resins (Botta *et al.*, 1981). Mills, White, and Gough (1984) have come to a different conclusion. In analysing the ether-soluble fraction of Baltic amber they identified a pattern of monoterpenoids similar to that encountered in samples of New Zealand Tertiary fossil resins of supposed *Agathis* origin. Analysis of modern *Agathis* resin also showed that the polymerisation and ageing process of fossil resins are very similar to those undergone by modern Araucariaceae resins which contain large proportions of labdatriene compounds. It was unlike that of modern *Pinus* resins which have a high abietadiene content.

The major problem here is that the plant fossils associated with the Baltic amber are generally called *Pinus succinifer*. This could easily influence the thinking behind any comparisons. Larsson (1978) however prefers to call the conifer by its original name of *Pinites*. As Mills *et al.* (1984) have pointed out, this name removes the impulse always to look for comparisons with *Pinus*. Its use would also allow the possibility that it was a primitive conifer representing an early stage in the developmental history of the Pinaceae which still retained archaic chemical characteristics in common with the Araucariaceae.

Analysis of fossil assemblages from the lignite beds of the Oligocene Bovey Formation in Devon has been taken to show that they were ultimately derived from plants belonging to the Cupressaceae (Grantham and Douglas 1980). This is based on the presence of retinellite and the montan wax of sesquiterpenoids typical of modern genera of the Cupressaceae. It is therefore in conflict with the palaeobotanical evidence which suggests that the lignite is composed mainly of *Sequoia* with no presence of the Cupressaceae.

Both of these examples indicate that resin analyses may be able to provide additional evidence for relating Tertiary woods and shoots to modern taxa. Modern woods yield recognizable resins so a biochemical survey of fossil woods and shoots should be expected to give valuable results. If it can provide a useful extra character for Tertiary conifers then perhaps it could also help in our understanding of the relationships of the earlier Mesozoic taxa.

The fossilized remains of non-articulated laticifers have been recovered from the Eocene brown coal deposits of the Geiseltal, near Halle in the DDR (Mahlberg, Field, and Frye 1984). Analysis with 13 NMR showed their contents to consist of cis-1,4 configuration rubber, very like that of modern *Hevea* rubber. This indicates that the mechanism for rubber biosynthesis is an ancient one that evolved with the nonarticulated laticifers as a cell type. The cis-1,4 configuration occurs in the non-articulated laticifers of *Euphorbia* as well as the articulated ones of *Hevea*. It does not occur in living gutta percha (*Pelaquin*, Sapotaceae) where the *trans*-1,4 configuration may represent a derived condition.

GLC analysis of the rubber showed the presence of hydrocarbons, possibly including triterpenoids. Similar profiles are known from the latex of living euphorbiaceous and moraceous species, but all are different from that of the Eocene fossil. Mahlberg *et al.* (1983) have shown the biochemical profiles of *Euphorbia* species to be distinctive, so there is clearly an opportunity to follow this through to distinguish the fossil species that produce laticifers.

Phenolic acids and Lignins

Some studies have been made of the phenolic constituents of living plants. There is a correlation between the presence of phenolic acids that can be obtained in ethanolic extracts and the type of lignin constituents that can be deduced from oxidation products (e.g. Ibrahim, Towers, and Gibbs 1962, Towers and Maass 1964). There have been no detailed attempts to look at the phenolic contents of fossil plants although some work has been published on lignin contents.

Lignin is a polymeric natural product arising from enzyme initiated dehydrogenated polymerization of three natural precursors: *trans* coniferyl, *trans* sinapyl and *trans*-p-coumaryl alcohols (Sarkanen and Ludwig 1971).

The lignin macromolecule can be oxidized to aromatic fragments (phenolic aldehydes) which can be used to characterise the lignin. The ratios of acetovanillone, p-hydroxybenzaldehyde, syringic aldehyde and vanillin prepared by alkaline oxidation can be used to distinguish gymnosperms from angiosperms (Logan and Thomas 1985). The major oxidation product of gymnosperms is vanillin and that of angiosperms is syringic aldehyde. The probable reason for this difference is because the fibres have relatively high proportions of syringic aldehyde in their walls. It is nothing to do with the presence of vessels as high proportions of syringic aldehyde were obtained from the fibres of *Agave* leaves and coconut drupes. Conversely the very small amounts of syringic aldehyde may be due to its presence in the walls of fibre-tracheids. Comparable analyses have been made of fossil plants (Logan and Thomas, in press). Quaternary, Tertiary, and Cretaceous woods showed the same range of oxidation products and a similar distinction between gymnosperms and angiosperms.

The same oxidation products can be obtained from compressions. Analyses of Mesozoic gymnosperms shoots and leaves gave p-hydroxybenzaldehyde, vanillin, and acetovanillone. The absence of syringic aldehyde may be explained by its demythoxylation and by its preferential loss during oxidative preparation (Logan and Thomas, in press). Palaeozoic arborescent lycophytes also yielded variable amounts of the oxidation products from their outermost stem tissues. *Sigillaria* gave much larger quantities than *Lepidodendron* and *Lepidophloios*, which probably came from the inner layer of thick walled periderm that in this genus is shed with the thinner walled sub-epidermal cells.

Lignin residues from living pteridophytes are very diverse and appear to be related as much to anatomy as to their taxonomic relationships. Syringic aldehyde is much more widespread than originally thought with large quantities being recovered from *Dennstaedtia bipinnata, Equisetum fluviatile* and three species of *Selaginella*. Logan and Thomas (1985) suggested that the syringic aldehyde might be coming from the well developed stelar sclerotic sheath in *Dennstaedtia*, the lignified cortical cells in *Equisetum* and the hypodermal sclerenchyma cells in *Selaginella*. Unfortunately no syringic aldehyde was recovered from *Lycopodiella cernua* which has a distinctive sclerenchyma sheath around its central vascular bundle, so the two characters are not firmly linked.

Unfortunately the picture becomes even more confusing when the bryophytes are examined. Both gametophyte and sporophyte tissues give the same four oxidation products, although once again there are major difference between genera (Logan and Thomas 1985). *Polytrichum commune* (sporophyte and gametophyte) and *Plagiochila aspleniodes* (gametophyte) have only a trace of syringic aldehyde while *Conocephalum conicum* (gametophyte) has large amounts. Interestingly, *Conocephalum* has thick-walled fibres (Watson 1967).

Although it is not certain if the oxidation products recovered from bryophyte are lignin precursors or phenolic compounds of cell walls, their identification from these plants means that they are not the exclusive products of vascular plants. Their identification cannot alone prove that a plant fossil was vascular.

This therefore negates the biochemical extrapolation put forward by Niklas and Pratt (1980). They suggested that the extraction of vanillin and p-hydroxybenzaldehyde from early Silurian (Llandoverian) plant fossils showed these plants to have had the capacity to synthesise a lignin-like constituent which was a necessary event prior to the appearance of vascular plants. The available evidence might equally be taken to suggest that they were bryophytes or even precursors to both bryophytes and pteridophytes.

Amino Acids

Amino acids can be recovered from fossil wood but there have been no surveys of their compositions or concentrations in living woods. There is scope for work here but the major problem will be the change in amino acid enantiomeric ratios that occurs through recemisation. This increases with the age of wood and can therefore be used for estimating the ages of woods (Lee, Bada, and Peterson 1976).

Conclusions

There is an increasing amount of biochemical information on fossil plants that is valuable for systematic and also phylogenetic studies. However, as yet the available information gives only useful hints and our state of palaeobiochemical knowledge is comparable to that of compressions and coal ball studies many decades ago.

Results all seem to be generally confirming the systematics based on more traditional reproductive and morphological data. Harborne and Turner (1984) wrote 'We do not believe that biochemistry represents a panacea for all systematic problems. It is clear from many experiments . . . that the available biochemical characters can often do no more than provide supplementary data for phylogenetic consideration'. While I fully appreciate that it must appear to be even truer for plant fossils where there is less recoverable biochemical evidence, it must be remembered that palaeobotanists have to make the best of fragmentary evidence. In our studies of plant fossils every little piece of information helps.

I do however urge caution for, although data from the analyses of single species are very interesting, if ideas are extrapolated to the generic or even family levels it could be as much a disservice in the long run as a service in the short term. It would appear better to first look for biochemical agreement and differences at the species level. Comparisons of different organs would also be of immense value here as would sampling from a wide range of specimens from different sites to detect any changes through either microbiological activity prior to fossilization or later diagenesis.

References

Botta, M., De Angelis, F., Nicoletti, R. and Tricarico, M. (1982) The role of acid catalysis in the genesis of amber. *Phytochem.* **21**, 381–4.

Bouska, V. (1981) Coal Science and Technology. Vol. 1. *Geochemistry of Coal.* Elsevier Scientific Publishing Company, Amsterdam.

Brassell, S.C. Eglinton, G. and Maxwell, J.R. (1983) The geochemistry of terpenoids and steroids. *Trans. biochem. Soc.* **11**, 575–81.

Clayton, J.L. and Swetland, P.J. (1978) Subaerial weathering of sedimentary organic matter. *Geochim. Cosmochim. Acta* **42**, 305–12.

Cooper-Driver, G.A. (1980) The role of flavonoids and related compounds in fern systematics. *Bull, Torey Bot. Club.*

Crawford, D. (1978) Flavonoid chemistry and angiosperm evolution. *Bot. Rev.* **44**. 431–56.

——and Levy, M. (1978) Flavonoid profile affinities and genetic similarity. *System. Bot.* **3**, 369–73.

Degens, E.T., Reuter, J.H. and Shaw, K.N.F. (1964) Biochemical compounds in offshore California sediments and seawater. *Geochim. Cosmochim. Acta* **28**, 45–66.

Dilcher, D.L. (1967) Chlorophyll in der Braunkohle des Geiseltales. *Natur und Museum* **97**, 124–30.

—— Pavlick, R.J. and Mitchell, J. (1970) Chlorophyll derivatives in Middle Eocene sediments. *Science* **168**, 1447–9.

Fox, D.L. and Anderson, L.J. (1941) Pigments from marine muds. *Proc. nat. Acad. Sci.* **27**, 333–7.

Gemmrich, D.R. (1977) Fatty acid composition of fern spore lipids. *Phytochemistry* **16**, 1044–6.

Giannasi, D.E. (1978) Generic relationships in the Ulmaceae based on flavonoid chemistry. *Taxon* **27**, 331–44.

—— and Niklas, K.J. (1981) Comparative paleobiochemistry of some fossil and extant Fagaceae. *Amer. J. Bot.* **68**, 762–70.

Grantham, A. and Douglas, A.G. (1980) The nature and origin of sesquiterpenoids in some tertiary fossil resins. *Geochim. Cosmochim. Acta.* **44**, 1801–10.

Harborne, J.B. (1970) *Phytochemical Phylogeny*. Academic Press, New York.

—— (1973) *Phytochemical Methods*. Chapman and Hall, London.

—— (1978) The Evolution of flavonoids in the flowering plants. *Pure Appl. Chem.* **39**,

—— and Mabry, T.J. (eds.) (1982) *The Flavonoids, Advances in Research*. Chapman and Hall, London.

—— Mabry, T.J. and Mabry, H. (1975) *The Flavonoids*. Chapman and Hall, London.

—— and Turner, B.L. (1984) *Plant Chemosystematics*. Academic Press, London.

Harris, T.M. (1969) Naming a fossil conifer. In *J. Sen Memorial Volume*, (ed. H. Santarau *et al.*) Calcutta Bot. Soc. Bengal, Calcutta.

—— (1979) *The Yorkshire Jurassic flora V. Coniferales*. British Museum (Natural History) London.

Ibrahim, R.K., Towers, G.H.N. and Gibbs, R.D. (1962) Syringic and sinapic acids as indicators of differences between major groups of vascular plants. *Bot. J. Linn. Soc.* **58**, 223–36.

Laflamme, R.E. and Hites, R.A. (1978) The global distribution of polycyclic aromatic hydrocarbons in recent sediments. *Geochim. Cosmochim. Acta* **42**, 289–303.

Lam, J. and Pedersen, K.R. (1978) Organic compounds from the Cretaceous coals of Nugssuaq, West Greenland. *Gronl. Geol. Und. Bull.* **130**.

Larsson, S.G. (1978) *Baltic Amber—A Palaeobiological study*. Scandinavian Science Press, Klampenborg.

Lee, G., Bada, J.L. and Peterson, E. (1976) Amino acids in modern and fossil woods. *Nature* **259**, 183–6.

Logan, K.J. and Thomas, B.A. (1985) Distribution of lignin derivatives in plants. *New Phytol.* **99**, 571–85.

—— and —— (in press) Distribution of lignin derivatives in fossil plants. *New Phytol.*

Mahlberg, P.G., Field, D.W. and Frye, J.S. (1984) Fossil laticifers from Eocene brown coal deposits of the Geiseltal. *Amer. J. Bot.* **71**, 1192–200.

—— Pleszczynskya, J., Rauh, W. and Schnepf, E. (1983) Evolution of succulent *Euphorbia* as interpreted from latex composition. *Bothalia* **14**, 533–39.

Mills, J.S., White, R. and Gough, L.J. (1984) The chemical composition of Baltic amber. *Chem. Geol.* **47**, 15–39.

Niklas, K.J. and Brown, R.M. (1981) Ultrastructural and palaeobiochemical correlations among fossil leaf tissues from the St. Maries River (Clarkia) area, Northern Idaho, U.S.A. *Amer. J. Bot.* **68**, 332–41.

—— and Gensel, P.G. (1976) Chemotaxonomy of some Paleozoic vascular plants. Part I: Chemical composition and primary cluster analyses. *Brittonia* **28**, 353–78.

—— and —— (1977) Chemotaxonomy of some Paleozoic vascular plants. Part II: Chemical characterization of major plant groups. *Brittonia* **29**, 100–11.

—— and —— (1978) Chemotaxonomy of some Paleozoic vascular plants. Part III: Cluster configuration and their bearing on taxonomic relationships. *Brittonia* **30**, 216–32.

—— and Giannasi, D.E. (1978) Angiosperm paleobiochemistry of the Succor Creek flora (Miocene) Oregon, U.S.A. *Amer. J. Bot.* **65**, 943–52.

—— and Pratt, L.M. (1980) Evidence for Lignin-like Constituents in Early Silurian (Llandoverian) Plant Fossils. *Science* **209**, 396–7.

—— Tiffney, B.H. and Leopold, A.C. (1982) Preservation of unsaturated fatty acids in Palaeogene angiosperm fruits and seeds. *Nature* **296**, 63–4.

Priestly, D.A. Galinat, W.C. and Leopold, A.C. (1981) Preservation of polyunsaturated fatty acid in ancient Anasazi maize seed. *Nature* **292**, 146–8.

Sanger, J.E. and Crowl, G.H. (1979) Fossil pigments as a Guide to the Paleolimnology of Browns Lake, Ohio. *Quat. Res.* **11**, 342–52.

Sarkanen, K.V. and Ludwig, C.H. (1971) Definition and Nomenclature. In *Definition and Nomenclature in Lignins: Occurrence, Formation, Structure and Reactions* (ed. K.V. Sarkanen and C.H. Ludwig) pp. 1–18, Wiley-Interscience, New York.

Smiley, G.J. and Huggins, L.M. (1981) *Pseudofagus idahoensis* N. gen., et sp. (Fagaceae) from the Miocene Clarkia Flora of Idaho. *Amer. J. Bot.* **68**, 741–61.

Stone, D.E., Adrouny, G.A. and Flake, R.H. (1969) New World Juglandaceae, II. Hickory nut oil, phenotic similarities and evolutionary implications in the genus *Carya*. *Amer. J. Bot.* **56**, 928–35.

Swain, F.M., Bratt, J.M. and Kirkwood, S. (1967) Carbohydrate components of some Palaeozoic plant fossils. *J. Paleont.* **41**, 1949–54.

——, —— and —— (1968) Possible biochemical evolution of carbohydrate of some Palaeozoic plants. *J. Paleont.* **42**, 1078–82.

Towers, G.H.N. and Maass, W.S.G. (1965) Phenolic acids and lignin in the Lycopodiales. *Phytochem.* **4**, 57–66.

Watson, E.V. (1967) *The Structure and Life of Bryophytes.* 2nd ed. Hutchinson University Library, London.

4. Identifying plant fragments

C.J. CLEAL

Abstract

Palaeobotanists, as students of fossil plants, are concerned with two types of taxonomy: identifying fossils and identifying plants. The plants are those reconstructed from the fossils, and the approach differs little from that used by neobotanists identifying extant plants. Ideally, the fossil would be given the same name as the parent plant, but this is often impossible, because the overall form of the parent is usually unknown. This would not be a problem if each species of whole plant was distinct in each of its organs, but many cases have been demonstrated of homologous organs in different species and heterogenous organs in the same species. It is therefore necessary to have a separate taxonomic scheme for identifying most fossil plants: the taxonomy of plant fragments. This can often be as 'natural' as the taxonomic scheme used for whole plants, in that every attempt is made to construct it to mirror the evolutionary pattern of the Plant Kingdom.

Introduction

The dictionary definition of palaeobotany is the science that deals with extinct or fossil plants (Little *et al.* 1975) and the final result of many paleobotanical studies is indeed the interpretation of a particular extinct plant. However, palaeobotanists usually spend most of their time studying fossils which, because of the disruptive influence of the fossilization process, represent only fragments of plants. Ideally, one would hope to reconstruct the original plant by piecing the fragments together, in which case the component parts might be given the same name as the parent plant. There have been several successful reconstructions of this type (e.g. Andrews 1948, Harris 1961, Eggert 1961, Barthel 1968) and the problem is discussed more fully by Chaloner (in this volume), but it has to be recognized that such cases are, and probably always will be, the exception rather than the rule. It might be argued (erroneously in my view) that it is better to make an intuitive guess as to the nature of the parent plant where a real reconstruction is unavailable. The danger of this can be seen in some of the early palaeobotanical studies (e.g. Martius 1825) where cacti, yuccas and bamboos have been recorded from the Carboniferous palaeoequatorial floras. Such major errors are perhaps less likely to occur today because of our much better understanding of the plant fossil record. However, one only has to remember the relatively recent records of cordaites from the middle and upper Devonian, based on misinterpretations of isolated progymnosperm wood, to realize the potential danger of trying to draw taxonomic conclusions from isolated plant structures.

How, therefore, should plant fragments be identified and named, where little is known of the whole parent plant? It might be argued that it is botanically meaningless to try to identify only a fragment of a plant, unless perhaps it is a fertile organ, and that it is better not to name such fragments at all. However, fossilized plant fragments have much more than just a narrow botanical significance, and it is important to be able to name them accurately so as to provide data for biostratigraphical and palaeoecological studies. Furthermore, it is possible to demonstrate important evolutionary trends within particular plant organs purely by studying the stratigraphical distribution of isolated plant fragments (the 'stratophenetic' approach discussed by Hill and Crane 1982), without understanding the overall form of the whole plant (e.g. Chandler 1923, Asama 1959, Chaloner and Sheerin 1979).

Clearly we need a means of identifying plant fragments, which defines exactly what we can see, and not what we assume. The introduction of speculative criteria is diametrically against the principles of biological and palaeontological taxonomy (see Matthews 1973). The basis of such a scheme is already available in the *International Code of Botanical Nomenclature* (hereafter referred to as the *ICBN*), and it can be readily extended, by the adoption of methodology used for naming plant microfossils, to form a fully nested hierarchy of taxonomic ranks for use with plant fragments. Such taxa can provide an accurate means of identifying fossil plants, since it takes into account the disarticulation and differential preservational states introduced by the fossilization process. This paper discusses the background to such a scheme of nomenclature for fossil plants, dealing particularly with those points whereby it differs from the nomenclature used for whole plants.

Different ranks of plant fragment taxa

At the rank of genus, the *ICBN* provides a type of taxon which allows the separation of palaeobotanical and neobotanical nomenclature: the form-genus (*forma-genera*). The definition of the form-genus in the *ICBN* has had a long and varied history, and still needs refinement before it becomes entirely satisfactory. The problems have mainly arisen because its definition has included a purely taxonomic criterion: the form-genus cannot be assigned to a family. This is against the principles of the *ICBN*, which is not supposed to have any effect on purely taxonomic problems. To avoid this difficulty, a reworded version of *Article 3* of the *ICBN* was proposed during the 1975 International Botanical Congress at Leningrad, which essentially freed the definition of form-genus from this constraint of family non-assignability. It was accepted by the Committee for Fossil Plants of the International Association of Plant Taxonomy (Traverse 1975, Voss 1976) but, for a combination of erroneous reasons (Meyen and Traverse 1979), the Editorial Committee rejected it and the published version of the *ICBN* (Voss *et al.* 1983) retained the lack of family assignability in the definition of form-genus. This remains the position of the *ICBN* on the form-genus.

In a subsequent attempt to resolve the problem, Meyen and Traverse (1979) proposed that the concept of form-genus should be entirely dispensed with. However, Meyen and Traverse themselves give five fundamental ways in which genera of fossils differ from genera of extant plants.

1. Different organs of a fossil plant can be assigned different generic names. This could never happen with living plants (except Fungi Imperfecti).
2. Different species belonging to a single genus of fossil plant can belong to different families. This could never happen with living plants.
3. A genus of fossil plant does not have to be assigned to suprageneric taxa, if the available information does not allow it. *Article 3* of the *ICBN* requires that genera of living plants *have* to be assigned to a full nested structure of suprageneric taxa.
4. A particular part of a plant, when preserved in different modes of fossilization, can be assigned to different genera. Such states of preservation do not influence neobotanical taxonomy.
5. Different ontogenetic phases of fossil plants may be assigned to different genera. This never happens (at least not by design) with living plants.

These are surely sufficient differences to justify retaining the concept of form-genus for identifying fossil fragments. In my opinion, it would be best if some sort of improved wording of the concept adopted by the Fossil Plant Committee in 1975 was reintroduced. It might perhaps be argued that it would be better for it to be renamed a 'fossil-genus' (Boulter 1979, Mosbrugger 1983) but this could be misleading, since reconstructed fossil plants may be named using 'ordinary' whole plant genera. In this paper, the name form-genus has been retained, but is used in the sense of the 1975 proposal.

The above five points can also be used to argue for separate taxa for fossil plant fragments at other ranks. If form-genera are acceptable, then the concept of form-species would seem both logical and relatively uncontroversial. Sylvester-Bradley (1954) defined 'true form-species' in a palaeozoological context as 'taxa within form-genera and themselves. . . based on populations', and this can surely be adapted to palaeobotany (his 'morphological form-species' are quite different in concept and are briefly covered later in the present paper). A widely quoted example of a palaeobotanical form-species is *Stigmaria ficoides* Brongniart, which is acknowledged as being the rootstock of a variety of different families of Palaeozoic arborescent lycophytes. This is perhaps an extreme example. However, any 'species' based on fossil plant fragments, can never be safely assumed to have the same circumscription as a whole plant species, unless large numbers of whole plant reconstructions are available for the group.

The separation at higher ranks might be considered unnecessary, since it is often possible to assign fossils, to say, a natural family. However, this is usually only completely successful when dealing with fossils belonging to an extant group of plants (e.g.the Tertiary angiosperms—Collinson 1983). In some extinct groups, sufficient progress has been made in producing whole or part plant reconstructions that at least an outline superstructure of suprageneric taxa is available. However, in such cases there is often a residual group of form-genera, usually of foliage and stems, which cannot be definitely accommodated within families. In discussing the classification of the lycophytes, Thomas and Brack-Hanes (1984) have proposed that this problem can be overcome by basing the suprageneric taxa exclusively around fructification form-genera, making the classification as near to the neobotanical concept of taxonomy as possible. The residual taxa are then referred to as satellite taxa (an extension of the satellite genus concept of Meyen 1978) which can be appended to lists of components of the suprageneric taxa, but which in no way are used to define them.

This simple method is a major step forward in helping rationalize the classification of the better understood plant groups, such as the lycophytes. However, other groups have seen far less progress in whole plant reconstruction, due either to natural disarticulation during life (e.g. the abscission of seeds at maturity), taphonomy, or scarcity of material. In such cases, suprageneric taxa (if they are to be used) have to be established on particular organs, and specialists working on different organs can produce conflicting schemes. An example is the Carboniferous pteridosperms, where different classifications have been established by workers dealing with stem anatomy (Barnard and Long 1975), seeds (Seward 1917), synangia (Halle 1933) and foliage (Corsin 1960). Such parallel schemes for different organs can become broadly correlatable as more whole plant reconstructions become available. As already pointed out, however, such reconstructions are unlikely ever to be made with the majority of plant fossils and so the complete integration of such systems will never be achieved. One answer is to formalize the concept of 'form-family', 'form-order', etc., so that different hierarchical schemes become available for the different organs. This is essentially the approach of the *turma* system introduced principally for micropalaeobotanical work (e.g. Erdtman 1947, Potonié and Kremp 1954, Roselt and Schneider 1969) but in recent years also adopted for macropalaeobotany (e.g. Meyen 1971, Maheshwari and Meyen 1975). The reluctance to adopt it more widely in macropalaeobotany is probably because the artificiality of the scheme tends to be rather overemphasized. However, the scheme needs to be no more or no less artificial than any suprageneric classification based on morphological criteria.

The value of the *turma* system is that it can be used to provide building blocks for constructing whole plant suprageneric taxa. In the early stages of the study of a group of plants (i.e. prior to the reconstruction of many whole plants) each organ can be classified within its own system of *turmae*. At first, these form–taxa will be largely artificial but, by judicious emendation, it should be possible to modify them so that they become more 'natural' in concept. As the study of the group progresses and as more whole plant reconstructions become available, it may be possible to construct (for example) a whole plant family from a *turma* of seeds, a *turma* of synangia, a *subturma* of foliage and a *superturma* of stems. At this stage, the satellite taxa method should come into its own, since the basic structure of the suprageneric taxa will be understood, but the taxonomic positions of some of the form-genera and -species might still be in doubt. Eventually, in the unlikely circumstance that the majority of the component species have been reconstructed as whole plants, then an entirely 'natural' (i.e. neobotanical style) classification can be adopted for the group. In this way, the style of suprageneric classification can evolve as the plants become better understood, so that at any one time it provides the maximum of verified information and the minimum of unverified speculation.

The circumscription of the form-species

The palaeobotanist is clearly unable to use features such as breeding patterns or chromosomal composition, by which the neobotanist may often define living plant species. Fossil plant species must therefore be circumscribed almost entirely by morphological features, whether or not one is dealing with whole plant or plant fragment taxa. In the latter case, particularly when the fragments belong to parts of the plant rarely used for neobotanical taxonomy (e.g. foliage, rooting structures—although see Just 1946 for a discussion on the subjectivity of such decisions on taxonomic value of different organs) it is tempting to propose an entirely 'artificial' classification, where the form-taxa are defined by very restricted morphological constraints. Such an approach has given rise to the widely used morpho-species of Carboniferous non-marine bivalves (e.g. Trueman and Weir 1945-1968) and the 'morphological form-species' used for conodonts (Sylvester-Bradley 1954), where the taxa are restricted to those specimens which are identical in form to the holotype. Specimens which differ from the various known holotypes must either themselves become holotypes of new morpho-species, or must be compared with one of the existing morpho-species with a 'cf.' or 'aff.'. The appeal of such a scheme is the ease with which individual specimens can be named, but it does give rise to rather unwieldy species lists for individual faunas and floras, far in excess of the natural species composition. This in itself is not necessarily an overriding objection and, in the case of the non-marine bivalves, the method has produced important biostratigraphical and palaeoecological results. However, if at all possible, it would seem preferable to adopt a more 'natural' classification, even if the goal of the study is not intrinsically botanical. Most of the conclusions which are drawn from even non-botanical studies are controlled by influences on the original plants: biostratigraphy reflects the evolutionary changes of the plants, whilst palaeoecology is dependant on the mode of life of the plants. It would, therefore, seem desirable to make the taxonomy of the fossil plant fragments mirror as closely as possible that of the parent whole plant.

How can fossil plant fragment taxonomy be made to become a closely comparable image of whole plant taxonomy? Stebbins (1950, p. 34) argued that living plant species must consist of systems of populations that are separate from each other by complete or at least sharp discontinuities in the variation pattern, and that these discontinuities must have a genetic basis. It seems reasonable, therefore, to base fossil species determinations on the assessment of morphological patterns in assemblages of fossil plant fragments rather than on individual specimens. There is, of course, the problem of the much greater degree of variation in plant fragments. The variation of the whole plant phenotype provides that variation which the botanists uses to recognize taxa in whole plants. In addition, the palaeobotanist has to take into account the range of variation of the organ *within* the individual plant. Furthermore, he may only be dealing with a part of an organ (e.g. a pinnule from a frond) and so another degree of variation is introduced (the variation of the suborgan within the organ). A further problem is whether or not the observed variation in fossil plant fragment assemblages is mainly genotypic. Without being able to observe directly the genetic structure of the plant fossils, such uncertainly can never be fully resolved. However, the careful recording of detailed variations in assemblages in different locations and sedimentary environments might help separate out at least some of the non-genotypic variation. Despite these additional difficulties, the essential problem is similar to that usually encountered by the neobotanist (i.e. the recogition of discontinuities within populations), the only difference being the amount and type of variation to be taken into account, and the lack of an absolute check of the nature of the genotype.

Taxonomies for different plant organs

If it is acknowledged that plant fragments should be classified in a different way to whole plants, then the problem arises as to where, within the plant, should the taxonomic boundaries be drawn. If parts of a plant do not normally become disarticulated during fossilization, there is clearly no need to assign them different taxonomic names. For instance, one would not normally have separate taxonomic schemes for the integuments and megaspores of gymnospermous seeds. However, if they do frequently become separated (e.g. the spores and sporophylls of arborescent lycophytes, the cyclopterid and pinnate segments of some pteridosperms fronds) then the problem arises as to whether or not they should be treated separately.

Since the necessity for having a separate set of taxa for plant fragments is based on the practical difficulty of finding articulated plant remains, then the decision as to whether different parts of the plant should be treated differently should also be one of practical (and thus to a certain extent subjective) choice. However, as a general guide the plant organ may be regarded as the minimum unit which should be given a separate taxonomic treatment. An organ generally tends to be relatively coherent entity within the plant, since it usually 'possesses a specific shape and performs a specific function' (Sinnott 1946, p. 29). It will probably, therefore, have less variation than a subdivision of an organ (e.g. a pinnule in a frond) and will thus be easier to classify in a more natural way. This is, in fact, the approach that most palaeobotanists have instinctively followed with the development of separate classifications for seeds, leaves, stems, etc.

A further complication is the presence of different preservational states for the same organs (a useful summary of the principal preservational states is found in Schopf 1975). This is particularly the case when compression/impression fossils are compared with permineralizations. The different types of data which each preservational state can supply are often mutually exclusive, and so the diagnostic criteria established for taxa in one state cannot be used in another. Very rarely, a specimen will show both states of preservation (an example of *Anabathra* showing just such a dual mode of preservation is being studied by H. Pearce, pers. comm.) and so a correlation is possible between the two sets of data. However, as with whole plant reconstructions from fossils, this is the exception rather than the rule, and one can never be certain whether all of the members of a taxon based on internal anatomy will show the same morphological features, and *vice versa*. At least for the foreseeable future, therefore, it will be necessary to have distinct but parallel taxonomic schemes for different preservational states as well as for different organs.

Assessing variation patterns in fossil plants

Traditionally, the variation of fossil plant fragments has been mainly assessed by making subjective visual comparisons. Clearly, a more objective method would be to adopt the morphometric approach, as has been done by Schiehing and Pfefferkorn (1980) and Mosbrugger and Vogellehner (1982). However, even when dealing with only one organ of a plant, there may be innumerable variables involved. Highly refined multivariate statistical tools have been developed in recent years (reviewed by Blackith and Reyment 1971) but the computing power that they require makes them impractical for day-to-day taxonomic applications. Even if this was not a serious problem, it would still be impossible to analyse literally every variable present in a plant fragment using a statistical approach. It is therefore necessary to select those characters for analysis which are believed to be taxonomically sensitive, and this immediately introduces a subjectivity into a method whose principal strength was supposed to be objectivity. Morphometrics will undoubtedly have a role to play in the future of palaeobotanical systematics, particularly when specific and very narrowly defined characters need to be analysed, but it is unlikely ever to supplant the traditional 'subjective' approach of visual comparisons, which can immediately assess the whole form of a plant.

The most important point is that a plant fragment taxon should only be defined using an assemblage from a single geographical and stratigraphical locality. This is essentially the 'palaeodeme' of Anderson (1977) and Anderson and Anderson (1983), which they define as a collection of specimens, representing a single breeding population, showing a normal distribution of variation, taken from a distinct lithological unit of restricted area and stratigraphical extent. In theory, their definition is perhaps looser than the one adopted here, since they only require the assemblage to be taken from a restricted area rather than from a single locality. However, since it is impossible to be certain that an assemblage represents a single breeding population unless it is collected from a single locality, Anderson's definition of palaeodeme really needs to be modified to that used here if it is to have any genetic significance.

The palaeobotanist, following the principles adopted by most taxonomists, is constrained (by the *ICBN*) to nominate a single specimen as holotype (or lectotype or neotype). However, it is also important to use the assemblage associated with the type to demonstrate the full range of morphological variation. This is clearly a desirable course when defining whole plant taxa, but is essential when dealing with the more variable plant fragment taxa. Even today, some palaeobotanical studies only describe and illustrate the holotype in the protologue of a new species. This usually confuses more than it clarifies, because it is impossible to be certain whether such an isolated specimen is typical or atypical of the parent species. In such cases, it becomes impossible to extrapolate the full range of infraspecific variation from the lone holotype, an essential requirement if the species is to be identifiable outside of the type palaeodeme. Such inadequately based species may seem acceptable from the personal view of the original author, since he now has a name that he can give to his unique specimen; but for other palaeobotanists the species is unusable, at least until it is revised using specimens from the type palaeodeme. Taxomony is concerned with the accurate communication of ideas, and taxa which cannot be reliably translated from one locality to another merely provide meaningless clutter for the literature. Many of the problems might be avoided if palaeobotanists could be persuaded that it is quite acceptable to describe a unique and interesting specimen without assigning it to a new form-species. The open nomenclature discussed by Matthews (1973) provides a more than adequate means of naming such material.

The number of specimens which need to be illustrated is, of course, a matter of opinion. Hughes and Moody-Stuart (1969) argue that at least 100 are required to document one of their biorecords, although they do not specify how many should actually be illustrated in the protologue. Different types of fossil will clearly have different requirements, but the rule should be that the number should demonstrate the full range of variation present in the collection made by the palaeobotanist, which in turn will hopefully provide a reasonable sample of the original population.

When attempting to identify species originally described elsewhere, essentially the same approach should be adopted, with the assessment of morphological variations in extensive palaeodemes. The determination of the full range of such variation is an essential prerequisite for reliable fossil plant fragment identification. Sometimes, however, there are practical difficulties and some sort of taxonomic work has to be achieved using a restricted amount of material. This can become particularly acute when dealing with borehole material, where the core may be only a few centimetres in diameter. Plant macrofossil determinations based on borehole material can clearly never be as robust as those based on material collected at outcrop, but the potential risk of misidentification can be reduced by being familiar with the coeval floras found in the same area. If a particular fragment found in a borehole can be assigned to any one of several different form-species, but only one of those form-species has been previously recorded from that area, there is clearly a much greater likelihood that the specimen belongs there than to the others. This is not an ideal state of affairs and the value of extensive collections from outcrop will never be supplanted. However, it is often necessary to work with borehole collections when making biostratigraphical studies and so such assumptions unfortunately have to be made. It should also be pointed out that this is not a special problem to palaeobotany, but is present wherever macrofossils are being studied from boreholes.

Concluding remarks

A system of fossil plant classification, in which unverifiable assumptions are introduced as an intrinsic part of that system, is totally alien to the principles of biological and palaeontological taxonomy. The latter is only of value as a scientific tool if it uses consistent, accurate, and verifiable taxa. Most classifications of fossil plants which are claimed to be based on whole plant taxa fall outside of these constraints, since such taxa are usually unverifiable, and thus inaccurate and inconsistent. It is therefore necessary to have a taxonomic scheme specifically designed for use with fossil plants, which allows the accurate identification of plant fragments. The *ICBN* takes a first tentative step in this direction, with the reference to the form-genus concept. However, this concept needs to be revised along the lines proposed during the 1975 Leningrad Botanical Congress, and also needs to be extended to include all other taxonomic ranks. Only when such a system becomes available will the palaeobotanist have a fully flexible scientific tool for identifying his fossils.

It is generally accepted that the hierarchical structure of biological taxonomy should reflect the genetic and evolutionary structure of the organic world, either by prior design or by result. However, this does not mean that taxonomy itself is a natural structure. It is simply an artefact of the biologist and/or palaeontologist which he uses to help analyse the organic world, and which should be constructed so as to provide the best available data base. The palaeobotanist is usually concerned with two different types of object: the individual fossil, which usually represents only part of a plant; and the whole plant which can sometimes be reconstructed from the fragments. The fact that objects belonging to the first set cannot always be related to the second set has meant that the palaeobotanist has had to establish two separate types of systematics. Whole fossil plant taxonomy lies nearest to neobotanical systematics, and is the one most often referred to in botanically orientated reviews of palaeobotany. Plant fragment taxonomy is that most usually used by geologists when studying fossil floras, since it provides a more comprehensive assessment of the fossil assemblage. However, neither scheme has better attributes or is 'more natural' than the other. Each is entirely artificial in that it is designed by and for the scientist to fit a particular set of facts. The palaeobotanist should always be aware of both systems, and should always be attempting to bring them together. However, because of the disruptive action of fossilization, a final unification of the two schemes is unfortunately never likely to be achieved.

Acknowledgements

This paper was written whilst on contract to the Nature Conservancy Council, Geological Conservation Review Unit, which is acknowledged for providing facilities to help in its preparation. Dr W. A. Wimbledon, Dr B.A. Thomas, Dr K.M. Evans, Mr C.H. Shute and Mr S.I. Jusypiw are thanked for reading the manuscript and making helpful suggestions. Thanks also go to Mr H. Pearce for discussions on the specimen of *Anabathra*. Finally, the author would like to thank colleagues for valuable discussions on the paper following its presentation at the meeting held at Goldsmith's College, London.

References

Anderson, J.M.(1977) The biostratigraphy of the Permian and Triassic. Part 3. A review of Gondwana Permian palynology with particular reference to the northern Karoo Basin, South Africa. *Mem. bot. Surv. S. Afr.* **41**, 1–65.

—— and Anderson, H.M.(1983) Palaeoflora of southern Africa. Molteno Formation (Triassic). Volume 1. Part 1. Introduction. Part 2. *Dicroidium*. A.A. Balkema, Rotterdam.

Andrews, H.N. (1948) Some evolutionary trends in the pteridosperms. *Bot. Gazette* **110**, 13–31.

Asama, K. (1959) Systematic study of so-called *Gigantopteris*. *Sci. Rep. Tohoku Univ.* (2nd ser., Geol.) **31**, 1–72.

Barnard, P.D.W. and Long, A.G. (1975) *Triradioxylon*—a new genus of Lower Carboniferous stems and petioles together with a review of the classification of early Pterophytina. *Trans. R. Soc. Edinb.* **69**, 231–49.

Barthel, M. (1968) *'Pecopteris' feminaeformis* (Schlotheim) Sterzel und *'Araucarites' spiciformis* Andrae in Germar.—Coenopteriden des Stephans und Unteren Perms. *Paläont. Abh.* **2**, 727–42.

Blackith, R.E. and Reyment, R.A. (1971) *Multivariate morphometrics*. Academic Press, London & New York.

Boulter, M.C. (1979) A proposal to emend Article 3.2 of ICBN. *Taxon* **28**, 598–600.

Chaloner, W.G. and Sheerin, A. (1979) Devonian macrofloras. *Spec. Pap. Palaeont.* **23**, 145–61.

Chandler, M.E.J. (1923) The geological history of the genus *Stratiotes*: an account of the evolutionary changes which have occurred within the genus during Tertiary and Quaternary times. *Q. Jl geol. Soc. Lond.* **79**, 117–38.

Collinson, M.E. (1983) Fossil plants of the London Clay. *Pal. Assoc. Fld Guide to Fossils* **1**, 121 pp.

Corsin, P. (1960) Classification des Ptéridophytes et des Ptéridospermophytes du Carbonifére. *Bull. Soc. géol. France* (2nd ser.) **2**, 566–72.

Eggert, D.A. (1961) The ontogeny of Carboniferous Lycopsida. *Palaeontographica* B **108**, 43–92.

Erdtman, G. (1947) Suggestions for the classfication of fossil and recent pollen grains and spores. *Svensk. bot. Tidskr.* **41**, 104–14.

Halle, T.G. (1933) The structure of certain fossil spore-bearing organs believed to belong to Pteridosperms. *Kungl. Svensk. Vetenskapakad. Handl.* (3rd ser.) **12**, no. 6, 103pp.

Harris, T.M. (1961) The fossil Cycads. *Palaeontology* **4**, 313–23.

Hill, C.R. and Crane, P.R. (1982) Evolutionary cladistics and the origin of the angiosperms. In *Problems of phylogenetic reconstruction* (K.A. Joysey and A.E. Friday, eds.), pp. 269–361. *Systematics Association Special Volume* 21.

Hughes, N.F. and Moody-Stuart, J.C. (1969) A method of stratigraphic correlation using early Cretaceous miospores. *Palaeontology* **12**, 84–111.

Just, T. (1946). The relative value of taxonomic characters. *Amer. midl. Natur.* **36**, 291–7.

Little, W., Fowler, H.W. Coulson, J., Onions, C.T. and Friedrichsen, G.W.S. (1975) *The shorter Oxford English dictionary*. 3rd edn (revised). 2 vols. Oxford University Press, Oxford.

Maheshwari, H.K. and Meyen, S.V. (1975) *Cladostrobus* and the systematic of cordaitalean leaves. *Lethaia* **8**, 103–23.

Martius, C.F.P. de (1825) On certain Antediluvian plants susceptible of being illustrated by means of species now living within the tropics. *Edin. phil. J.* **12**, 47–56, 270–81.

Matthews, S.C. (1973) Notes on open nomenclature and on synonymy lists. *Palaeontology* **16**, 713–9.

Meyen, S.V. (1971) *Phyllotheca*-like plants from the Upper palaeozoic flora of

Angaraland. *Palaeontographica* B **133**, 1–33.

—— (1978) Systematics, phlogeny and ecology of Protopteridophytes. *Byull. mosk. Obshch. Ispyt. Prir. (Biol.)* **83**, 72–84 [in Russian].

—— Traverse, A. (1979) Proposal on Article 3: remove 'form-genus' too! *Taxon* **28**, 251–4.

Mosbrugger, V. (1983) Organische Zusammengehörigkeit zweier Fossil-Taxa als taxonomisches Problem am Beispiel der jungpaläozoischen Farnfruktificationen *Scolecopteris* und *Acitheca*. *Rev. Palaeobot. Palynol.* **40**, 191–206.

—— and —— Vogellehner, D. (1982) Zur Sippenabgrenzung in der Paläobotanik mit hilfe mathematisch-statistischer Methoden am Beispiel der jungpaläozoischen Pteridophylle *Pecopteris arborescens*—*cyathea*. *Palaeontographica* B**184**, 107–26.

Potonié, R. and Kremp. G. (1954) Die Gattungen der paläozoischen *Sporae dispersae* und ihre Stratigraphie. *Geol. Jb.* **69**, 111–94.

Roselt, G. and Schneider, W. (1969) *Cuticulae dispersae*, ihre Merkmale, Nomenklatur und Klassifikation. *Paläontol. Abh.* B 3, 1–128.

Scheihing, M.H. and Pfefferkorn, H.W. (1980) Morphologic variation in *Alethopteris* (Pteridosperms, Carboniferous) from St. Clair, Pennsylvania, USA. *Palaeontographica* B **172**, 1–9.

Schopf, J.M. (1975) Modes of fossil preservation. *Rev. Palaeobot. Palynol.* **20**, 27–53.

Seward, A.C. (1917) *Fossil plants*. Vol. 3. Cambridge Univ. Press, Cambridge.

Sinnott, E.W. (1946) *Botany principles and problems*. 4th edn. McGraw-Hill Inc., London and New York.

Stebbins, G.L. (1950) *Variation and evolution in plants*. Oxford University Press, London.

Sylvester-Bradley, P.C. (1954) Form-genera in paleontology. *J. Paleont.* **28**, 333–6.

Thomas, B.A. and Brack-Hanes, S.D. (1984) A new approach to family groupings in the lycophytes. *Taxon* **33**, 247–55.

Traverse, A. (1975) Report of the committee for fossil plants. *Taxon* **24**, 690.

Trueman, A.E. & Weir, J. (1946–1968) A monograph of British Carboniferous non-marine Lamellibranchia. Parts 1–9. *Pal. Soc. Monogr.*, 449 pp.

Voss, E.G. (1976) XII International Botanical Congress mail vote and final congress action on nomenclatural proposals. *Taxon* **25**, 169.

—— *et al.* (1983) *International Code of Botanical Nomenclature*: Thirteenth International Botanical Congress, Sydney, 1981. *Regn. Veg.* **111**, 487 pp.

5. Reassembling the whole fossil plant, and naming it

W.G. CHALONER

Abstract

In dealing with fossil plants the distinction between nomenclature and taxonomy is sometimes less clear-cut than with living organisms. The relationship between *Equisetum* (non-fossil genus) and *Equisetites* (fossil genus) exemplifies this. The controversy that has surrounded the concepts of form-genus, organ-genus and genus, as applied to fossil plants, is briefly reviewed. The nomenclatural problems that ensue when parts of fossil plants become linked are complex, and do not lead to a simple synonymy between the genera concerned. Six different ways in which fossil genera may be shown to be conceptually linked are enumerated. In such cases we must use great caution in attempting to extend the application of a name based on one organ to the whole plant.

Introduction

The problems of naming fossil plants are as old as the science of palaeobotany. Many features of fossil plants cause them to present particular nomenclatural problems. In this article I attempt to review some of these problems, and especially to consider whether, as we reconstruct whole fossil plants from fragments, we can—or should—adjust our nomenclatural treatment to take account of this.

The distinction between nomenclature and taxonomy

A concept dear to most systematists is that the activities of nomenclature (naming) and taxonomy (classifying) are totally separate. It is obviously possible to use a set of names (e.g. colloquial ones) for plants without considering their classification at all. Equally, one could designate a series of fossil spore types, for example, by numbers, and then proceed to group them into a hierarchy of levels of similarity (i.e. classifying them) without ever naming them. In one of his many contributions on spore nomenclature Traverse (1961) suggests optimistically that 'neophytes' may be confused about the distinction between nomenclature and classification. Alas, the confusion is not restricted to 'neophytes'; in dealing with fossil plants the two compartments of 'naming' and 'classifying' are not always as water-tight as we like to imagine, and the processes of nomenclature and taxonomy sometimes overlap.

Before pursuing this theme, it is important to remember that even in living plants there are plenty of circumstances in which one of two alternative generic names may be applied to the same plant. *Lycopodium selago* L. may equally correctly be referred to as *Huperzia selago* (L). In the first case one is taking a broad sense of *Lycopodium*, and in the latter, acknowledging its division into several distinct genera, of which *Huperzia* is one. Both usages are perfectly acceptable, and two different binomials may be correctly applied to the same plant. What is in a superficial sense a nomenclatural matter, relating to two alternative names, is of course reflecting an underlying taxonomic decision as to whether *Lycopodium s. l.* includes such diverse species that it needs to be split.

A comparable situation occurs in palaeobotany, but here the distinction between a purely nomenclatural matter and one of taxonomic judgement is rather different. The living genus *Equisetum* L. (the horsetails) is seen by many as being recognizable in fossils at least as old as the Triassic. Others are more cautious in this extrapolation (mainly based on vegetative material alone) and assign at least the Mesozoic species to a different genus, *Equisetites* Sternberg (based on the Triassic fossil, *E. muensteri*). Tom Harris, who described a number of species of fossil *Equisetum*, adopts the latter procedure in his earliest work (Harris 1926), but in his later publications he places all the Yorkshire Jurassic species in the living genus. He writes (1961): 'I have previously followed Seward and used *Equisetites* . . . latterly however, several authors have deliberately used the modern names of living genera for Mesozoic plants and as the convention is not kept it is best dropped . . . No morphological distinction has ever been proved between *Equisetum* and *Equisetites*, indeed progress in knowledge adds new points of agreement'.

In the current edition of the *International Code of Botanical Nomenclature* (Voss *et al.* 1983) the distinction is drawn (p. 1) between a 'fossil taxon' (of which the name is based on a fossil type) and 'a non-fossil taxon' when its name is based on a non-fossil type. In this usage, *Equisetum columnare* and *Equisetites columnare* are both fossil taxa (fossil species); but while *Equisetites* is clearly a fossil taxon (genus), *Equisetum* is a non-fossil taxon, even though many authors include fossil species within it. I labour this point to emphasize that the relationship between *Equisetum* and *Equisetites* is of a very different character from that between *Lycopodium* L. (*s. l.*) and *Huperzia*. The species included in the latter are simply a sub-set of those included in the former, and the boundary is clear-cut. The distinction between *Equisetum* and *Equisetites* (for those who wish to sustain it) is less well-defined.

I bring up this matter at the outset merely to illustrate how the use of names for fossil plants can involve uncertainties as to where objective nomenclatural matters end and subjective taxonomic considerations take over. Principle IV in the *International Code* (Voss *et. al.* 1983) stipulates that 'each taxonomic group with a particular circumscription can bear only one correct name . . . except in specified cases'. The text of the Code is perhaps not as helpful as it might be in specifying cases involving fossils, such as those just cited, where two alternative—and correct—names are applicable to one species.

Form-Genera, Organ-Genera and Genera

Over the last forty years a somewhat unfruitful debate has taken place over the concept of a genus as it may be applied to fossil plants. I will not review these issues as they have been amply explored elsewhere (e.g. by Beck 1969, Boulter 1979, Faegri 1963, Jansonius 1974, Meyen 1975, Meyen and Traverse 1979, Mosbrugger 1983, and Schopf, 1969, 1978).

In brief, three kinds of genera were acknowledged in the Codes of Nomenclature up to and including that published from the Seattle Botanical Congress of 1975. The first was essentially the genus as it is applied to living plants, and those fossils where supposedly the 'whole plant' was the basis for the name—perhaps *Rhynia* may be seen as an instance of this. Then, two kinds of genera based on less complete fossils were recognized—an organ-genus, which was based on a fossil organ showing enough detail and character that it could be assigned to a family, (and thence to higher taxa); and a form-genus, which lacked sufficient distinctive characters for family assignment. Examples of different situations involving these concepts are shown in Table 5.1. The genera given there are chosen to show a range of situations in terms of the nature of the type specimen and, in the case of fossils, the part of the plant on which the name was based, its state of preservation and its assignability at family level.

The first five are genera based on living (extant) plants, and hence are technically non-fossil taxa, column A. Within these five, *Volvox* represents a genus of very perishable algal coenobia, unlikely ever to be preserved as a fossil, and unknown as such. At the opposite extreme, among living plants, we have *Ginkgo* which is based on the single living species, *Ginkgo biloba* but which has many extinct fossil species assigned to the

Table 5.1. Summary diagram of changing concepts of the genus, form-genus and organ-genus.

Examples of Different Kinds of Genera:	Types of Genus:	A Fossil (F) or non-fossil (NF) taxon	B Fossil species known for the genus	C Genus ''assignable to a family''	D ICBN of Montreal, Seattle etc. (pre-1975)	E ICBN of Leningrad (Stafleu *et al.* 1978)	F Proposal *36* (Meyen & Traverse 1979)
Volvox	Living plant	NF	—	✓			
Bellis	Living plant	NF	—	✓			
Pinus	Living plant	NF	✓	✓	Genus		
Ginkgo	Living plant	NF	✓	✓			
Metasequoia	Living plant	NF	✓	✓		Genus	
Siltaria	Fossil pollen exine	F	✓	✓			
Lagenostoma	Fossil seed	F	✓	✓			Genus
Lepidocarpon	Fossil seed	F	✓	✓	Organ-Genus		
Lepidodendron	Fossil stem impression	F	✓	✓			
Stigmaria	Fossil rhizophore cast	F	✓	—			
Pecopteris	Fossil leaf impression	F	✓	—	Form-Genus	Form-Genus	
Dadoxylon	Fossil, permineralised wood	F	✓	—			
Leiotriletes	Fossil spore exine	F	✓	—			

genus. *Metasequoia* is noteworthy as having been initially described on the basis of a fossil species; the subsequent discovery of the living (non-fossil) species, *M. glyptostroboides*, meant that the genus came to be regarded as being typified by that species in accordance with the *ICBN Article 58*. The generic name *Metasequoia* Miki ex Hu et Cheng is accordingly actually conserved, as based on the non-fossil species.

The remaining eight genera in Table 5.1 are all fossil taxa, based on various fossil organs. *Siltaria* Traverse (1955) is an instance of a genus based on fossil pollen, which its author believed could be assigned to the family Fagaceae. It is therefore, at least on this interpretation, an organ-genus. The following three genera have been assigned to (extinct) families by at least some authors, so that technically they too are organ–genera. However, in the sense of the *ICBN* approved in Leningrad (Staffleu *et al.* 1978), sustained in the current revision of the *ICBN* (Voss *et al.* 1983) they are not differentiated from other plant genera in the application of the Code (column E). The last four genera are based on—and include—species which lack sufficient character for assignment to a family. In each of these cases there is indeed evidence of association or connection with plants known to represent more than one family or even order. The genus *Pecopteris*, for example, includes foliage of both true ferns, Marattiales, and seed ferns, Pteridospermales. *Dadoxylon* includes species some of which are of the Cordaitales, others probably of the Coniferales. There is therefore, in broad terms a gradient of 'naturalness' of generic concept down the Table. This goes from living genera, probably of closely related species, in the first two named, descending through taxa which are in varying degree progressively more remote from the concept of a genus of living plants. The last three columns illustrate how palaeobotanists have either sought to recognize this gradation in our taxonomic concepts (column D) or to reject it (column F).

The relationship between the four generic concepts referred to above may perhaps usefully be expressed in the form of a key, as follows:-

Key to generic concepts involving plant fossils:-

1.	Genus based on an extant ('non-fossil') type species.	..... *non-fossil genus*
	Genus based on a fossil type species.	..... (*fossil genus*).....2.
2.	Genus based on a fossil species (or including fossil species) which show sufficient characters to make them assignable at family level	..... *organ-genus*
	Genus based on a fossil species lacking sufficient characters for family assignment to be possible	..... *form-genus.*

Reassembling the whole plant

Most palaeobotanists would rate the reassembly of whole plants (or at least the linking of detached fossil organs) as one of the prime goals in fossil plant research. There are several rather different circumstances in which two or more parts of a plant may be shown to have been parts of a single species to which different names have come to be applied. They may, for instance, represent different organs of a single plant (leaves, seeds), preserved in the same kind of way, which after the setting up of names for the parts, are then found connected to each other. In other instances, two seemingly different structures are shown to probably represent different growth stages of a single species (or perhaps, at least, of the same genus). A further possibility is that organs in different states of preservation (one perhaps permineralized, one as a compression fossil), to which different names have been applied, are shown to represent the same organ. Some of the instances which have been documented are set out below to illustrate the diversity of cases in which different taxa may come to be linked:-

A. Linking of two genera based on different organs in the same preservation state:

1. *Cladophlebis denticulatus* later found in organic connection with the sporangium-bearing leaves *Todites undans* (Harris 1961).
2. The Jurassic female Bennettite fructification *Williamsonia gigas* and the leaf *Zamites gigas* associated with one another in many localities, and having significant agreement of cuticle characters, are regarded on these grounds as parts of a single species (Harris 1969).
3. The Carboniferous megaspores *Legenicula rugosa*, extracted from sporangia of the lycopod cone *Lepidostrobus russellianus* (Chaloner 1953a); this cone has recently been reassigned as *Flemingites russellianus*, see Brack-Hanes and Thomas 1983 (See also Fig. 5.2.).

B. Linking of two genera based on the same type of organ in two different preservation states

4. Stem impressions with distinctive and characteristic vertical seriation of leaf cushions, showing a more or less pronounced 'ribbed' stem surface formed the basis of the Carboniferous lycopod genus *Sigillaria*, (e.g. *S. scutellata*); permineralized stems, showing the same idiosyncratic leaf cushion arrangement as species preserved as impressions have long been included in the same genus (e.g. *S. elegans*). The peculiarity of leaf cushion and leaf scar character form the basis of correlating the two preservation states, and assigning both to a single genus.

5. Spores of the type called *Tuberculatisporites mamillarius* occur in the pedunculate lepidodendralean cone compressions called *Sigillariostrobus rhombibracteatus*; the same type of spore is present in the permineralized cone *Mazocarpon shorense* (Chaloner 1953b). On the basis of the distinctive type of spore present, and other circumstantial evidence (Schopf 1941), it is believed that the latter two cone genera represent different preservation states (compression, permineralization) of the cones of *Sigillaria*, and that these two species (*S. rhombibracteatus, M. shorense*) probably represent the same species of cone.

C. Linking of two genera, based on different organs, using a specimen combining different preservation states:

6. A compression specimen of *Archaeopteris* foliage (a Devonian progymnosperm) bearing sporangia had local pyrite permineralization of part of the leaf-bearing axis. The distinctive character of the xylem of the structurally preserved axis agreed with the stem genus *Callixylon* (Beck 1960). This established a link between the two genera, one based on fertile leaf compressions, one on permineralized axes.

These six examples demonstrate that links between fossil taxa are not all of comparable character or of comparable security. The circumstances are significantly different in each case.

The generic concept in fossil plants

Organ- and form-genera, being based on detached organs of fossil plants inevitably differ conceptually from genera based on whole living plants, which have at least the potential of yielding all available information concerning every phase of their life cycle. A genus based on dispersed spores is inevitably defined in terms of spore characters. Clearly there was a lot more to the plant than its spores, but our concept of the genus is perforce limited to spore features, and our sense of its time range must be founded on records of spores showing those characters. For me, a genus such as *Lycospora* Schopf Wilson and Bental is indeed a 'spore genus'—a genus based on a species typified by a holotype consisting solely of a spore, and defined in terms of spore morphology. There has long been evidence from spores found *in situ* in cones connected to leafy shoots (see e.g. Chaloner 1953a, and Brack-Hanes and Thomas 1983) that *Lycospora* was borne in cones comformable with those in the genus *Lepidostrobus*, and that cones of this character were borne by trees of the general character of *Lepidodendron (s. l.)*. I construe the relationship of these three genera (Fig. 5.1, left side) as that of a spore genus (*Lycospora*) some species of which, at least, were borne by cones assignable to the cone genus *Lepidostrobus*, some species of which were borne on stems having the character of the stem genus *Lepidodendron*.

I accept that this is a 'simplistic' treatment with all the derogatory implications that that term implies, for I am aware of the vulnerability of this interpretation of fossil taxa. Others, notably the late James Schopf, took a rather different view. His case is succinctly set out in his article on 'systematics and nomenclature in palynology' (Schopf 1969). Referring to the specimens on which fossil taxa may be based, he writes (p. 56) that, 'a single specimen. . .obviously can represent only a momentary

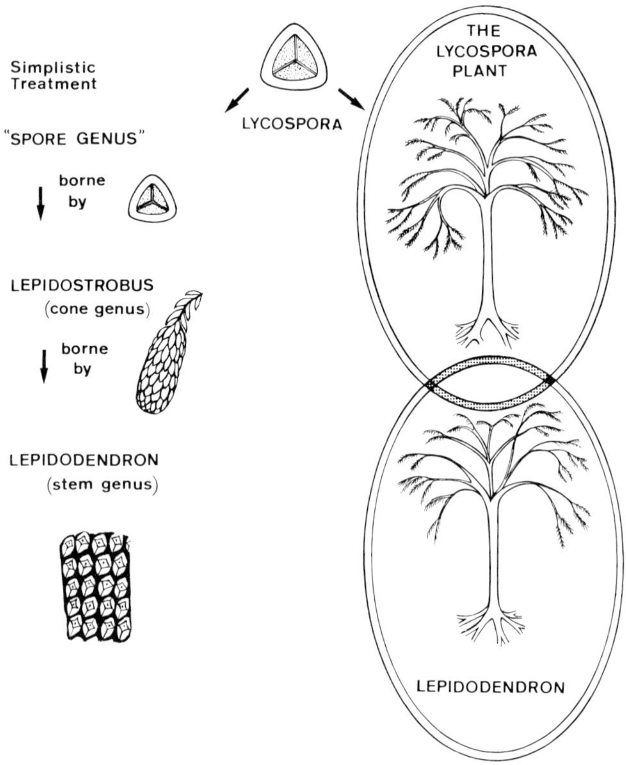

Fig. 5.1. A diagram to illustrate two concepts of fossil plant genera derived from the spore genus *Lycospora*. That advocated here is the empirical treatment shown at left: *Lycospora* is regarded as a spore genus, some species of which were borne by the cone genus *Lepidostrobus*. Species of this genus are known in connection with the stem genus *Lepidodendron*, based on an impression of a cortical surface with leaf cushions. Other authors (notably Schopf 1969) have advanced the view that *Lycospora* is to be regarded as a genus representing a whole plant of which we have only the spores. It is accepted that this plant is probably close to our concept of *Lepidodendron* as a whole plant, and that as shown in the diagram the two generic concepts probably overlap (stippled) to an unknown extent.

part of a contemporary stage. A species concept, however involves the theoretical recognition of the sequence of all life cycle stages, many of which must be generalised for want of specific information. A species represents a taxon of plants even though the species may be primarily defined only on the restricted basis of biocharacters morphologically represented in customary specimens'. He carries this theme to its logical conclusion in his discussion of organ-genera (p. 60), where he writes, 'organ genera are not genera of organs—such an expression is a contradiction in terms and is nonsense. Genera are taxa, or groups of allied species of plants; organs do not have any independent organic existence'. For Schopf there was therefore a *Lycospora* plant in its totality, and this was implicit in the genus *Lycospora* which for him referred to that whole plant, of which we have only the spore. For him, *Lycospora* was just as much a complete plant as *Lepidodendron*, based on an impression fossil of the trunk of that tree. For Schopf (personal communication ca. 1975) there was of course an overlapping relationship between the *Lycospora* plant and the *Lepidodendron* plant. I represent this by the overlap of the frames around these two theoretical concepts, at the right-hand side of the pictorial representation of Fig. 5.1. While I well understand the somewhat rarified concept of fossil taxa that Schopf subscribed to, I believe that the empirical treatment (left-hand side of Fig. 5.1) is much closer to the reality that we have to work with in applying names to fossil plants.

Some authors have regarded organ-genera as being essentially like non-fossil genera in this respect, and have synonymized their names, as the Code requires when two living plant (non-fossil) genera are shown to have been based on the same species. I cite one example, not wishing to condemn it as unacceptable practice, but simply to indicate that it can raise more problems than it solves. Stewart (1983), referring to the correlation of *Archaeopteris* and *Callixylon* mentioned above (p. 73) states that, 'correctly using nomenclatural priority, the earlier name first used by Dawson in 1871 was selected by Beck for the whole plant'. If this procedure is followed, it raises two types of problem. Firstly, did *all* the species of *Callixylon* trunk (with its characteristic anatomy) bear foliage conforming to the leaf genus *Archaeopteris?* It seems possible—indeed likely—that the evolution of foliage and stem anatomy were largely independent. The plant which first evolved a stem anatomy that palaeobotanists would assign to *Callixylon* may not have borne leaves of *Archaeopteris* type, and vice versa. To synonymize these two generic names (granting priority to *Archaeopteris*) denies the independent evolution of different organs of a plant. There is a further, separate but related problem here. If, as Stewart suggests, it is proper to now regard *Archaeopteris* as the name of the whole plant (including all *Callixylon* stem species) a report from a given locality of *Archaeopteris* may mean a leafy

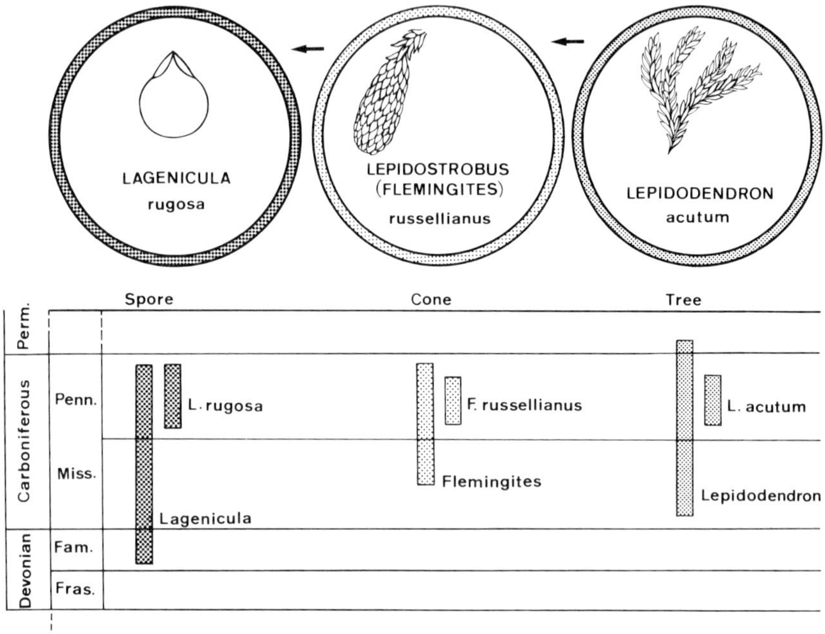

Fig. 5.2. Diagram to show the relationship between three genera based on isolated organs of Palaeozoic lycopods. The three species shown are believed to represent parts of a single species, in the upper part of the Pennsylvanian. However the three genera have a longer time range than the three species, and that of the megaspore *Lagenicula* extends into the Devonian. Despite the supposed connection of the three species illustrated, the three genera cannot be regarded as synonyms. The abbreviated designations at left refer to the Permian, Pennsylvanian and Mississippian, and within the late Devonian to the Famennian and Frasnian stages.

shoot with particular leaf character, or a fragment of permineralized wood. This ambiguity of use blurs our capacity to use names to specify the biological entities with which we are actually dealing.

A more serious—and interesting—problem arises in connection with the time dimension of our nomenclatural concepts. I illustrate this with the relationship between three genera of Carboniferous lycopods shown in Fig. 5.2. The cone believed to have been borne by the stem known as *Lepidodendron acutum* is *Lepidostrobus russellianus* (recently reassigned as *Flemingites russellianus* by Brack-Hanes and Thomas 1983). This cone is known to have borne megaspores of the type *Lagenicula rugosa* (Chaloner 1953a). The ranges of each of these three (Upper Carboniferous, Pennsylvanian) species lies, obviously, within the greater time range of each of these three genera. Further, within the limits of our knowledge, the three generic ranges are not coincident. Spores perfectly conformable with *Lagenicula* are known from Late Devonian rocks in Britain (Mortimer and Chaloner 1967). We have at present no knowledge of cones conforming to *Flemingites* sensu Brack–Hanes and Thomas, nor of lycopod stems conforming to *Lepidodendron s. s.* in rocks of Devonian age. It is obviously likely that some similar lycopod was producing the Devonian *Lagenicula* megaspores, but we know nothing at present of its cone structure or its stem. We cannot therefore simply treat *Lepidodendron, Flemingites* and *Lagenicula* as synonyms, much less assume that all species of all three genera were borne on the same type of whole plant. (For a more fully documented exploration of the problem touched on here, see Mosbrugger 1983.)

It is of course more cumbersome to retain separate names for fossil organs, even when connection is suspected or near-proven. None the less, if we accept that different organs could have evolved independently, it is inevitable. A recent instance of the practice advocated here is in Crane's (1984) treatment of *Cercidiphyllum*-like leaves, fruit and seeds from the British Tertiary. He uses the genus *Nyssidium* for the fruit and seeds of his fossils, based on *N. arcticum,* and *Trochodendroides prestwichii* for the leaves. The evidence for the connection between these two is strong, but Crane only refers informally to the '*Nyssidium* plant'. It is clear that this was similar to the living genus *Cercidiphyllum* but also shows significant differences. He acknowledges the possibility that, 'some *Trochodendroides*-like leaves. . .may actually represent other families'. This adds strength to the need to treat the leaf and fructifications fossils under separate generic names.

In conclusion, I simply renew the plea that fossil taxa, particularly genera, with their special problems of fragmentary types, varied preservation, and above all their time dimension, require that we treat them differently from living (non-fossil) taxa. Even when we establish the most secure connection between species of two fossil genera, those genera do no necessarily become synonyms, as they could if we were dealing with extant plants.

References

Beck, C.B. (1960) The identity of *Archaeopteris* and *Callixylon. Brittonia* **12**, 351–68.

—— (1969) Problems of Generic Delimitation in Paleobotany. *Proc. North Amer. Paleontological Conv.* **C**, 173–93.

Boulter, M.C. (1979) A proposal to emend Article 3.2 of ICBN *Taxon* 28(5/6), 598–600.

Brack-Hanes, S.D. and Thomas, B.A. (1983) A re-examination of *Lepidostrobus* Brongniart. *Bot. J. Linn. Soc.* **86**, 125–33.

Chaloner, W.G. (1953a) On the megaspores of four species of *Lepidostrobus. Ann. Bot.* **17**, 264–73.

—— (1953b) On the megaspores of *Sigillaria. Ann. mag. nat. hist.* **12**(6), 881–97.

Crane, P.R. (1984) A re-evaluation of *Cercidiphyllum*-like plant fossils from the British early Tertiary. *Bot. J. Linn. Soc.* **89**, 199–230.

Faegri, K. (1963) Organ and form genera: significance and nomenclatural treatment. *Taxon* **12**, 20–8.

Harris, T.M. (1926) The Rhaetic Flora of Scoresby Sound, East Greenland. *Meddr. Gronland* **68**, 45–148.

—— (1961) *The Yorkshire Jurassic Flora I. Thallophyta-Pteridophyta.* British Museum (Natural History), London. 212pp.

—— (1969) *The Yorkshire Jurassic Flora III. Bennettitales.* British Museum (Natural History), London. 186pp.

Jansonius, J. (1974) Form-genera versus organ-genera; a proposal. *Taxon* **23**, 867–68.

Meyen, S.V. (1975) Palaeobotanical taxonomy and nomenclature: the need for a new approach. *Taxon* **12**, 251–4.

—— and Traverse, A. (1979) Proposal on Article 3: Remove 'Form-Genus' too! *Taxon* **28**(5/6), 595–8.

Mortimer, M. and Chaloner, W.G. (1967) Devonian megaspores from the Wyboston borehole, Bedfordshire, England. *Palaeontology* **10**, 189–213.

Mosbrugger, V. (1983) Organische zusammengehörigkeit zweier fossil-taxa als taxonomisches problem am beispiel der jungpaläozoischen farnfruktifikationen *Scolecopteris* und *Acitheca. Rev. Palaeobot. Palynol.* **40**, 191–206.

Schopf, J.M. (1941) Contributions to Pennsylvanian paleobotany: *Mazocarpon oedipternum* sp. nov. and Sigillarian relationships. *Illinois Geol. Surv. Rept. Inv.* **75**, 1–40.

—— (1969) Systematics and nomenclature in palynology. In *Aspects of Palynology* (eds. R.H. Tschudy and R.A. Scott), pp. 49–78, Wiley, New York.

—— (1978) Unstated requirements in nomenclature for plant fossils. *Taxon* **27**(5), 485–8.

Stafleu, F.A. *et al.* (eds.) (1978) *International Code of Botanical Nomenclature*, 457pp., W. Junk, The Hague, Netherlands.

Stewart, W.N. (1983) *Paleobotany and the evolution of plants*, 405pp., Cambridge University Press, Cambridge.

Traverse, A.F. (1955) Pollen analysis of the Brandon Lignite of Vermont. *U.S. Bur. Miner. Rept. Inv.* **5151**, 1–107.

—— (1961) Effect of the 1959 International Botanical Congress on Nomenclature of fossil spores and pollen. *Micropaleontology* **7**(4), 485–9.

Voss, E. G. *et al.* (eds.) (1983) *International Code of Botanical Nomenclature*, 472pp., W. Junk, The Hague, Netherlands.

6. The role of reproductive organs in the classification of fossil plants

T.N. TAYLOR AND S.P. STUBBLEFIELD

Introduction

The reproductive parts of fossil plants have played a pivotal role in palaeobotany, being instrumental not only in characterizing major groups of plants, but also in providing a basic framework in which to consider their taxonomy, biology, and evolution. Because of their potential value, reproductive structures are perhaps the most prized fossil specimens. Their significance is evident in many groups. For example, the organization of the cone with adaxially positioned sporangia and the number and types of spores are perhaps the most useful characters in lycopod taxonomy (Thomas 1978, Thomas and Brack-Hanes 1984), and the distinctive features of the spores have been especially useful in plotting the stratigraphic and ecological distribution of lycopods (e.g. Peppers 1970). Furthermore, with fertile specimens it has been possible to characterize the reproductive strategies of many taxa of Carboniferous lycophytes, and recent information regarding the embryogeny of some forms has raised questions regarding evolutionary relationships among fossil and living lycopods (Stubblefield and Rothwell 1981, Rothwell and Erwin 1985).

Nevertheless, the value of reproductive organs in the classification of fossil plants is limited, and their discovery inevitably raises additional questions. Many groups of plants illustrate both the value and the limitations of reproductive material. Prior to about 1966, the coenopterid ferns were separated into four families based on the anatomy of the phyllophore, and various taxa were assembled into evolutionary lineages based on anatomical features and stratigraphic data. Because of the unusual anatomical organization of these plants when compared to extant ferns, the coenopterids were considered to be insignificant in the evolution of modern ferns (Eggert 1964). However, as sporangia were found in attachment to fronds it became clear that some coenopterid taxa were members of filicalean families while the affinities of others, also known with sporangia in organic attachment, remained questionable. The sporangia of *Sermaya*, for example, possess features suggestive of the Gleicheniaceae (Eggert and Delevoryas 1967). Those of *Tedelea* are most similar to sporangia of the Schizeaceae and Osmundaceae (Eggert and Taylor 1966), while the massive sporangia of the zygopterid ferns are unlike the sporangia of any living filicalean taxa (Galtier and Scott 1979).

The form genus *Callixylon* provides a similar example. Prior to 1960 *Callixylon* was believed to be the wood of a large Devonian gymnosperm while *Archaeopteris* was regarded as the foliage of a fern. So basic was this understanding that the vegetative foliage was described in pteridophytic terms. When these taxa were demonstrated to be parts of the same plant palaeobotanists were forced to modify long standing taxonomic concepts, and recognize an additional, major group of vascular plants (Beck 1960). However, further information concerning reproductive organs has raised additional questions. Recent discoveries of cupulate seeds from contemporaneous strata (Pettitt and Beck 1968; Gillespie *et al.* 1981) and the presence of both homosporous and heterosporous species of *Archaeopteris*, raise questions concerning the homogeneity of the genus and the taxonomic position of the group (Taylor 1981).

In other instances, reproductive material provides little, if any, insight into the natural affinities of a fossil plant. In the genera *Barinophyton* (Brauer 1980) and *Protobarinophyton* (Brauer 1981, Scheckler 1979) for example, it is the reproductive organs that provide the diagnostic features of the taxa. Although the reproductive structures are known in detail, their organization provides little information about the natural affinities of the plants. However, recent information about the fine structure of the spores suggests some relationship with members of the Lycophyta (Taylor and Brauer 1983, Cichan *et al.* 1984).

Other examples of plants whose affinities are uncertain despite knowledge of attached reproductive organs are the early land plants *Tortilicaulis* and *Horneophyton*. Although attached sporangia are known, it is difficult to determine whether *Tortilicaulis* is a bryophyte or vascular plant (Edwards 1979), largely because of inadequate preservation. *Horneophyton lignieri*, on the other hand, is an exceptionally well-preserved permineralization with attached sporangia with spores, but it possesses certain features which make it difficult to position systematically (Eggert 1974).

The above examples represent but a few of the many instances in which reproductive organs in fossil plants are relevant to classification. To underscore some of the problems and important taxonomic contributions that have come from the study of reproductive structures, we have chosen to highlight two widely differing groups—pteridosperms and Paleozoic fungi.

Pteridosperms

Our understanding of seed ferns has greatly expanded since the initial characterization of the group by Potonié based on vegetative features, and later by Oliver and Scott (1904) based on reproductive parts. In recent years palaeobotanists have amassed detailed information about the pteridosperms, recognizing almost all stages of the life histories of certain taxa, and have paid particular attention to the seeds.

Historically, Palaeozoic seeds were assembled into three orders: Lagenostomales, Trigonocarpales and Cardiocarpales. Lagenostomalean seeds are small (generally several millimeters), radially symmetrical and possess an integument fused with the nucellus along most of its length, a single vascular system and a complex pollen-receiving mechanism (Taylor and Millay 1981). In addition, the small size of these seeds suggests they may have been borne in a cupulate organ. In contrast, seeds of the Medullosales are radial and large (some up to 13cm long) with a histologically complex integument attached to the nucellus only in the chalazal region of the seed, vascular tissue in both the nucellus and integument, and a simple, bell-shaped pollen chamber (Taylor 1965). They were borne singly. Ovules of the Cardiocarpales are distinguished principally by their bilateral symmetry and features of the seed coat and vascular system. Today, these artifical orders based only on features of isolated seeds are roughly equivalent to the Lyginopteridales, Medullosales and Cordaitales respectively. However, as we have learned more about the Palaeozoic pteridosperms other seed ferns have been recognized whose features encompass more than one of these groups. Perhaps the most well known Carboniferous pteridosperms are the Callistophytales (Rothwell 1975; 1980). Many aspects of the life history of *Callistophyton* are known in great detail, including such ephemeral components as stages in the development of the microgametophyte and aspects of pollination. However, the identity of the Callistophytales was masked for many years by and artifical system of classification in which organs were placed in different groups. The stems were believed to be monostelic and were, therefore, placed in the Lyginopteridales; the seeds (*Callospermarion*) were bilateral and regarded as cordaitean; the pollen was saccate and considered to be cordaitean or coniferophytic, and some pollen organs were thought to be synangia of homosporous ferns. As various organs were either found in attachment or demonstrated to be isolated parts of the same plant, a biological concept of the plant *Callistophyton* emerged. Today it is the most well known seed fern, and there is a basis for understanding not only its reproductive history, but also its evolutionary position with respect to other major groups of seed plants.

At the present time the Carboniferous pteridosperms may seem, at least superficially, to constitute a well-defined group whose features provide a framework for interpreting both earlier and later forms. This, however, is not the case even among the late Palaeozoic genera. For example, among the seeds that would be included in the Trigonocarpales (= Medullosales) are seeds like *Hexapterospermum* that are radial and large, but whose pollen-receiving mechanism is more like that found in the seeds of some members of the Lyginopteridales (Taylor 1966). Moreover, among the medullosans there are many more types of seeds than stems. Although this may reflect, in part, developmental variation and even preservational modifications, ovule types still markedly outnumber types of stems. Thus, there is a continuing need for an artifical system in which to assemble isolated seeds, and despite the inadequacies of the traditional categories, it provides a general framework in which to compare and evaluate seeds whose affinities are unknown.

The morphology and anatomy of the pollen organs of the Carboniferous seed ferns have been less useful than seeds in the calssification of pteridosperms. Although many types of pollen organs have been described, until recently they have been classified principally on the basis of the pollen. Organs containing small, trilete grains were generally associated with the lyginopterid seed ferns (Taylor and Millay 1981), while synangia containing large monolete grains are associated with the medullosan pteridosperms (Dennis and Eggert 1978, Millay and Taylor 1979, Millay *et al.* 1978). The utilization of pollen in the classification of seed fern reproductive organs was advantageous in allowing the correlation of different preservational types. In recent years the classification of seed fern pollen organs has been refined through the discovery of several structurally preserved synangia which demonstrated the fundamental pattern of organization. Among the medullosan pollen organs the single ring of elongate sporangia is the basic type, with structurally more complex forms (e.g. *Dolerotheca*) resulting from aggregation and plication. An informal classification of these organs based on their level of complexity includes the categories simple, aggregate and compound (Millay and Taylor 1979). At this time, information about pollen organs is more useful in tracing the evolution of the organ than in contributing to the systematics of the group. Moreover, there are other Carboniferous pollen organs constructed of numerous elongate sporangia whose organization is still poorly understood. For example, the genus *Potoniea*, initially known from compression specimens, is now known to consist of several rings of clustered sporangia (Stidd 1978). In this case the organ, although structurally complex and similar to medullosan forms, does not reflect the basic organizational pattern seen in other medullosans, nor does it contain the expected type of pollen.

The late Palaeozoic and Mesozoic seed ferns pose even greater taxonomic problems than the Carboniferous pteridosperms. This, in part, reflects their preservation. The reproductive organs of these seed ferns generally occur as impression/compression fossils so that almost nothing is known about their internal structure (Thomas 1925, 1933). A notable exception is the order Glossopteridales. These plants, known only from Gondwana continents, were a dominant element of the Permian flora, but are also known from sediments of the Triassic. The leaf type, *Glossopteris*, is a widely distributed but relatively homogeneous form genus. The seed bearing organs, on the other hand, are morphologically diverse with some compressed forms apparently cupulate and others considered to be more strobilar in organization (Surange and Chandra 1976). The recent discovery (Gould and Delevoryas 1977) of structurally preserved *Glossopteris* remains from the Bowen Basin of Australia indicates that at least one of the reproductive organs of this order consisted of an infolded megasporophyll bearing sessile non-cupulate gymnospermous seeds on its inner surface. Although a sporophyll with this organization may assume various configurations when compressed, Rex (1985) has demonstrated using modelling experiments that not all of the compressed *Glossopteris* reproductive organs can be explained as a variation of the infolded megasporophyll. A similar degree of variability is present in the pollen organs and pollen (Lacey *et al.* 1975, Surange and Chandra 1976). The variability of the reproductive organs implies that there are several unrelated groups of plants within this order of pteridosperms, and the recent discovery of a cycad stem from the Triassic of Antarctica underscores the fact that there were other seed plants with reproductive organs that may have been morphologically similar to some of those produced by associated glossopterids (Smoot *et al.* 1985).

A second problem in the systematics of the late Palaeozoic and Mesozoic seed ferns is their ill-defined relationship with earlier pteridosperms. The apparent cupulate nature of some of the seed-bearing organs and the organization of permineralized glossopterid megasporophylls appear to be the principal reasons they are considered seed ferns, but they differ significantly from their supposed Carboniferous counterparts. Furthermore, the small size of the plants, with the exception of some glossopterids, also increases the difficulty of relating these seed plants to geologically older taxa. In dealing with the seed ferns, it would appear that, as in most groups of fossil plants, the reproductive organs by themselves provide an insufficient basis for taxonomy, and that the recognition of natural affinities will require the reconstruction of entire plants.

Palaeozoic fungi

Although the fungi have an extensive fossil history, we know relatively little about their diversity and structure through geologic time. This may reflect their delicacy and the general perception that fungi are not well preserved in the fossil record. The group may also have received little concentrated attention since there are few palaeobotanists sufficiently trained in mycology to deal with the complexity of these organisms. Finally, fossil fungi are often difficult to interpret because they consist mostly of vegetative remains, whereas the classification of extant fungi is based largely on reproductive structures. Nevertheless, fungal remains with diagnostic features are present in the fossil record. In many cases these are reproductive structures that provide sufficient evidence to classify a fossil fungus to class. However, as in the vascular plants, even well-preserved reproductive bodies may be subject to different interpretations and the affinities of a fungus may be unclear. In instances where reproductive structures are unavailable, classification may still be possible. Vegetative characters and evidence of fungal activity may also be taxonomically significant.

One group of fungi whose geologic history has been disputed is the Ascomycota. Until recently the fossil history of the group was not thought to predate the Cretaceous (Pirozynski and Weresub 1979), although possible imperfect fungal remains have been described from earlier strata (e.g. Pratt *et al.* 1978). The Carboniferous reproductive structure *Protoascon* is an example of a fungus that was initially thought to represent an ascomycete (Batra *et al.* 1964). This fungus is known from structurally preserved material and consists of a swollen hypha with a terminal pore. Surrounding the vesicle are a whorl of nonseptate hyphae. In the original report the vesicle was described as a multicellular perithecium. However, a re-examination of the material suggests that the vesicle is more similar to an oogonium surrounded by antheridia of some species of water molds (Baxter 1975). In this example, even though a reproductive structure was permineralized it was still difficult to determine the systematic position of the fungus.

Another Carboniferous fungus whose taxonomic position has been debated is *Traquairia*. This ornate spherical body has been classified with various groups of plants and animals. Initially, *Traquairia* was thought to be a radiolarian test (Carruthers 1873, Scott 1911) while other authorities regarded the spheres as megaspores of a lycopod (Williamson 1880) or the massulae of a heterosporous fern (Solms-Laubach in McLean 1912). The use of additional specimens and different techniques, especially critical point drying and scanning electron microscopy has resolved this taxonomic confusion (Stubblefield and Taylor 1983). The outer, ornate wall of *Traquairia* consists of interwoven, highly divided hyphae, and the discovery of sac-like structures containing spores within the larger sphere suggests that *Traquairia* is the cleistothecium of an ascomycete containing asci and ascospores. The spiny nature of the cleistothecium of *Traquairia* is reminiscent of erysiphalean cleistothecia (powdery mildews), but the absence of a hymenial layer suggests that these structures are more closely related to the Eurotiales. Included within this order are the sexual stages of *Penicillium* and *Aspergillus*.

Vegetative reproductive remains as well as vegetative hyphae may also be useful in the recogniton of ascomycetes. A recent report by Sherwood-Pike and Gray (1985) extends the probable geologic history of the ascomycetes into the Silurian based on multiseptate spores, ovate unicellular spore, filaments with flask-shaped appendages resembling phialides and branching hyphae with perforate septae.

Reproductive structures have also been of use in identifying several groups of lower fungi. An aquatic fungus that may be classified based on its reproductive structures is a Pennsylvanian chytrid-like fossil (Millay and Taylor 1978). Attached to the central body of saccate pollen grains are several sporangia with characteristic discharge papillae. Contained within the sporangia are numerous dark bodies up to 2.5 μm in diameter that may represent zoospores. To a large degree the classification of extant chytrids is based on features of the zoospores, especially the type of flagella. If present, the flagella of the fossil cannot be resolved. However, it is interesting that the dark bodies in the Pennsylvanian specimens contain what appear to be refractile bodies that are a feature of modern chytrids. The occurrence of these specimens on pollen grains, a common substrate for extant chytrids, also suggests that they are chytrids.

Other lower fungi known from reproductive structures are found among the Zygomycetes. Fossil *Glomus*-like chlamydospores from the Devonian to the Triassic occur as spherical, globose spores that possess a multilayered wall and are attached to hyphae by a narrow stalk (Wagner and Taylor 1982). These structures resemble the chlamydospores of modern endomycorrhizal fungi in morphology when viewed with transmitted light and scanning electron microscopy (Pirozynski and Malloch 1975, Stubblefield *et al.* 1985a). In living plants they are associated with roots and in the fossil record they have been found in the underground organs of vascular plants. Although these bodies resemble extant *Glomus* even in the nature of their attachment to hyphae, a particular physiological relationship cannot necessarily be inferred from morphological similarity. While their association with underground plant parts would imply the possibility of endomycorrhizal relationships, the absence of arbuscules and differences in the ultrastructural organization of the wall leave the question of an endomycorrhizal association open at the present time (Stubblefield *et al.* 1985b).

Reproductive structures have been particularly useless in the recognition of Palaeozoic basidiomycetes because fossil fruiting bodies are exceedingly rare. Nevertheless, features of the vegetative mycelium are sometimes useful in the recognition of this group. The most obvious are the diagnostic clamp connections often associated with vegetative hyphae. An excellent example of Palaeozoic clamp connections was reported by Dennis (1970) from a fungus associated with the Pennsylvanian fern *Zygopteris*. Additional evidence of basidiomycetes in the fossil record comes from the interaction of a fungus and its host. Specimens of the Devonian genus *Callixylon* were found infected with a fungus consisting of branching, septate hyphae and terminal and intercalary chlamydospores (Stubblefield *et al.* 1985a). Although clamp connections are not evident, the wood shows many characteristics of fungal white rot that is caused today by a number of modern basidiomycetes. A similar infection occurs in specimens of *Araucarioxylon* collected from the Permian of Antarctica (Stubblefield and Taylor 1985). Transverse sections of these woody stems show numerous areas of decayed tracheids. The decay in these specimens shows features of both white rot and white-pocket rot, two types of decay caused today by basidiomycetes. Where reproductive structures are not available, the affinities of a fossil fungus may determined by more indirect evidence.

Summary

The examples cited above illustrate some of the uses and limitations of reproductive parts in the classification of fossil plants. Many of the problems we have described arise because of the fragmentary nature of the fossils under study. However, even in instances where reproductive organs are exceptionally well-preserved, both the identity of their parent plants and their natural affinities may remain problematical. In the calssification of modern plants reproductive characteristics are usually augmented by suites of vegetative features which are often helpful in refining the classification. Likewise, reproductive parts rarely provide a complete answer to taxonomic questions involving fossil plants and must be used in conjunction with vegetative features.

In recent years our knowledge of fossil plants has greatly expanded as increasing numbers of plants have been reconstructed in part or in full. This is a primary goal of modern palaeobotany and is necessary to developing an understanding of the organism. This in turn provides the basis for considering many biological parameters including palaeoecology, the evolution of reproductive strategies, and development. As these areas become more fully understood, palaeobotany will be able to move toward a more natural classification. Until the entire organism is more completely known our classification schemes, whether they be based on vegetative or reproductive parts, represent only a convenient method of treating fragmentary remains of plants through geologic time. 'No system should ever be accorded the veneration of a creed nor should we ever regard as heresy efforts to change it. Classification systems should always reflect current knowledge, never perfect, but constantly being improved' (Arnold, 1959).

Acknowledgements

This study was supported in part by funds from the National Science Foundation (BSR-8402813 and BSR-8313786).

References

Arnold, C.A. (1959) *Paleobotany and plant classification*. Sixth Sir Albert Charles Seward Memorial Lecture, Birbal Sahni Institute of Palaeobotany, Lucknow, pp. 3–7.

Batra, L. R., Segal, R.H. and Baxter, R.W. (1964) A new middle Pennsylvanian fossil fungus. *Amer. J. Bot.* **51**, 991–5.

Baxter, R.W. (1975) Fossil fungi from American Pennsylvanian coal balls. *Univ. Kansas Contrib.* Paper **77**, 1–6.

Beck, C.B. (1960) The identity of *Archaeopteris* and *Callixylon*. *Brittonia* **12**, 351–68.

Brauer, D.F. (1980) *Barinophyton citrulliforme* (Barinophytales *incertae sedis*, Barinophytaceae) from the Upper Devonian of Pennsylvania. *Amer. J. Bot.* **67**, 1186–206.

—— (1981) Heterosporous, barinophytacean plants from the Upper Devonian of North America and a discussion of the possible affinities of the Barinophytaceae. *Rev. Palaeobot. Palynol.* **33**, 347–62.

Carruthers, W. (1873) On *Traquairia*, a radiolarian rhizopod from the Coal-measures. *Rept. British Assoc. Brighton,* p. 126

Cichan, M.A., Taylor, T.N. and Brauer, D.F. (1984) Ultrastructural studies of *in situ* Devonian spores: *Protobarinophyton pennsylvanicum* Brauer. *Rev. Palaeobot. Palynol.* **41**, 167–75.

Dennis, R.L. (1970) A middle Pennsylvanian basidiomycete mycelium with clamp connections. *Mycologia* **62**, 578–84.

Dennis, R.L. and Eggert, D.A. (1978) *Parasporotheca* gen. nov. and its bearing on the interpretation of the morphology of permineralized medullosan pollen organs. *Bot. Gaz.* **139**, 117–39.

Edwards, D. (1979). A late Silurian flora from the Lower Old Red Sandstone of south-west Dyfed. *Palaeontology* **22**, 23–52.

Eggert, D.A. (1964) The question of the phylogenetic position of the Coenopteridales. *Torrey Bot. Club* **21**, 38–57.

—— (1974) The sporangium of *Horneophyton lignieri* (Rhyniophytina). *Amer. J. Bot.* **61**, 405–13.

—— and T. Delevoryas. (1967) Studies of Palaeozoic ferns: *Sermaya*, gen. nov. and its bearing on filicalean evolution in the Palaeozoic. *Palaeontographica* **120B**, 169–80.

—— and Taylor T.N. (1966) Studies of Palaeozoic ferns: on the genus *Tedelea* gen. nov. *Palaeontographica* **118B**, 52–73.

Galtier, J. and Scott, A.C. (1979) Studies of Palaeozoic ferns: on the genus *Corynepteris*. A redescription of the type and some other European species. *Palaeontographica* **170**, 81–125.

Gillespie, W.H., Rothwell G.W. and Scheckler, S.E. (1981) The earliest seeds. *Nature* **293**, 462–4.

Gould, R.E. and Delevoryas, T. (1977) The biology of *Glossopteris*: evidence for petrified seed-bearing and pollen-bearing organs. *Alcheringa* **1**, 387–99.

Lacey, W.S., Van Dijk, D.E. and Gordon-Gray, K.D. (1975) Fossil plants from the Upper Permian in the Mooi River district of Natal, South Africa. *Ann. Natal. Mus.* **22**, 349–420.

McLean, R.C. (1912) A group of rhizopods from the Carboniferous Period. *Proc. Camb. Philo. Soc.* **16**, 493–513.

Millay, M.A., Eggert, D.A. and Dennis, R.L. (1978) Morphology and ultrastructure of four Pennsylvanian prepollen types. *Micropaleontology* **24**, 303–15.

Millay, M.A. and Taylor, T.N. (1978) Chytrid-like fossils of Pennsylvanian age. *Science* **200**, 1147–9.

—— and Taylor, T.N. (1979) Palaeozoic seed fern pollen organs. *Bot. Rev.* **45**, 301–75.

Oliver, F.W. and Scott, D.H. (1904) On the structure of the Palaeozoic seed *Langenostoma lomaxi*, with a statement of the evidence upon which it is referred to *Lyginodendron*. *Phil. Trans. Roy. Soc. London* **197**, 193–247.

Peppers, R.A. (1970) Correlation and palynology of coals in the Carbondale and Spoon Formations (Pennsylvanian) of the northeastern part of the Illinois Basin. *Ill. Geol. Surv. Bull.* **93**, 1–89.

Pettitt, J.M. and Beck, C.B. (1968) *Archaeosperma arnoldii*—a cupulate seed from the Upper Devonian of North America. *Cont. Mus. Paleo. Univ. Mich.* **22**, 139–54.

Pirozynski, K.A. and Malloch, D.W. (1975) The origin of land plants; a matter of mycotrophism. *BioSystems* **6**, 153–64.

—— and Weresub, L.K. (1979) A biogeographic view of the history of Ascomycetes and the development of their pleomorphism. *IN*: Kendrick, B. (ed): *The Whole Fungus: the sexual-asexual synthesis*, 93–123. Second International Mycological Conference, Calgary 1979, Proceedings Vol. 1 (Co-published by National Museum of Natural Sciences, National Museums of Canada and the Kananaskis Foundation).

Pratt, L.M., Phillips, T.L. and Dennison, J.M. (1978) Evidence of non-vascular land plants from the early Silurian (Llandoverian) of Virginia, U.S.A. *Rev. Palaeobot. Palynol.* **25**, 121–49.

Rex, G.M. (1986) Experimental modelling as an aid to interpreting the original three-dimensional structures of compressions. (*This volume*).

Rothwell, G.W. (1975) The Callistophytaceae (Pteridospermopsida) I. Vegetative structures. *Palaeontographica* **151B**, 171–96.

—— (1980) The Callistophytaceae (Pteridospermopsida) II. Reproductive features. *Palaeontographica* **173B**, 85–106.

Rothwell, G.W. and Erwin, D.M. (1985) The rhizomorph of *Paurodendron*:

implications for homologies among rooting organs of Lycopsida. *Amer. J. Bot.* **72**, 86–98.

Scheckler, S.A. (1979) *Protobarinophyton*: a link between the Zosterophyllales and the Barinophytales: *Bot. Soc. Am. Misc. Publ.*, **157**, 36–7.

Scott, R. (1911) On *Trquairia*. *Ann Bot.* **25**, 459–67.

Sherwood-Pike, M.A. and Gray, J. (1985) Silurian fungal remains: probable records of the class Ascomycetes. *Lethaia* **18**, 1-20.

Smoot, E.L., Taylor, T.N. and Delevoryas, T. (1985) Structurally preserved fossil plants from Antarctica. I. *Antarcticycas*, gen. n., a Triassic cycad stem from the Beardmore Glacier area. *Amer. J. Bot.* **71**, 410–23.

Stidd, B.M. (1978) An anatomically preserved *Potoniea* with in situ spores from the Pennsylvanian of Illinois. *Amer. J. Bot.* **65**, 677–93.

Stubblefield, S.P. and Rothwell, G.W. (1981) Embryology and reproductive biology of *Bothrodendrostrobus mundus* (Lycopsida). *Amer. J. Bot.* **68**, 625–34.

—— and Taylor, T.N. (1983) Studies of Palaeozoic fungi. I. The structure and organization of *Traquairia* (Ascomycota). *Amer. J. Bot.* **70**, 387–99.

—— and Taylor, T.N., (1985) Wood decay in silicified gymnosperms from Antarctica. *Bot. Gaz.* (in press).

—— Taylor, T.N. and Beck, C.B. (1985a) Studies of Palaeozoic fungi V. Wood decay in the Upper Devonian progymnosperm *Callixylon newberryi*. *Amer J. Bot.* (in press).

—— Taylor, T.N. and Miller, C.E. (1985b) Studies of Palaeozoic fungi IV: Wall ultrastructure of fossil endogonaceous chlamydospores. *Mycologia* **77**, 83–96.

Surange, K.R. and Chandra, S. (1976) Morphology and affinities of *Glossopteris*. *Palaeobotanist* **25**, 509–24.

Taylor, T.N. (1965) Palaeozoic seed studies: a monograph of the American species of *Pachytesta*. *Palaeontographica* **117B**, 1–46.

—— (1966) Palaeozoic seed studies: on the genus *Hexapterospermum*. *Amer. J. Bot.* **53**, 185–92.

—— (1981) *Paleobotany: An introduction to fossil plant biology*. McGraw-Hill, New York p. 589.

—— and Brauer, D.F. (1983) Ultrastructural studies of *in situ* Devonian spores: *Barinophyton citrulliforme*. *Amer. J. Bot.* **70**, 106–12.

—— and Millay, M.A. (1981) Morphologic variability of Pennsylvanian lyginopterid seed ferns. *Rev. Palaeobot. Palynol.* **32**, 27–62.

Thomas, B.A. (1978) Carboniferous Lepidodendraceae and Lepidocarpaceae. *Bot. Rev.* **44**, 321–64.

—— and Brack-Hanes, S.D. (1984) A new approach to family groupings in the lycophytes. *Taxon* **33**, 247–55.

Thomas, H.H. (1925) The Caytoniales, a new group of angiospermous plants from the Jurassic rocks of Yorkshire. *Phil. Trans. Roy. Soc. London*, Ser. B, **213**, 299-363.

—— (1933) On some pteridospermous plants from the Mesozoic rocks of South Africa. *Phil. Trans. Roy. Soc. London,* Ser. B. **222**, 193-265.

Wagner, C.A. and Taylor, T.N. (1982) Fungal chlamydospores from the Pennsylvanian of North America. *Rev. Paleobot. Palynol.* **37**, 317–28.

Williamson, W.C. (1880) On the organization of fossil plants of the Coal Measures X. *Phil. Trans. Roy. Soc. London.* **171**, 493–539.

7. Use of modern generic names for plant fossils

M.E. COLLINSON

Abstract

Some plant fossils exhibit diagnostic suites of characters all of which fall within the range of variation of comparable parts of a modern genus (or species) and are also unique to that genus (or species). Whereas the purpose of a name is solely as a means of reference and communication and usually any validly published name is appropriate, in these cases if the taxonomic information is to be recognized then use of the modern generic (or specific) name is required. The modern name (i.e. Linnean binomial), however, carries implications beyond those of the taxonomic information available on the fossil to include the whole plants and the various biological species concepts. Rejection of the modern name however ignores taxonomic information and infers that modern plants have no history in past time. Furthermore, there is a limit to the taxonomic information available on modern plant nomenclatural types and also there is a gradation from modern through subfossil to fossil within which no obvious boundary can be drawn where modern ceases and fossil begins. Use of various open nomenclature methods to indicate limitation of available taxonomic information on fossils have been considered but none are without practical problems. Communication of taxonomic information and hence the use of modern names is preferred here but emphasis must be placed on the limitation of this information. Unsubstantiated extrapolations via whole living plants of past phylogeny, ecology and environmental parameters must be discouraged. Reid and Chandler (1933) have pointed out,

> 'It is undoubtedly desirable and necessary to name well defined fossil genera of which the relationship is questionable but it can be done without blurring incomplete evidence and confusing the boundaries of living generic names. Having entered this plea for caution we would enter a complementary one for courage. If a species can definitely be proved to

belong to a living genus then let the name of the genus be used, and discard such half-hearted names as *Nipadites* which imply a lack of evidence that is not warranted by facts. We could wish for an altogether stricter usage of names in Tertiary Palaeobotany. Let the degree of uncertainty [and certainty—author's insertion] be reflected in the published lists and not be found merely in the discussion of species'.

Introduction

It might well once have been true to say that authors studying angiosperm dominated fossil floras attempted to identify all their specimens to modern genera. On occasion such identifications were achieved at the expense of critical assessment, specimens being assigned to modern genera or species based upon superficial resemblance, often also after examination of very limited comparative material. Such attitudes no longer prevail amongst the critical students of such floras.

This paper concerns only the use of modern generic (and specific) names in instances where a given plant fossil exhibits a diagnostic suite of characters all of which fall within the range of variation of comparable parts of the modern genus (or species) and are also unique to that genus (or species).

I am not concerned here with any continued 'casual' assignment of fossils to modern taxa. This is a matter for peer review and subsequent confirmation or revision. I am also not concerned with 'alternative names' in cases where the criterion above does not apply e.g. due to non-diagnostic characters, widely variable characters etc. Such material, which might once have been assigned casually to modern taxa or roughly grouped and held unnamed in collections, can now be assessed and categorised by computing technology. Such assessments (see the separate papers by Burnham, Hughes, and Spicer in this volume) may require alternative nomenclature systems. There is in my opinion no conflict between such other systems and the issues considered in this paper.

The time continuum—the transition from modern to fossil

New collections of Recent plant material are identified by botanists via specimens in reference herbaria though these latter may be one to two hundred (or more) years older. Fragmentary plant material for forensic, agricultural, or ecological (e.g. diet) assessment is also dealt with in this manner. Sub-fossils or potential fossils are examined by taphonomists. These are compared with Recent material growing around depositional sites and with reference herbarium material. Again a time-scale of one to two hundred (or more) years is involved. Plant material from archaeological sites (up to several thousand years old), Quaternary plant material (up to 2 million years old) and Pliocene and older material may be treated in a similar manner.

Although geological time is subdivided into units for reference purposes and hence plant material may be categorised as above, the boundaries between these categories are gradual, not sharp. In all the instances the suite of characters on the 'unknown' is compared with that available on reference material. If the two coincide (to the exclusion of any other) then identity is assumed.

There is no a priori reason to consider this procedure less appropriate as the time difference between reference and unknown increases and no sharp boundary or boundaries at which to impose any limitation. Furthermore, stratigraphic morphoclines (e.g. for *Stratiotes* seeds, Cook and Urmi-König 1983 and references there cited) are often documented in the fossil record. There are also numerous instances where a given fossil organ (or organs) with a diagnostic character suite is indistinguishable from the comparable organ (or organs) of modern plants (e.g. Collinson 1983a, b, Collinson and Crane 1978, Friis 1985, Mai and Walther 1978, Takhtajan 1974, 1982, Truchanowiczowna 1973).

If a thorough and critical comparison is made and the diagnostic suite of characters exhibited by a plant fossil organ falls within the range of those of a certain taxon of modern material (and not within those of any other modern material) then there is no reason to assume the two are different but good reason to assume that they can be identified to the same taxon (see also Crane 1984).

The critical point—nomenclature and taxonomy intertwined

All palaeobotanists follow the above uniformitarian comparative approach in their study of fossils, commencing at the recognition of plant (i.e Kingdom Plantae) and proceeding through phylal, ordinal familial, generic and specific level comparisons. (Ramifications at the structural and functional level are all-encompassing, ranging for example, through recognition of cells, spores, leaves, seeds, the tree stature, hydrophytic strategy, pollination biology etc.—all being based upon similarity between ancient and modern examples i.e. the uniformitarian concept.)

A critical point however occurs when assessment of similarity involves the generic (or lower) taxonomic rank. At this point application of the above approach requires the use of the modern name for the compared material. This then intertwines taxonomy and nomenclature, two classically distinct procedures; the former involving study, description and classification along with subsequent identification of unknowns within the classification and the latter merely involving the name. Again there is no a priori reason for ceasing to make use of the uniformitarian approach at the generic (and lower) level because of this problem. However the difficulties imposed by the implications conveyed by the name have given rise to alternative suggestions. Neobotanists are apparently content to apply existing names to newly collected material perhaps at least two hundred years younger than the nomenclatural type. Palaeobotanists show more concern about this procedure in spite of the time continuum because of the increasing periods of time involved and the consequent evolutionary implications (e.g. Hughes 1976).

Abuse of the modern name

Much of this concern arises in my opinion from the casual or indiscriminate use of modern names by authors who have followed one of three inappropriate procedures. The first involves inadequate comparative study of reference material and the second reliance on inadequate data on the fossil e.g. impressions lacking critical cellular detail. The third involves a thorough and critical comparison using a full suite of diagnostic information available on the fossil. However, inspite of recognition of one or several features which fall outside the range of variation of the modern taxon, the modern name is still applied to the fossil.

Of these three circumstances the first is possible to remedy and equally applicable in neo-and palaeobotany as plants may be inadequately known (or not even collected yet) and some groups offer huge numbers of taxa with which unknowns must be compared. In the second and third categories however some degree of subjectivity arises. In the second case the situation is again quite comparable for neo-and palaeobotanists and subsequent peer review must assess the adequacy of the data.

Situation three also applies in both neo-and palaeobotany but is much more acute, indeed critical, for the palaeobotanist. Two perhaps contrasting aims of palaeobotany conflict here; the first, to recognise first occurrences of the combinations of characters defining living plants and the second, to recognise the ancestors from which the living plants may have acquired these characters. Any palaeobotanist more interested in the former is likely to include a slightly expanded character suite in a modern group whereas one interested in the latter is likely to exclude it.

The criterion stated on p. 92 must be critically applied. Even with such rigour there remain two problems concerned with characters, problems which again apply equally to neo-and palaeobotany. Firstly features regarded as characters by one author may not be by another, several characters may be combined as one or indeed characters may be split to infinite degree. Such character definition is one of the fundamental difficulties in taxonomy. Furthermore new characters are continually being created, especially recently biochemical, chromosomal, and ultrastructural characters. For these reasons the 'all' in the criterion (p. 92) cannot be rigorously defined. For neo-and palaeobotanist alike 'all characters' tends towards the meaning 'all features currently used by other workers plus those additional in this new work'.

Rejection of the modern name

The other main source of concern however argues much more fundamentally against the entire methodology of applying modern names to fossils. In an alternative system modern nomenclature is applied to modern time and another nomenclature to past time, the element of time being emphasized as requiring that modern names cannot be used for fossils (Hughes, this volume).

Whilst such emphasis on the valuable additional evidence from the time sequence available from the fossil record is laudable there are several practical difficulties with this approach.

Firstly authors emphasizing the time element and the alternative system find no difficulty in applying comparisons with modern material at taxonomic ranks above the genus. That certain fossil pollen are angiosperm for example is acceptable. These then presumably possess a suite of diagnostic characters unique to modern angiosperm pollen. It is therefore also possible for certain fossil pollen to possess characters diagnostic of pollen of a named modern angiosperm order, family, genus or species. This does not of course mean that all will, merely that some might. If so then accepting the taxonomic information of angiosperm, angiosperm order, angiosperm family but rejecting that of angiosperm genus or species simply because nomenclature impinges on taxonomy seems inappropriate.

Clearly the intertwining of nomenclature and taxonomy is a problem as the implications carried with the name can be of whole plants, interbreeding populations or monophyletic lineages. Perhaps however a means of reducing these implications can be found which doesn't require the a priori rejection of a taxonomic decision simply because of the name.

Secondly, given the time continuum referred to above, it is not possible to locate a boundary, potentially acceptable to workers from all points in the continuum, where modern time ceases and past time begins. A possible boundary would be at the point where an organ is shed or removed from the plant or the plant removed from the population as occurs with fossils. However this would obviously also include all modern herbarium specimens. Modern could be taken from the time of collection of the first herbarium specimen or as with nomenclatural priority for modern plants at Linnaeus 1753, extrapolated to include specimens from sediments collected post that time. Whilst this would produce an operable arbitrary boundary it would probably be impossible to convince archaeologists e.g. those working on material from the sixteenth century warship *Mary Rose* or on human remains such as Neolithic peat burials e.g. 'Pete Marsh' to conform to such a boundary. Such difficulty is not in itself a reason for rejecting a useful step forward. However any new system must be applicable to all students of fossil material so that we can once and for all shed the stigma of 'Quaternary and Holocene fossils placed in modern species, Tertiary fossils in modern genera and Cretaceous fossils in modern families' which has consistently affected the credence given, especially to the angiosperm component, of the fossil record. Obviously a boundary at Recent (Holocene) time is also unsuitable for this reason.

Another possible boundary would be to include all 'buried' or 'dead and buried' objects as past time. Death would be a requirement as numerous small organisms often inhabit sediments (diatoms etc.). However death in the case of a seed or a resting cyst could be difficult to establish. Dead and buried objects, that is, those covered with sediments, would thus have a small grey area of doubt but might still prove an operable boundary. Objections raised above would also apply here.

The most significant difficulty with rejection of modern names hangs upon loss of information and rejection of a taxonomic statement. If the criterion on page 92 has been established and met, then rejection of the modern name becomes a rejection of a taxonomic statement directly ignoring known evidence to the contrary. Whilst use of the modern name carries implication of whole plants in interbreeding populations from monophyletic lineages (or some combination of these to different authors), rejection of the modern name implies certainly that these never existed, and more precisely that modern plants have no history in past time (whatever that has been designated to be).

Both are equally unsatisfactory. Possible solutions are considered in the final part of this paper.

Incomplete information on plant fossils?

The maximum available information for any plant fossil taxon will probably always be less then a similar maximum for a modern equivalent. The full suite of taxonomic information including structural, morphological, biochemical, chromosomal, biogeographical, ecological, overall population and interbreeding characteristics, evolutionary lineage and so on can never be known for ancient plants. It is pertinent to note that very frequently some of this information is also lacking for modern examples. The contrast, it may be claimed, is that the potential is available for the modern but not for the fossil.

However in nomenclatural terms modern taxa are frequently based on highly incomplete material i.e. the type specimen—which may for example be a sterile leafy shoot to represent a tree species. In these terms the information available to support the nomenclatural attribution is as limited as that usually available for fossils.

Indeed sometimes the information available on fossils surpasses that on modern material due to preservation states, fossilization processes and so forth. For example fruit stones (endocarps) have proved highly diagnostic of modern angiosperm taxa but are scarcely mentioned by taxonomists studying modern material because of the difficulties of observing features which are only expressed clearly following decay. The dearth of information on modern fern stelar anatomy compared with that available for fossil permineralizations, readily sectionable, is another example.

Furthermore as many modern type specimens were collected over one hundred, sometimes several hundred, years ago aspects of the taxonomic information, (mentioned above) may well have modified since that time. This will be especially true for species with rapid generation times. Aspects of the populations genetics have probably changed for all or most 'species' since their types were collected. Some of course are now extinct and thus have 'transferred to fossil status' since deposition of their types! Some taxonomic information must therefore be considered as totally lost for modern taxa as it is for fossil examples.

Especially where fossils consist of organs in attachment (or less ideally in recurrent and consistent association) the specimens cannot be considered any more or less complete than their equivalent modern herbarium specimens. Many plant fossils are however represented only by isolated organs for which the remainder of the plant is unknown. It may be argued that some of these, for example, a seed, a leaf, a pollen grain, a megaspore, convey much less taxonomic information than any type specimen from a modern species. Whilst this is arguable it is also important to note that some of these isolated organs (such as seeds) do actually represent the whole plant at one stage of the life cycle. Furthermore many of them can be clearly shown to fulfill the criterion established on page 92 (see page 93).

What name to use

Whereas a name can be anything at all, provided it conforms to the *International Code of Botanical Nomenclature* rules, this is not a simple situation (as discussed above) where rejection of the name also implies rejection of a taxonomic statement. In practice a diversity of information may be conveyed in the names of fossils, much of which is of taxonomic nature. Perhaps therefore it is pertinent to note here name categories, and the taxonomic information they convey, which are currently applied to fossils in an example of an angiosperm dominated fossil flora.

Categories of generic names used for Tertiary plant fossils

General

Names used for various fossils may often be categorised in terms of authors opinions regarding the diagnostic value of the organ (or organs) amongst modern plants. This, at least partially, reflects the attention (in the history of the subject) paid to comparative studies and also the role of the fossil in applied areas such as biostratigraphy, palaeoclimatology, palaeoecology and so on.

Isolated woods, for example, are frequently identified to modern families and not to lower rank modern taxa. Their names often reflect the family stem with an ending *-oxylon* for wood (cf. category (4) below). Isolated leaves include name categories similar to those listed below for fruits and seeds with *Dicotylophyllum* and *Monocotylophyllum* substituted for *Carpolithes* of category (5b). below. Isolated pollen however has two fundamentally different name categories, one as in category (1) below and the other reflecting the morphology of the grain. These convey very different taxonomic information e.g. *Tilia* versus *Intratriporopollenites*, one to modern plants, the other to a morphological classification only. In terms of nomenclatural category the *PDHC (Palaeobotanical data handling code*; Hughes, this volume) and it's variations are equivalent to the latter. A more detailed explanation is given below for names of fruits and seeds only.

Fruits and seeds

There are several categories of information conveyed by the names currently used for Tertiary fruits and seeds, although, of course, according to the *ICBN* these are actually only either modern (Category (1)) or form generic (Categories (2)–(5)) names. Also, incorrectly according to ICBN , most of these are considered to be assigned to modern plant families. Examples below are drawn from the London Clay Flora unless otherwise stated (see summary Collinson 1983a).

(1) Fossils whose diagnostic suite of characters falls within the range of variation of (and is unique for) that of a certain modern genus. The

name is that of the modern genus, e.g. species of *Magnolia, Dracontomelon* and so on.

(2) Fossils whose suite of characters falls close to, but not within, the range of variation of a certain modern genus; but is not as close to any other modern genus. The name is often that of the modern genus with a prefix, for example, *Eo, Palaeo* etc., e.g. *Eohypserpa, Palaeobruguiera, Pseudosclerocarya*.

(3) Fossils whose suite of characters falls within the range of variation of a certain modern family, but is not particularly similar to any modern genus or combines characters now segregated in two or several modern genera. Names may reflect the section of the family concerned but more usually have some other derivation, for example, from places, collectors, former names and so on, e.g. *Cupanoides, Wardensheppeya, Minsterocarpum Langtonia* and also *Sabrenia* (Collinson 1980).

(4) Fossils whose suite of characters falls within the range of variation of a certain modern family, but whose characters are general for that family. Names usually have the family stem with various suffixes such as *-spermum, -carpum*, e.g. *Anonaspermum, Bursericarpum, Laurocalyx, Icacinicarya, Euphorbiotheca*.

(5) Fossils whose suite of characters has no close living analogue.
  (a) those given distinctive names (due to abundance, highly distinctive features etc) even though their affinities are enigmatic e.g. *Wetherellia, Hightea, Lagenella*.
  (b) those assigned to species of *Carpolithes* or *Carpolithus*—both names erected for fruits and seeds of unknown affinity, the former having priority.

Alternative Approaches (Concerning category (1) above only)

Given that 'a plant fossil organ, (or combination of organs) exhibits a diagnostic suite of characters, all of which fall within the range of variation of comparable organs of the modern genus (or species) and are also unique to that genus (or species)' then, if this taxonomic information is to be recognized, the modern name must be applied to the fossil.

Some implications of this procedure may be inappropriate, for example, that whole plants in interbreeding populations or from monophyletic lineages are represented. Furthermore, with increasing time comes the increasing likelihood that other aspects of the taxonomic information might differ from those conveyed by the organ or organs available. Rejection of the modern name does not however seem to be a solution as, firstly, this could be equally applicable to new material of modern plants and secondly, it carries an equally unjustified assertion that modern plants did not have any history in past time (whenever that is designated to have begun).

One possible solution would be to reach a compromise in which the name also conveyed some aspects of the limitations. Such nomenclatural compromises are available in open nomenclature. Matthews (1973) states 'open nomenclature was developed as a remedy . . . if identity cannot be exactly determined if it [material] is too hastily referred to a known species or genus a previously clear taxonomic concept may be diminished. If one refuses altogether to identify it, potentially useful information may be left unemployed. . . The method offers a clear expression of the fact of uncertainty and also some of the degree of uncertainty involved'.

Possible methods

Method 1

Prefixes or suffixes attached to the name as a part of it, e.g. *-oxylon*, *-phylum*, *-spermum* indicating the nature of the record, such as wood, leaf or seed. Smith (1984) after detailed discussion (ibid. pp. 35–82) of the problems involved with fossil fungal material and the solutions proposed by Pirozynski and Weresub (1979) suggested (Smith 1984, p. 64) that a suffix . . . '*-sporites*' should be used appended to the modern generic name for fossil fungal spores.

Problems and comments

(a) Certain examples are in widespread use for fossils as mentioned above e.g. *Palmoxylon*, *Anonaspermum*, *Dicotylophyllum*. However these have been used largely to indicate general family or higher rank affinity and not the modern genus itself. Confusion would always be possible as to whether *Anonaspermum* implied seeds of *Anona* of Anonaceae unless the latter became '*Anonaceaespermum*'. Even if emendations were internationally agreed by *ICBN* to make this possible there is still the problem with the suffix '*-spermum*' where some modern names already include this e.g. *Menispermum*, *Lithospermum*. There may well be similar problems with other suffixes. If the terms were added as prefixes it is possible that these problems could be overcome.

(b) The suffixes of prefixes used for each organ might need to be internationally agreed, or at least the language (ideally Latin) on which they would be based could be internationally agreed. Furthermore there are a large number of potential organs involved e.g. for angiosperms- flower, anther, pollen grain, seed, fruit, leaf, mature wood, twig wood, leafy shoot, stem, root, inflorescence, infructescence, stipule, thorn and so on. The list could reach at least 100 including all plant groups. Obviously organic connection between organs would need to be considered (see also Chaloner, this volume). Flowers are an obvious example which are likely to be found in in dispersed parts as well as intact. However this is not a severe problem as a name incorporating, for example, *flos* could be used for the entire flower, or *anther* for anthers and so on for each part found isolated.

(c) There will be room for argument about how to apply certain prefixes and suffixes: for example is the suffix *-oxylon*, for wood, appropriate in the case of *Palmoxylon* where this is actually a primary stem?

Method 2

A term indicating the organs could be added after the name, but always cited with it as is the author, for example *Genus species* author seed. In effect this is merely moving the first citation of the organ as the first word in the diagnosis forwards to become a component of the full citation of the name.

Problems and Comments

(a) Severe risk of confusion with authors such as Dr Seed, Mr Leaf, Ms Wood. Could be avoided by use of italics or by latinising term e.g. *Genus species* author *seed,* or *Genus species* author *spermum*. International agreement would be necessary.

(b) Ease of removal by printers, editors and typesetters and potential confusion with other words in the text.

(c) The final term would be dropped, as are authors names in the general text.

Method 3

Use of open nomenclature symbols. Collinson (1980) proposed the use of *?* as the first 'letter' of the name where a fossil organ agreed in all available information with the comparable modern organ, but where certain information, shown to be diagnostic in the modern organs, was lacking on the fossil. This could easily be extrapolated to any situation where any aspect of the taxonomic information is lacking (although as remarked earlier in reality this could mean all taxonomy!). Other symbols could perhaps also be used.

Problems and comments

(a) Possible confusion with open nomenclature already operating.

(b) Cannot be pronounced with the name and therefore not useful in conversation or verbal presentations.

(c) Does not indicate anything about the organs involved and is therefore much less informative.

(d) Might be misunderstood and removed, for example, by editors of typesetters.

Method 4

Use of *fossilis*. Frequently in earlier literature the term *fossilis* was added e.g. *Genus species* author *fossilis*, conveying obviously that the material was fossil.

Problems and comments

(a) Could be unclear whether applicable at the generic or specific level.

(b) As for 2(b) and (c) and 3(c) and (d) mentioned above.

Conclusions

Casual use of modern names for fossils and assumptions that, for example, Cretaceous fossils can be placed in modern plant families, Tertiary fossils in modern plant genera and Quaternary and Holocene fossils in modern plant species should be eradicated from palaeontological persuit. Each unknown must be individually investigated and compared with known material, any taxonomic assessments being based upon this thorough comparison.

Unfortunately when taxonomic assessment indicates that a fossil exhibits a diagnostic suite of characters all of which fall within the range of variation of comparable parts of a modern genus (or species) and are also unique to that genus (or species) : then this necessitates either the use of the modern name or the rejection of this taxonomic information. Use of the modern name (i.e. Linnean binomial) carries the implication of additional aspects of taxonomic information not available on the fossil and also of the various biological species concepts involving interbreeding populations or monophyletic lineages.

Whereas one suggestion is therefore to reject the use of modern names for fossils this produces incongruities. Firstly much taxonomic information is also often lacking on the nomenclatural type specimens of modern species and also especially at the population level much of this information will have changed since the type specimen was collected. Secondly there is a continuum both in time and morphology between 'modern' and 'fossil' in-spite of the regular use of these terms. This continuum may be exemplified by; a fruit on a tree, a fruit on an herbarium sheet, a fruit on the ground beneath the tree, a fruit buried in the leaf litter beneath the tree, a fruit blown away from the tree, a fruit in an ancient compost heap on an archaeological site, a fruit in a Pleistocene river gravel, a fruit in a Tertiary clay etc., etc., back in time.

The only point at which a boundary can be placed in this continuum occurs at the biological level of removal from the plant or removal of the plant from the population or lineage—this of course applies equally to entry to the potential fossil record or entry to the herbarium. Very frequently plant fossils are indistinguishable from comparable organs of modern plants and others occur in stratigraphic morphoclines within which boundaries cannot be drawn.

Whilst it must be recognised that modern and fossil nomenclature cannot be segregated without the taxonomic implication of rejection of taxonomic information it must also be recognised that with increasing time comes increasing likelihood that other aspects of the taxonomic information might differ from those conveyed by the available material. Thus some compromise retaining the taxonomic information that all available data on the plant fossil falls within the range of variation of that on a certain modern taxon, but emphasizing the recognition that other features, if available, might not, would be a possible solution to the dilemma.

Various 'open nomenclature' methods have been considered. The use of a suffix composed of the Latinised term for the taxonomic information (i.e. the organ or organs represented) is probably most appropriate and informative, but is not without practical problems.

Until such a system is perfected use of modern names must be accompanied by emphasis on the limitations of the information. The *Stratiotes* fossil record for example, consists (largely) of lineages of *Stratiotes* seeds, not whole plants. (It is my opinion that such distinctive and diagnostic fossil seeds do indicate that the plants which bore them were involved in the ancestral lineage of modern *Stratiotes* plants. However such opinions must not of course, influence, or be influenced by, nomenclature.)

Similarly when, palaeoecological or palaeoclimatic inferences are drawn from floral lists involving modern names these must be supported by independent data. Key points may be established where for example facies or diverse biota may be combined to support ecological inferences concerning plant fossils (e.g. Collinson 1983b). Isolated occurrences of such fossils may then be more reasonably assumed to carry the same implications. Similarly palaeoclimatic inferences may be supported by data from a diversity of plant and animal groups and from oxygen isotope studies, rather than relying solely on the distribution of near living relatives inferred via the use of modern names for fossils.

Acknowledgements

Participation in numerous lively discussions on the matters raised in this paper, along with others raised in this volume, have greatly enhanced my awareness of the problems considered. I should like to thank the following for their part in these: Dr M. Boulter, Professor W.G. Chaloner, Mr C. Cleal, Dr W. Diver, Dr J. Hooker, the late Ms M. Jones, Dr P. Smith, Mr T.M.F. Windle and Dr G. Wilkinson. This work was carried out under the tenure of a Royal Society 1983 University Research Fellowship which is gratefully acknowledged.

References

Burnham, R.J. (this volume) Morphological systematics and the Ulmoideae.

Chaloner, W.G. (this volume) Reassembling the whole fossil plant, and naming it.

Collinson, M.E. (1980) Recent and Tertiary seeds of the Nymphaeaceae *sensu lato* with a revision of *Brasenia ovula* (Brong.) Reid and Chandler. *Ann. Bot.* **46**, 603–32.

—— (1983a) *Fossil Plants of the London Clay*. Palaeontological Association Field Guides to Fossils No. 1. Palaeontological Association, London.

—— (1983b) Palaeofloristic assemblages and palaeoecology of the Lower Oligocene Bembridge Marls, Hamstead Ledge, Isle of Wight. *Bot. J. Linn. Soc.* **86**, 177–225.

—— and Crane, P.R. (1978) *Rhododendron* seeds from the Palaeocene of southern England. *Bot. J. Linn. Soc.* **76**, 195–205.

Cook, C.D.K. and Urmi-König, K. (1983) A revision of the genus *Stratiotes* (Hydrocharitaceae). *Aquatic Botany* **16**, 213–49.

Crane, P.R. (1984) Misplaced pessimism and misguided optimism : a reply to Mabberley. *Taxon*, **33**, 79–82.

Friis, E.M. (1985) Angiosperm Fruits and Seeds from the Middle Miocene of Jutland (Denmark). *Biol. Skr.* **24 (3)**, 165pp.

Hughes, N.F. (1976) *Palaeobiology of Angiosperm Origins*. Cambridge University Press, Cambridge.

—— (this volume) The problems of data-handling for early angiosperm-like pollen.

Mai, D.H. and Walther, H. (1983) Die Floren der Haselbacher Serie im Weisselster-Beckens (bezirk Leipzig, DDR). *Abh. Staatl. Mus. Dresd. Mineral. Geol.* **28**, 1–200.

Matthews, S.C. (1973) Notes on open nomenclature and on synonomy lists. *Palaeontology*, **16**, 713–19.

Pirozynski, K.A. and Weresub, L.K. (1979). The classification and nomenclature of fossil fungi. In *The Whole Fungus* Vol II (ed. B. Kendrich), National Museums of Canada, Ottowa.

Reid, E.M. and Chandler, M.E.J. (1933) *The London Clay Flora.* Brit. Mus. (Nat. Hist.), London.

Smith, P.H. (1984) *Fossil fungi from early Tertiary deposits of southern England.* Unpublished PhD thesis, University of London.

Spicer, R.A. (this volume) Comparative leaf architectural analysis of Cretaceous radiating angiosperms.

Takhtajan, A. (Ed) (1974) *Magnoliophyta Fossilis URSS vol. 1, Magnoliaceae—Eucommiaceae* (in Russian). Nauka, Leningrad.

—— (1982) *Magnoliophyta Fossilia URSS vol. 11 Ulmaceae—Betulaceae* (in Russian). Nauka, Leningrad.

Truchanowiczowna, J. (1973) Variability of the recent and fossil fruits of the genus *Dulchium*. *Acta palaeobot., Cracov* **14**, 119–43.

8. Morphological systematics and the Ulmoideae

R.J. BURNHAM

Abstract

Morphology is the basis for classification of fossil angiosperm leaves. The taxonomic utility of characters recognizable in fossils can be evaluated best by reference to the pattern and range of variation of these same characters in their modern descendants. Multivariate analyses of the modern taxa provide the basis to determine critical characters of the foliage useful in distinguishing extant taxa. These characters then can be used at an equivalent rank in classifying fossil leaves.

Classification of fossil leaves of the Ulmaceae from the Palaeogene of western North America serves as an example of these alternative systematic methods. Generic affinities of the fossil specimens were determined using discriminant functions based on the modern genera of the Ulmoideae. Intra-generic groups were determined by ordination in combination with subjective considerations. Of the seven genera in the extant Ulmoideae, three, *Chaetoptelea*, *Ulmus*, and *Zelkova*, are recognized in Paleogene collections. Each can be subdivided into two or more morphotypes. Circumscription of morphotypes has advantages over application of binomial names to fossil forms. The morphotypes are considered to be population complexes that may be equivalent to more or less than one species. In contrast, the implications of defining a binomial species in the fossil record include its treatment as an interbreeding, monophyletic taxon with a unique biogeographic history.

A statistical approach to modern and fossil organs also allows evaluation of the distribution and variability of individual characters. The mean and variance of metric characters of modern organs may be compared to those of ancestors in the fossil record. Sampling of many modern taxonomic groups that also occur in the Tertiary fossil record should provide a first approximation to the range of character variability expected in fossil taxa. The application of this technique may extend by inference to the expected variance of characters found in fossil plants with no modern descendants.

Morphological systematics and the Ulmoideae

Introduction

Palaeobiologists are unique in the biological sciences in their ability to incorporate geologic time as a dimension of working hypotheses. The pattern of morphological change and the pattern of response to ecological change may be documented over a time scale relevant to significant evolutionary questions. The fossil record is the sole source of data on long-range ecosystem stability and on the pertinence of modern ecological principles to ancient ecosystems. Palaeobotanical data are applicable to questions concerning the tempo and mode of evolution because of the sessile nature, rare behavioural complications, and relatively abundant fossil record of plants. Fragmentation of plant remains, both pre- and post-burial, need not impose as severe restrictions on palaeobotanical applications as is commonly thought (e.g. Briggs and Walters 1984, Beck 1969). Once assemblages of organs derived from the same parent-plant species are determined, the predictable disaggregation and differential deposition of different organs of the same plant prior to fossilization (e.g. leaves and pollen or spores) permit stratigraphic, biogeographic, and ecological interpretations. In the case of angiosperm fossils, fragmentation in the form of abscission results in preservation of many individual plant organs from which the morphological variability of characters can be estimated once the organs are determined to belong to the same taxon.

The usefulness of fossil taxa increases dramatically as the phylogenetic affinities and ecological ranges are determined. Taxonomic treatments of specimens should reflect our confidence in their placement within a phylogenetic hierarchy. At the present time, regardless of the confidence in the naturalness of the taxon to which a fossil specimen is assigned, all specimens are given binomials in accordance with the *International Code of Botanical Nomenclature*.

The aim of this contribution is to present an alternative for infrageneric classification and to discuss the implication of this alternative in light of systematic investigations in palaeobotany. For angiosperms in particular, organs of extant taxa can be used as the basis of a strictly morphological classification scheme in which characters useful for distinguishing taxa of different rank can be determined. Characters selected must be those commonly preserved in the fossil record and may be micro- or macro-morphological features. Multivariate techniques are used to classify fossil specimens, and are objective and repeatable in their circumscription of the taxa. Such methods permit a more rigorous definition of fossil taxa and enhance our confidence in phylogenetic analyses, which always are constrained by our confidence in the taxa under scrutiny. Knowledge of the phylogenetic relationships of taxa will strengthen the role of palaeobotany in evolutionary biology.

The Ulmoideae as a case history

Fossil and extant members of the subfamily Ulmoideae (Ulmaceae) have been investigated in a taxonomic revision of early Tertiary fossil specimens from the western United States. The study serves as an example of the methods proposed for morphological systematics (Burham, in press). The subfamily was chosen for study because it is a group of deciduous trees with extant representatives occupying a variety of lowland habitats: conditions that provide abundant potential specimens for the fossil record. Extensive morphological work on the modern Ulmoideae has centred on foliage characters, reflecting the tendency of taxonomists to rely heavily on uniquely derived vegetative characters in their classifications (Melville 1978, Melville and Heybroek 1971, Richens 1983). Therefore, the Ulmoideae is a group for which the bases to classify fossil and extant plants are similar.

The extant Ulmoideae is composed of seven genera, *Ulmus*, *Zelkova*, *Chaetoptelea*, *Planera*, *Hemiptelea*, *Holoptelea* and *Phyllostylon* (Grudinskaya 1967, Gianassi 1978). Of these, only the first four have been reported previously from the fossil record. The Ulmoideae probably represent a monophyletic group, although *Zelkova* and *Hemiptelea* have been placed in the Celtoideae in the past (Elias 1970, Sweitzer 1971).

In all, 256 modern leaf specimens were examined, representing 48 species in the seven genera of the Ulmoideae. Twenty metric and discontinuous morphological characters, commonly preserved in fossil leaves of the Ulmoideae, were scored in the extant leaves. Eight additional characters, designed to express aspects of shape and vein density, were made from ratios of the simple characters. Some characters were eliminated from the analyses because they reflected size, which was measured by other characters, or because there was too little character variation among genera to allow separation. Nine characters remained after screening and these were used for discriminant analysis (Nie *et al.* 1975):

(1) mean number of teeth per secondary vein in the lower half of the lamina
(2) apex shape
(3) degree of base obliquity
(4) difference between basal and medial secondary vein angles
(5) length to width ratio
(6) linear proportion of leaf below the widest point on the lamina (proximal to petiole)

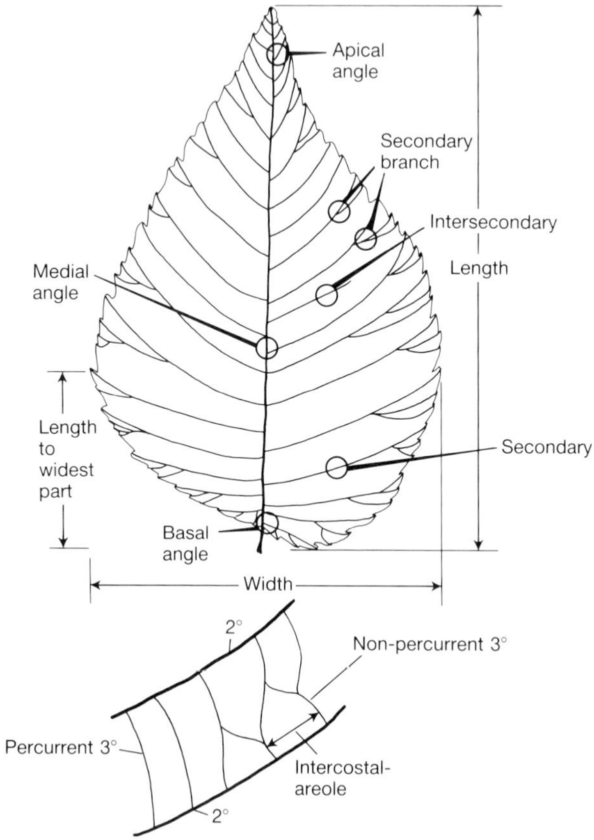

Fig. 8.1. Morphological features measured on modern and fossil Ulmoideae specimens.

(7) number of secondary veins per centimetre
(8) proportion of secondary veins with prominent branching point(s)
(9) number of intersecondary veins/ number of secondary veins (see Fig. 8.1).

Three genera, were eliminated from the analysis, *Phyllostylon*, *Holoptelea*, and *Hemiptelea*. These genera are distinct morphologically from the other four and none of the fossils from the Paleogene resemble them.

Discriminant analyses were used to distinguish genera of the Ulmoideae based on characters of the leaves. This type of analysis is employed in other fields (Sokal and Rohlf 1981, Davis 1973) but rarely is applied in palaeobotany. The analysis determines an algebraic equation that maximizes the differences between the mean character values of pre-defined groups (Davis 1973). A discriminant function is determined for each pair of groups. The simple case for two genera scored on two characters will serve as illustration. Each genus may be plotted as a frequency

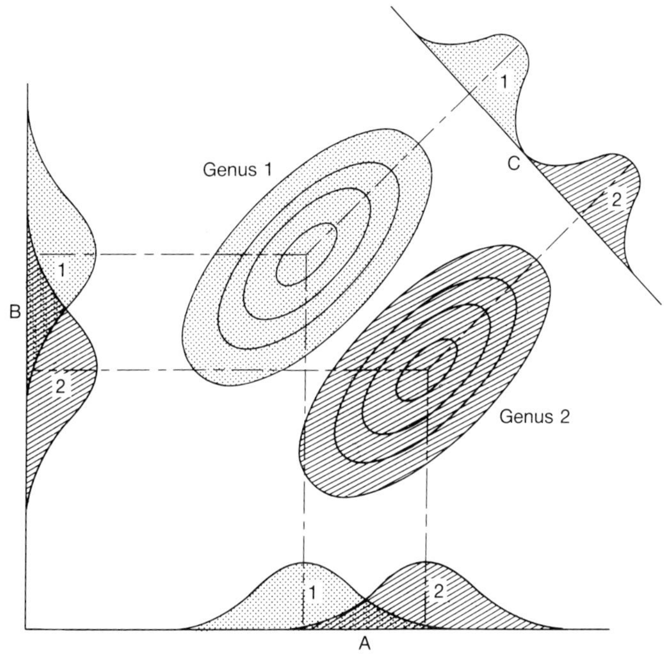

Fig. 8.2. Bivariate distributions overlap on characters A and B but are distinguished by projecting specimens of the two genera onto the discriminant function C (redrawn from Davis 1973).

distribution based on the two characters, A and B (Fig. 8.2). In two dimensions these distributions may overlap to some degree. As a result, neither character A nor B adequately distinguishes the two genera. Discriminant analysis makes a specific combination of the values of A and B as a new variable, C, that best distinguishes the genera. The new variable provides an axis onto which the distributions are projected and easily distinguished. The genera must be known a priori; the resulting discriminant functions then can be used to classify unknowns according to the morphological criteria that will separate the known genera.

Discriminant functions were made that best distinguished among the four genera *Chaetoptelea*, *Ulmus*, *Zelkova*, and *Planera*. The functions were tested for accuracy by treating known specimens as unknowns and determining the percentage of correct classification. Then, discriminant functions were made using each combination of two genera in paired comparisons. The resulting functions also were tested for accuracy treating known specimens as unknowns. Accuracy in determining the generic identity of individual specimens increased dramatically with the following use of the discriminant functions:

(1) a known modern specimen, treated as unknown, is classified by a discriminant function based on all four genera
(2) the two genera most similar to the 'unknown' are determined
(3) the 'unknown' specimen then is classified using the pairwise discriminant function based on these two most similar genera. The accuracy in determining the correct genus using this procedure ranges from 85 to 100 per cent (see Burnham, in press).

Fossil specimens, as true unknowns, are classified using these same discriminant functions, following the procedure outlined above. Over 500 fossil specimens from the Paleocene and Eocene of western North America were classified to genus using this procedure. The generic placement of fossil specimens therefore is determined on the basis of the range of morphological characters of extant genera. Fossil specimens were not found that showed character states intermediate between two modern genera or beyond the limits of the modern genera. Thus, establishment of new genera was not warranted in the Ulmoideae. The genus *Chaetoptelea*, containing a single species often placed in the genus *Ulmus*, has a distinct morphology that is recognizable early in the Tertiary and is justified as a distinct genus. No fossil specimens were placed in the genus *Planera* by the discriminant analysis, a condition that supports recent authors who have transferred fossil specimens previously assigned to *Planera* to other genera (Hickey 1977, MacGinitie 1941, 1969). The present analysis indicates that *Planera* is not present in the early Tertiary specimens from the western United States.

The facility with which genera are distinguished based on leaves alone indicates that the modern genera within the Ulmoideae are distinct entities with few intermediate forms. Metric characters distinguishing the genera intergrade, but modern specimens rarely are misclassified because the combination of characters is unique for each genus. For fossil specimens previously examined by other authors, there is an 80 per cent coincidence between the multivariate and traditional methods (*sensu* previous authors: MacGinitie 1941, 1969, Hickey 1977, Wolfe 1977, Wing 1981) for identification. Thus, fossil genera within the Ulmoideae are distinct and apparently have remained so since the early Tertiary.

Subdivision of fossil genera proceeded using principal components analysis (Nie *et al.* 1975). This ordination technique is a multivariate method used to represent many characters and individual specimens in one to three dimensions with as little loss of information as possible (Gauch 1982). A plot is produced in which similar specimens are close to each other and dissimilar ones are separated. All fossil specimens of a given genus, as determined by the discriminant analysis, were entered in an ordination. The same characters were used for ordination as were used in the discriminant analysis. Sub-groups are recognized within the genera *Ulmus* and *Zelkova*. The characters unique to the subgroups were largely the same ones useful in distinguishing modern species as well as those that best grouped specimens from the same locality. The genus *Chaetoptelea* was difficult to subdivide by ordination. In this case, the subjective characters of size, shape and distribution of marginal teeth, and the density of tertiary venation were used to partition the genus into two groups.

The infrageneric groups are designated as *morphotypes* in the resulting classification. Morphotypes, as used in this study, are groups of specimens that share morphological features and that may be supra- or sub-specific in taxonomic rank. A morphotype can be characterized by the mean and variation of all specimens assigned to it in the fossil collections. A specimen, close to the mean in many diagnostic characters, may be designated as a specimen characteristic for the morphotype, analogous to the type specimen of a palaeontological species. There are fundamental differences between the classical methodology for designating palaeontological species and that described here for morphotypes within the Ulmoideae. First, the genera are defined by a repeatable multivariate technique, using all characters from modern genera. Second, morphotypes within genera are designated by ordination techniques based on characters used to distinguish modern species and are determined using all Paleocene and Eocene fossil specimens available from the western United States. Third, the morphotypes convey only the implications of morphological difference existing among morphotypes. A range of variation may be assigned to each morphotype and thus provide a means to assess the probability that specimens found subsequently belong to the morphotype.

Species within the genera of the extant Ulmoideae often are difficult to distinguish due to frequent interspecific hybridization and to a high degree of variability, both among leaves produced early and late in the season on the same tree, and among different populations (Melville 1938, 1978, Richens 1983). Consequently, species in the fossil record of the Ulmoideae should be accepted with great caution unless they are defined on the criteria used to distinguish extant modern species (fruit morphology, distribution and number of leaves on short shoots, and inflorescence morphology—Melville and Heybroek 1978). I have chosen to characterize all fossil specimens as morphotypes (not bearing binomials) within the clearly defined genera. The logic of this approach and the implications for evolutionary studies will be discussed in the following sections.

Linnean Binomials

Species named using binomial nomenclature imply a biological species concept, requiring that the specimens belong to genetically cohesive populations that are potentially interbreeding and reproductively isolated from other such populations (Mayr 1965, Ghiselin 1975). Most biologists would add that species are monophyletic (Wiley 1978). These implications, hard enough to justify for modern taxa, are essentially impossible to defend in palaeontology. Riedel (1978) states that binomials indicate: (1) reproductive isolation of a group of populations from other such groups (2) interbreeding within the group of populations (3) morphological description of specimens belonging to the group and (4) phylogenetic placement. Palaeontologists can only hope to fulfill his third and fourth objectives. I propose that these objectives are not realized adequately by palaeontological species assigned Linnean binomials.

Species defined based on specimens from the fossil record often are described from localities of limited exposure, providing few individuals on which a characterization of the species is based. The species normally is known only to the collector and/or author of the flora and it is to this authority that new collections will be referred. Unless a thorough review of the modern and fossil relatives of specimens has been undertaken in defining each new species from a flora, species are based on the limited experience of the author with the related taxa. This problem is prevalent in fossil angiosperm taxonomy, perhaps because until recently the standard approach was to describe floras, rather than to monograph taxa. In recent years many researchers have turned to monographic treatments of related taxa (Manchester and Dilcher 1982, Crane 1981, Schwarzwalder 1984, Wing and Hickey 1984). Even with large numbers of fossils representing a few taxa, the taxonomy of some groups may not be resolved clearly to the specific level. Numerous factors contribute to this difficulty: use of one organ for taxonomy (e.g. leaves, fruits), hybridization among species (see Roth and Dilcher 1979), and continuous variation in many characters. The question becomes whether each apparently different morphological type is to be described as a species, thus *implying* biological integrity at a specific taxonomic level. A better method may exist for conveying this palaeobotanical information.

I do not propose that palaeontological taxonomy should be independent of modern taxonomy. Instead we must recognize the limitations of the fossil record and work within the bounds of the data available. Morphotypes may be a form of taxonomic expression that reflects our level of confidence, yet does not compromise integration with modern evolutionary biology.

The designation of morphotypes has advantages over the use of binomials in the following ways.

(A) The extent of confidence in characters used in circumscribing groups is stated explicitly.

(B) Attention is focused on those taxonomic ranks where data from modern taxa indicates confidence in taxonomic segregation based on morphology.

(C) Even though species are not named as such, estimates of the diversity of a fossil flora may be based on morphotypes, thus one important reason for recognizing species is handled adequately using morphotypes.

(D) Once defined, morphotypes may be used stratigraphically in the same way that species are, without the added implication that the taxa are individual, monophyletic species.

(E) The designation of angiosperm fossil genera by reference to modern genera or families requires a phylogenetic approach, even if the goal of the research is not discovering ancestor–descendant relationships. This is equally true for designating morphotypes. A fossil group, by this approach, bears the name of the genus to which it is most similar and most likely belongs. At lower taxonomic levels, fossil taxonomy is a convention and only conveys morphological description in the sense of Riedel (1978).

(F) Characters of the plants are studied individually, providing a data base that is ideal for investigating the change in the mean and variance of characters through time. Patterns emerging from these studies will complement hypotheses of functional adaptation and phylogenetic constraint as controls on character evolution.

Terminal taxa

The biological species concept is an idea familiar to modern biologists. Acceptance of the concept is not wholehearted (Shaw 1969, Farris 1975, Levin 1979, Cronquist 1978). Problems arise in applying the concept to modern taxa, especially plants, as well as to fossil taxa (Burger 1975, Scheihing and Pfefferkorn 1980, Crane 1984). Consequently there remains considerable dispute over how a species concept should be applied, if at all (Raven 1976, Shaw 1969, Haldane 1956). Underlying this controversy is the idea that the basic unit of evolution should be the unit of intense focus for taxonomists as well as evolutionary biologists. Løvtrup's (1979) terminal taxa are groups of individuals that have not left the 'arena of evolution'.

> 'The only taxa in the phylogenetic hierarchy which can undergo further change are the terminal ones. Individuals therefore participate in evolution *qua* members of a terminal taxon. When the reproduction is sexual, it usually, but not always occurs between members of the same terminal taxon. The unit of evolution is a population which may range from all to a single member of a terminal taxon.' (Løvtrup 1979).

Wiley (1978, 1981) has revised the evolutionary species concept of Simpson (1961), considering species as temporal and spatial entities.

> 'A species is a single lineage of ancestral descendant populations of organisms which maintains its identity from other such lineages and which has its own evolutionary tendencies and historical fate.' (Wiley 1978).

The most significant difference between Løvtrup's terminal taxon and the evolutionary species concept of Wiley (1978, 1981) is that Løvtrup allows taxa other than species to participate in evolutionary processes. Certainly modern work on natural interspecific hybrids within the genus *Ulmus* (Santamour 1970, 1972, Bate-Smith and Richens 1973, Townsend 1975) indicate that species in *Ulmus* are not taxa 'bounded in time and space with a unique history' (Wiley 1980). Analysis of the species concept in *Quercus* by Burger (1975) may be a similar case. The difference between these two otherwise very similar philosophies underscores the diversity of opinions as to the position of species as a taxonomic rank. The idea of a terminal taxon emphasizes that the unit of evolutionary importance need not be equivalent from taxon to taxon. This may account for the observation that some genera are species rich while others are depauperate.

The biological species concept is based solely on sexual reproduction (Mayr 1969). The concept states that species are 'groups of actually or potentially interbreeding natural populations which are reproductively isolated from other such groups' (Mayr 1965). Populations of individuals that cannot interbreed are designated as independent species because they have independent evolutionary trajectories. However, closely related 'species' of a genus may interbreed and thus not qualify under Mayr's definition. A supraspecific group to which they belong represents the unit of evolution in this case. The potential to interbreed is a character retained during evolution by each species as they have diverged on other features and is no more than an ancestral character common to the genus (Wiley 1981). In the situation where all species of a genus are interfertile, the biological species concept requires that genera be monotypic, and that what may amount to large morphological divergence be masked because of retention of one, albeit significant, ancestral trait.

A conception of the species as the only lineage with unique evolutionary tendencies may be restricting our view of evolution to patterns present only at lower taxonomic levels. With an adequate fossil record, palaeontologists may be able to determine the evolutionary tendencies of a taxon at any one point in its past history by studying diversification within the taxon in successively younger strata. The most direct method for palaeontologists to determine both the history and evolutionary tendencies of taxa is to identify a morphological hierarchy. The stratigraphic and aereal distribution of each entity within the morphological hierarchy can be estimated directly using the fossil record. The evolutionary species or terminal taxon may be defined based on this trajectory of the morphological entity.

Shaw observed that the use of lower-level taxonomy and binomial nomenclature by palaeontologists was completely controlled by the personal objectives of the researcher (Shaw 1969). For palaeobotany, applications in stratigraphy, systematics, and morphology have been the major focal points. Farris (1975) has suggested some solutions to the problems of adherence to classical taxonomy by palaeontologists. Implicit in his idea is the assumption that palaeontology should strive to reconstruct phylogeny. To make the most of the interdisciplinary nature of palaeobotany, the basic unit of organization should be agreed upon.

Morphological characters

Fossil taxa, characterized by only a few specimens from a single locality, are known from virtually every geological epoch. 'Overspeciation' of this sort vexes stratigraphers and biologists alike because the species cannot be used reliably as age-indicators nor trusted as biological entities. A method is proposed here that may be capable of predicting the amplitude of character values expected in fossil taxa.

The mean and variance of twenty foliage characters measured from mature specimens of three genera of the extant Ulmoideae were computed. In order to compare the variation of characters in one genus with supposedly homologous characters of other genera and with other characters within the same genus, a coefficient of variation was calculated. Coefficient of variation (c.v. = population standard deviation divided by population mean) is a measure *not* dependent on scale (Sokal and Braumann 1980) and thus may be used to compare variability of characters measured at different scales, for example leaf length (in tens of millimetres) and angle of the basal secondary vein (from 60 to 140 degrees). The statistic does not reflect absolute values of the characters measured, rather it is a measure of the breadth of the distribution.

The genera are plotted by character in terms of coefficient of variation (Fig. 8.3). Variation in foliage characters, measured by c.v., generally is of the magnitude of 0.2—0.6, with 6 of 20 characters tending to exceed this value. Those characters that do exceed the value, i.e. those that are most variable, generally are consistently more variable in all genera. Likewise characters with low variability are consistently low in all genera. The genus *Ulmus* is composed of 15–30 species, *Zelkova* of approximately five, and *Chaetoptelea* one, so it is to be expected that those with more species (*Ulmus*) normally would be more variable than those

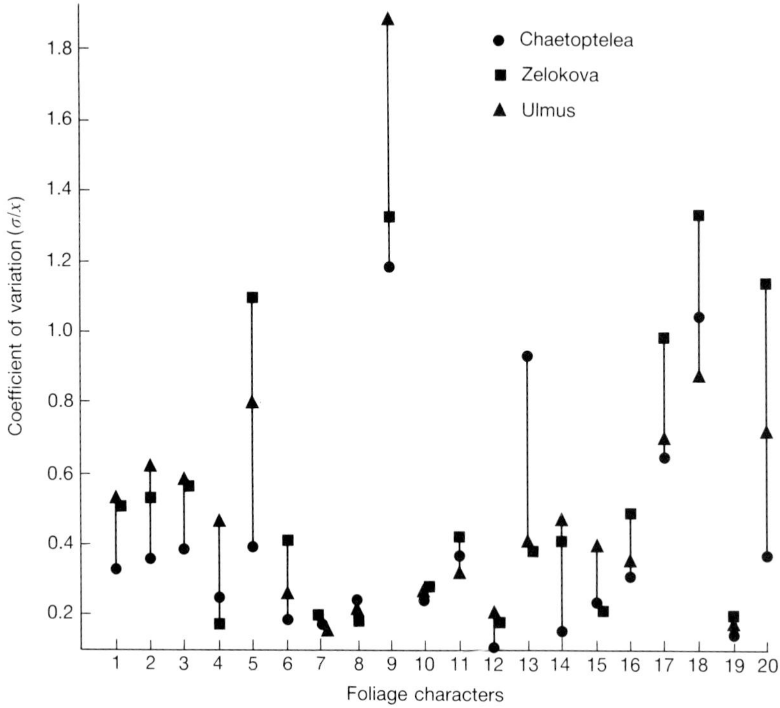

Fig. 8.3. Twenty foliage characters of extant specimens of three genera of the Ulmoideae. Variation profiles are similar for all genera.

with few species (*Zelkova*, *Chaetoptelea*). These similar variance profiles allow us to predict that unmeasured extant genera of the Ulmoideae would have variability profiles similar to those of these three genera. The prediction is tested using a sample of specimens of one species of *Planera*, *P. aquatica*. The variance profile of *Planera* is expected to follow a pattern similar to the other three genera of the Ulmoideae, the greatest congruence being with *Chaetoptelea*, also represented by one species. A t-test designed to test for differences in c.v. between two independent samples (Sokal and Braumann 1980) indicates that 17 of 20 of the characters have equivalent c.v. ($\alpha = 0.05$). This indicates that the method may be used to predict c.v. for characters of taxa related to those taxa measured.

The method may be applied to other groups, particularly fossils, if a large data base is constructed based on coefficient of variation of characters of many extant taxa. Thus, an estimate of the amount of variation in characters of fossil genera or species may be made. The method is applicable to organs other than leaves, e.g. fruits and seeds, and is applicable to modern groups of gymnosperms and pteridophytes as well as angiosperms. The technique provides an objective method for determining the status of species-rich fossil groups in which many species are represented by a few specimens. Alternatively, it may provide a basis for subdividing morphologically variable groups such as the *Alethopteris* complex studied by Scheihing and Pfefferkorn (1980). Once a large data base is compiled, including characters from gymnosperms and pteridophytes as well as angiosperms, the technique may be extrapolated to fossil taxa that are not represented in the modern flora.

Morphological Systematics and Evolutionary Patterns

I have advocated delimitation of fossil taxa based on morphological features of modern taxa at the same hierarchical rank. Within the Ulmoideae, morphotypes are designated as terminal taxa of the Palaeocene and Eocene. These are characterized by features of fossil leaves alone, probably insufficient for constructing a phylogenetic history of the Ulmoideae. Studying the modern descendants of the Palaeogene taxa that share these foliage characteristics will permit inference regarding the physiological and reproductive traits in the fossils. A phylogenetic tree then may be proposed using this data from fossil and modern specimens.

In the Ulmoideae there are no smooth morphological transitions between genera, either in modern or fossil representatives. However, modern species within the genus *Ulmus* are highly interfertile and do intergrade morphologically, raising the probability of intrageneric reticulate evolution. This suggests that modern genera arose as single species, bearing most if not all of the traits that characterize the genus. The stem species of modern genera must have been isolated rapidly to maintain such distinctly different morphologies. Genera are artificial constructs, yet in this instance the genera are distinct as ecological and morphological groups. Thus, in the Ulmoideae, the ecological conditions under which the stem species of a genus evolved or the genetic basis of the evolutionary change may be different than those under which subsequent species within genera formed. Similar patterns have been observed in characters of vertebrate genera and species (Lemen and Freeman 1984), in which species within genera are distinguished by size differences and genera are distinguished by shape differences.

Valentine's (1980) model for the origination of higher taxonomic categories proposes that under normal conditions, the morphological changes leading to stem species of genera are of a fundamentally larger size and lower frequency than those leading to species within genera. Genera would be recognized by their distinct morphologies, which should be present in the stem species of the genera. Subsequent speciation within the genus is on a smaller morphological scale. The Palaeogene data on the Ulmoideae suggests this pattern of generic appearance and diversification.

Carson (1982) has suggested that a relaxation in some aspect of the selective environment may allow the establishment of different morphologies, should they arise. He proposes a disruption of the genetic integrity of the population, leading to genomic reorganization and the origination of new forms. If this occurs under relaxed selective conditions, new forms have the opportunity to expand. Under resumption of normal selective conditions, new morphologies are maintained during reorganization of polygenic systems. In this proposal, large step-sizes may occur at higher frequencies under certain genetic conditions and the maintenance of larger steps may be favoured in altered selective environments. Carson's open and closed genome model really is not necessary to account for the pattern described here. A model of stochastically occurring changes producing different morphologies, via a variety of known developmental mechanisms, can produce the pattern if new forms encounter relaxed selective environments. The problem becomes one of coincidence of a variant form within a low-selective (low competition) setting.

The lowland habitats of the modern genera of the Ulmoideae do not appear to offer relaxed selective conditions. However, modern habitats in which the Ulmoideae live may not be similar to those of the late Cretaceous because the diversification of significant grassland and herbaceous families did not occur until the Eocene (Muller 1980, Axelrod 1981). Alternatively, short periods of relaxed competition may have occurred as a result of volcanic or climatic events of the late Cretaceous and early Tertiary. Intensive investigations of the palaeoenvironments within which genera of the Ulmoideae first appear would provide an opportunity to test the pattern of appearance as predicted by Carson's model. The morphological stability of ulmaceous genera since their appearance in the Palaeocene and Eocene may be due to the nature of the generic morphologies. The genera represent groups of populations that share a similar morphology. All new forms are relatively similar to the stem species. If a new form of a radically different type had appeared, it would have been produced in a lowland environment in which its ancestors had lived previously and to which the ancestors were well adapted. If a radically different form had survived, we would probably have placed it in a new genus. This could be demonstrated with a rigorous phylogenetic analysis in which the paraphyletic groups are detected. It seems likely that within the modern Ulmoideae today, each generic morphology is an integrated system from which radically new morphologies have a low probability of origination and maintenance.

Acknowledgements

Thanks are extended to Jack A. Wolfe of U.S. Geological Survey, Francis Hueber of Smithsonian Institution, the Oregon Museum of Science and Industry, Steve R. Manchester of Indiana University, Howard M. Schorn of University of California at Berkeley, Wesley Wehr of the Washington State Burke Museum, and Estella B. Leopold of the University of Washington for gracious loans of fossil specimens used during this research.

For helpful comments on the manuscript, I sincerely thank Alan J. Kohn, Melinda F. Denton, Richard G. Olmstead, and especially William A. DiMichele.

The work was undertaken while the author was a recipient of a National Science Foundation Graduate Fellowship.

References

Axelrod, D.I. (1981) Role of volcanism in climate and evolution. *Geol. Soc. Am. Spec. Pap.* **185**, 1–59.

Bate-Smith, E.C. and Richens, R.H. (1973) Flavonoid chemistry and taxonomy in *Ulmus*. *Biochem. Syst. Ecol.* **1**, 141–46.

Beck, C.B. (1969) Problems of generic delimitation in paleobotany. *Proc. North Am. Paleo. Conv.* pp. 173–93.

Briggs, D. and Walters, S.M. (1984) *Plant variation and evolution*. Cambridge University Press, Cambridge.

Burger, W.C. (1975) The species concept in *Quercus*. *Taxon* **24**, 45–50.

Burnham, R.J. (in press) Foliar morphological analysis of the Ulmoideae (Ulmaceae) from the early Tertiary of western North America. *Palaeontographica Abt. B.*

Carson, H.L. (1982) Speciation as a major reorganization of polygenic balance. In *Mechanisms of speciation* (ed. C. Barigozzi), pp. 411–433, A. R. Liss, New York.

Crane, P.R. (1981) Betulaceous leaves and fruits from the British Upper Paleocene. *Bot. J. Linn. Soc.* **83**, 103–136.

—— (1984) A re-evaluation of *Cercidiphyllum*-like plant fossils from the British early Tertiary. *Bot. J. Linn. Soc.* **89**, 199–230.

Cronquist, A.R. (1978) Once again, what is a species? In *Biosystematics in Agriculture.* (ed J.A. Romberger) pp 3–20, Montclair, Allanheld, Osmun & Co.

Davis, J.C. (1973) *Statistics and data analysis in geology*. John Wiley and Sons, New York.

Elias, T.S. (1970) The genera of Ulmaceae in the southeastern United States. *J. Arnold Arb.* **51**, 18–40.

Farris, J.S. (1975) Phylogenetic classification of fossils with recent species. *Syst. Zool.* **25**, 271–82.

Gauch, H.G., Jr. (1982) *Multivariate analysis of community ecology*. Cambridge Univ. Press, Cambridge.
Ghiselin, M.T. (1975) A radical solution to the species problem. *Syst. Zool.* **23**, 536–44.
Gianassi, D.E. (1978) Generic relationships in the Ulmaceae based on flavonoid chemistry, *Taxon* **27**, 331–44.
Grudinskaya, I.A. (1967) The Ulmaceae and reasons for distinguishing the Celtoideae as a separate family, Celtidaceae Link. *Bot. Zhur. Leningr.* **52**, 1723–49.
Haldane, J.B.S. (1956) Can a species concept be justified? *Systematics Association Special Volume* **2**, 95–6.
Hickey, L.J. (1977) Stratigraphy and paleobotany of the Golden Valley Formation (early Tertiary) of western North Dakota. *Geol. Soc. Am. Mem.* **150**.
Lemen, C.A. and Freeman, P.W. (1984) The genus: a macroevolutionary problem. *Evolution*. **38**, 1219–37.
Levin, D.A. (1979) The nature of plant species. *Science*. **204**, 381–4.
Løvtrup, S. (1979). The evolutionary species: fact or friction? *Syst. Zool.* **28**, 386–92.
MacGinitie, H.D. (1941) A Middle Eocene flora from the central Sierra Nevada. *Carn. Inst. Wash. Publ.* **599**.
—— (1969) The Eocene Green River flora of northwestern Colorado and northeastern Utah. *Univ. Calif. Publ. Geol. Sci.* **83**.
Manchester, S.R. and Dilcher, D.L. (1982) Pterocaryoid fruits (Juglandaceae) in the Paleogene of North America and their evolutionary and biogeographic significance. *Amer. J. Bot.* **69**, 275–86.
Mayr, E. (1965) *Animal Species and Evolution*. Belknap, Cambridge.
—— (1969) The biological meaning of species. *Biol. J. Linn. Soc.* **1**, 311–20.
Melville, R. (1938) The application of biometrical methods to the study of elms. *Proc. Linn. Soc. Lond.* **151**, 152–9.
—— (1978) On the discrimination of species in hybrid swarms with special reference to *Ulmus* and the nomenclature of *U. minor* Gled. *Taxon* **27**, 345–351.
—— and Heybroek, H. (1971) The elms of the Himalaya. *Kew Bull.* **26**, 5–28.
Muller, J. (1981) Fossil pollen records of extant angiosperms. *Bot. Rev.* **47**, 1–142.
Nie, N.H., Hull, C.H., Jenkins, J.G., Steinbrenner, K. and Bent, D.H. (1975) *Statistical Package for the Social Sciences*, McGraw Hill, New York.
Raven, P.H. (1976) Systematics and plant population biology. *Syst. Bot.* 1, 284–316.
Richens, R.H. (1983) *Elm*. Cambridge University Press, Cambridge.
Riedel, W.R. (1978) Systems of morphologic descriptions in paleontology. *J. Paleontol.* **52**, 1–7.
Roth, J. and Dilcher, D.L. (1979) Investigations of angiosperms from the Eocene of North America: stipulate leaves of the Rubiaceae including probable polyploid population. *Am. J. Bot.* **66**, 1194–207.
Santamour, F.S. (1970) Natural hybrids between Chinese and cedar elms from Texas. *Bull. Torrey Bot. Club*. **97**, 72–3.
—— (1972) Interspecific hybridization with fall and spring flowering elms. *For. Sci.* **18**, 283–9.

Scheihing, M.H. and Pfefferkorn, H.W. (1980) Morphologic variation in *Alethopteris* (Pteridosperms, Carboniferous) from St. Clair, Pennsylvania, U.S.A. *Palaeontogr. Abt. B.* **172**, 1-9.

Schwarzwalder, R.N. (1984) Aspects of the evolution of the Platanaceae and its relation to the Hamamelidaceae. (Abst.) *Am. J. Bot.* **71**, 111.

Shaw, A.B. (1969) Adam and Eve, paleontology and the non-objective arts. *J. Paleontol.* **43**, 1085–98.

Simpson, G.G. (1961) *Principles of Animal Taxonomy*. Columbia University Press, New York.

Sokal, R.R. and Braumann, C.A. (1980) Significance tests for coefficients of variation and variability profiles. *Syst. Zool.* **29**, 50–66.

—— and Rohlf, F.J. 1981. *Biometry*. Freeman and Co., New York.

Sweitzer, E.M. (1971) Comparative anatomy of the Ulmaceae. *J. Arnold Arb.* **52**, 523–73.

Townsend, A.M. (1975) Crossability patterns and morphological variation among elm species and hybrids. *Silvae Genet.* **24**, 18–23.

Valentine, J.W. (1980) Determinants of diversity in higher taxonomic categories. *Paleobiology* **6**, 444–50.

Wiley, E.O. (1978) The evolutionary species concept reconsidered. *Syst. Zool.* **27**, 17–26.

—— (1980) Is the evolutionary species fiction? A consideration of classes, individuals, and historical entities. *Syst. Zool.* **29**, 76–80.

—— (1981) *Phylogenetics*. John Wiley and Sons, New York.

Wing, S.L. (1981) A study of paleoecology and paleobotany in the Wilwood Formation (Early Eocene, WY) PhD. Dissertation, Yale University.

—— and Hickey, L.J. (1984) The *Platycarya* perplex and the evolution of the Juglandaceae. *Am. J. Bot.* **71**, 388–411.

Wolfe, J.A. (1977) Paleogene floras from the Gulf of Alaska region. *U.S. Geol. Surv. Prof. Pap.* 997.

9. The epidermis/cuticle and *in situ* spores and pollen in fossil plant taxonomy

C.R. HILL

Abstract

Palaeobotanical research in recent years has concentrated increasingly on the cuticle/epidermis and on *in situ* spores and pollen of fossil plants. Because of their wide diversity amongst taxa, these suites of characters provide valuable sources of taxonomic information, particularly (though by no means exclusively) in compresssion/impression fossils. The purpose of this introductory review is to explore briefly the role that cuticles/epidermises and *in situ* spores and pollen play in fossil plant systematics, to provide a lead-in to the literature, and to focus on some recent developments of interest.

Role of cuticles and spores in perspective of the systematics of whole plants, extant and fossil

Exhaustive reviews of the epidermis and cuticle, of excellent quality, are already available in the literature. Therefore no attempt is made here to cover the same ground. Beginning students are referred particularly to Cutler, Alvin and Price (1982), Martin and Juniper (1970), Stace (1965), Van Cotthem (1970), and Wilkinson (1979). These investigations make it clear that the structure of the epidermis, and hence the cuticle, of living higher plants shows considerable diversity, though within the limits of biological variation it is more or less constant for a given species or higher taxon. Equivalent reviews in the literature for *in situ* pollen and spores (Erdtman 1971, Erdtman and Sorsa 1971) make it clear that they too exhibit a comparable range and type of variation in characters.

So far this states nothing about cuticles or spores that makes them appear different in kind from any other sort of character useful in plant systematics, such as for example wood anatomy or strobilar structure.

Indeed within a purely systematic, i.e. comparative, context involving equal weighting of characters there is nothing necessarily special about these suites of characters, particularly the epidermis and cuticle. This is because of the heterobathmy (mozaic distribution) of character–states generally in biological material (see Hill and Crane 1982, pp. 297–9 for an explanation of the term heterobathmy and the conclusions of Stace (1965, pp 68–71) which are particularly appropriate in this context). Thus it is worth remembering that in a systematic study of extant plants (Hill and Camus 1986), the epidermis/cuticle and spores or pollen can be investigated together with many other characters, and that in context of overall character comparison of the group being studied, these particular sources of characters may have greater or lesser taxonomic value compared to other characters. What I am trying to emphasize is that such characters are always potentially useful in a comparative context for delimiting species or higher taxa, indeed in many cases they are of actual use, but are not necessarily universally useful. Thus there are occasions when gross morphology may be equally or even more useful than cuticular or spore characters (for example in Czekanowskiales, see Harris *et al.* 1974). There is therefore a potential limitation of becoming preoccupied with a particular group of taxa, such as Czekanowskiales or Ginkgoales, looking for minute differences in cuticle or pollen/spore characters that may turn out to be of equal or even less value than certain other characters. The peculiar paracytic stomata of Bennettitales, *Welwitschia*, perhaps of *Pentoxylon* and of some Angiosperms, for instance, provide a satisfactory character for potentially defining a group composed collectively of these taxa, but the paracytic character in itself is of little proven use for defining groups at a narrower level other than phenetically. In contrast, the very possession of an acid-resistant cuticle (Alvin 1971) is a unique character that discriminates the fossil matoniaceous fern *Weichselia* from most other ferns and, very distinctly, from its nearest relative *Piazopteris*. Similarly, in extant Marattiales (e.g. Hill and Camus, 1986) the huge stomata of *Christensenia*, reputed to be the largest of any vascular plant, are diagnostic of that genus, whereas stomatal size does not separate out any of the other five extant genera of the group except as uniformly different from *Christensenia*.

A major reason that palaeobotanists spend so much effort on the epidermis/cuticle (and on *in situ* spores and pollen) is in fact practical rather than biological. Cuticles, as well as spores and pollen, particularly in compressions/impressions, are often the major or even the only source of useful preserved characters available apart from those of gross morphology. This is because of their resistance to decay before and during preservation and their resistance to destruction by acid-based preparation techniques. In addition to this practical reason, pollen and spores are of particular biological interest since they represent a discrete, integral phase in the life-cycle of whole plants. (It should be added that this is the position as viewed within the context of systematics: in other biological perspectives, such as those of reproductive physiology or ecology, both the epidermis/cuticle and pollen/spores have additional biological significance.)

Recent developments

An important conceptual development in recent years has been increasing recognition of the distinction between the structure of mature stomata and the pattern of their ontogenetic development. It is therefore considered no longer appropriate, for instance, to use the term 'syndetocheilic' to describe the mature stomata of Bennettitales. The term 'paracytic' is more appropriate (Wilkinson 1979). There is nothing particularly special about Bennettitalean stomata (Alvin, pers. comm., Krasilov 1978) other than within the context of 'gymnospermae'. An additional development has been the extension of application of cuticular and otherwise epidermal studies, from the pioneering work (done mainly on Mesozoic 'gymnosperms') to studies of dicotyledonous Angiosperms (Dilcher 1973, followed by many studies), and to ferns (Harris 1961, p. 105, Fig. 34D, Appert 1973, Hill, Wagner and El-Khayal 1986, Fig. 1), as well as to Palaeozoic material.

In terms of laboratory techniques, the basic preparation methods utilized both for cuticles and for fossil pollen and spores are essentially similar. With emphasis on dispersed spores and pollen they have been comprehensively reviewed by Gray (1965), Doher (1980) and by Phipps and Playford (1984). (The use of phenol in glycerin jelly, recommended by Doher, is not endorsed here. Thymol is preferred.) See also Wesley (1954) and Lacey (1968).

Electron microscopy

Undoubtedly by far the most significant advances in taxonomic research on fossil cuticles/epidermises and on *in situ* pollen and spores during the past 20 years have resulted from the application of electron microscopy: scanning electron microscopy (SEM) of cuticles (Alvin 1970) and of pollen and spores (Taylor and Rothwell 1982), as well as transmission electron microscopy (TEM) of cuticles (Archangelsky *et al.* in press) and of pollen and spores (Taylor and Rothwell 1982, Taylor and Alvin 1984).

TEM work on fossil spores and pollen was pioneered in the 1960s by J. Pettit in the UK and by T. Taylor in the USA, who, through his co-workers and students, has subsequently continued to play a prominent pioneering role in the application of TEM work generally in palaeobotany. An overview of the most suitable preparation and staining techniques, as well as a synthesis of the results obtained to date, would be of much value to other workers. Use of the TEM has been of particular value in characterizing the internal architecture of spore and pollen grain walls. Its initial recent application to fossil cuticles (Archangelsky *et al.* 1986) also appears promising in this respect (Fig. 9.2).

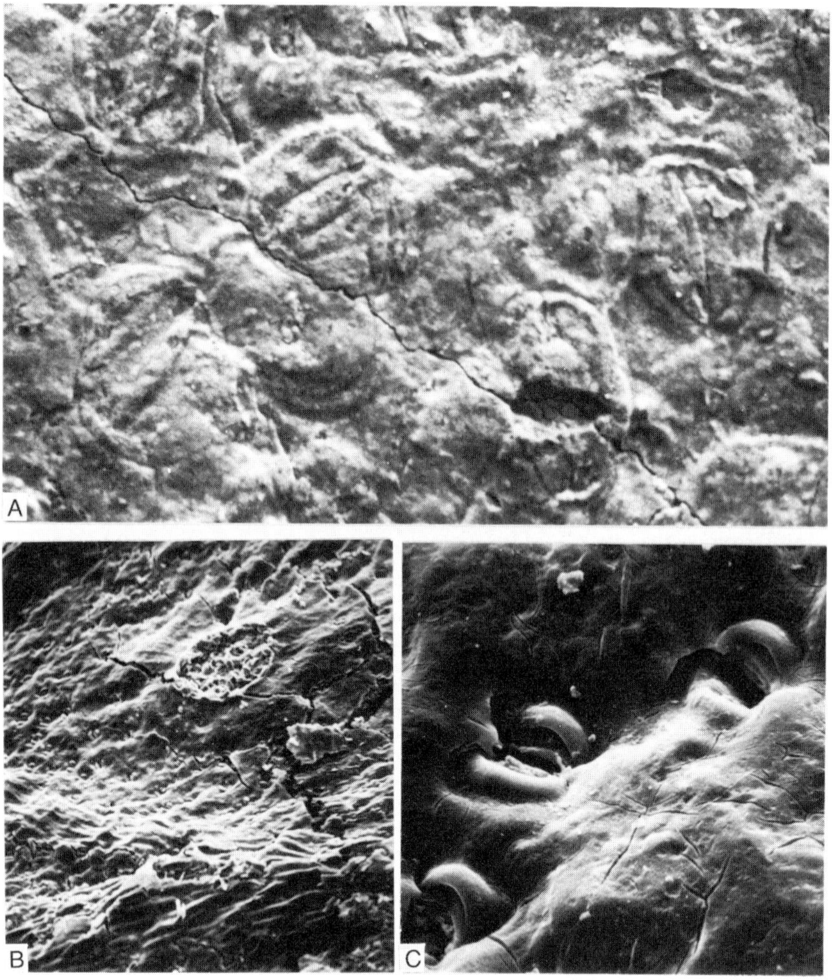

Fig. 9.1. Epidermal structure of fossil fern pinnules observed by direct examination with SEM. A. *Qasimia schyfsmae* (Lemoigne) Hill, Wagner and El-Khayal, Permian, Saudi Arabia (Marattiales). Impression material showing stomata, x 750. B, C. *Phlebopteris woodwardii* Leckenby, Middle Jurassic, Yorkshire, UK. Fusainised material, B approx. x 65, C approx. x 650.

Fig. 9.2. Transmission electron micrograph of a leaf cuticle, showing lamellated outer region and inner homogeneous sub-layer of cuticle proper, x72000. *Ticoa harrisii* Archangelsky, L. Cretaceous, Argentina.

Although TEM work has shown much promise, SEM has in fact been more widely used in palaeobotany than TEM. Utilizing suitable freeze-fracturing techniques in conjunction with high resolution field-emission SE microscopes (Barnes and Blackmore 1984, Blackmore and Barnes 1985, Barnes 1985), SEM may even come to displace TEM for comparative palaeobotanical studies in the future. In particular, SEM has facilitated study of epidermal characters in material such as compressions/impressions of ferns, which do not normally yield an acid resistant cuticle. This is illustrated in Fig. 9.1 by Permian-age fern pinnules preserved as impressions (*Qasimia*, Fig. 9.1A) and fusainised compressions (*Phlebopteris*, Fig. 9.1B,C). In the impression material of *Qasimia* (Fig. 9.1A), the intervenal region of the abaxial epidermis shows several cyclocytic stomata, which are characteristic of Marattiales and therefore support assignment of *Qasimia* to that group. Numerous stomata in the intervenal region, as well as the elongate cells of the epidermis over the veins of *Phlebopteris* can be recognized in Fig. 9.1B (the circular structure towards the top of the micrograph is the broken placenta of a sorus). At a higher magnification (Fig. 9.1C) details of the stomata in exterior view of the pinnule epidermis can be seen.

SEM has also facilitated study of cuticles/epidermises and spores/pollen in impression remains generally of higher plants, which constitute

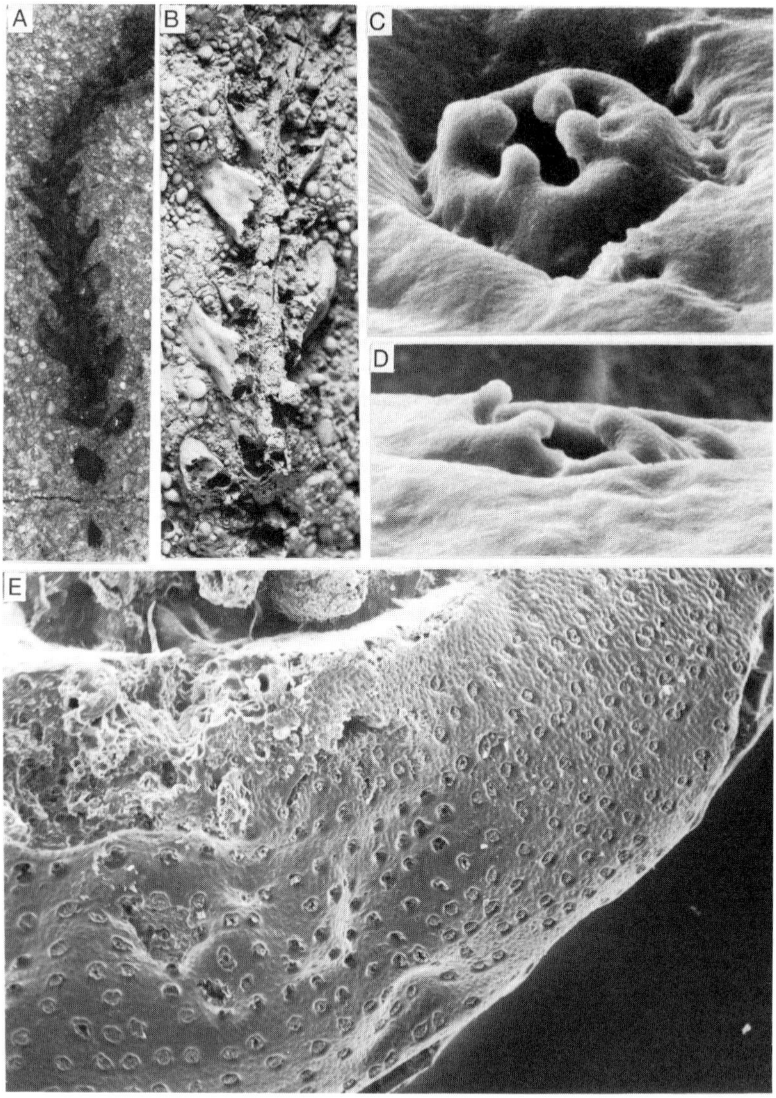

Fig. 9.3. Use of latex replicas for examining surface features of impression material at low magnifications. *Pagiophyllum araucarinum* (Pomel) Saporta, Middle Jurassic, Oxfordshire, UK (Coniferales), BM(NH) 38932. A. actual specimen, xl.5. B. latex replica of specimen in A, x3. C–E. SEM photos. of latex replica in B: C, D × 900, E × 40.

by far the major proportion of plant fossil material available. Since such impressions are merely imprints of surface features on the rock matrix they are unsuitable for acid-preparation treatment and would formerly have been considered of little value for examination of microscopic features. In practice, SEM may be used either for direct views of such an impression surface (Figs. 9.1A, 5A) or indirectly by use of replicas (Figs. 9.3,4). These may be in latex (Chaloner and Gay 1973), silicone rubber (Watson and Alvin 1976) or in other materials (see Hamilton-Waters (1983) for a general review of replication methods and materials, and for reviews with special reference to SEM see Pfefferkorn and Boyde (1974), Pameijer (1978), and Crankshaw (1984)). Such replicas have the advantage of 're-creating' the original surface rather than viewing a 'negative' imprint of it.

At low magnifications, c.x500 to x1000, as are used for observation of most epidermal/cuticular characters, latex and silicone rubber are adequate replicators (Figs. 9.3,4) though they tend to round off from the template. This leads to some loss of definition but is usually (though not always) unimportant at these low magnifications. The use of latex replicas for examining surface features of impression material of a coniferous

Fig. 9.4. Use of latex replicas for examining surface features of impression material at low magnifications. Replica of portion of a synangium of the marattialean fern *Qasimia schyfsmae* (Lemoigne) Hill, Wagner and El-Khayal, showing spore-mass *in situ* within a locule of the synangium, x200. Replica examined with the Environmental Chamber facility of an ISI 60A SEM.

shoot (*Pagiophyllum*) of Middle Jurassic age from Oxfordshire, England, is illustrated in Fig. 9.3. The specimen itself (Fig. 9.3A) appears at first sight poorly preserved and unpromising, occurring in a coarsely oolitic sandstone matrix. Figure 3B shows a latex replica of the same shoot coated with NH_4C1 sublimate and photographed optically. The replica was then cut up and mounted on stubs for examination with SEM (Figs. 9.3C–E). Figure 3E illustrates part of a leaf, showing the regular longitudinal files of stomata, with individual stomata in side and oblique views in Figures 3C, D. These show subsidiary cell papillae surmounting a rampart, typical of the species *Pagiophyllum araucarinum* (Pomel) Saporta. Critical identification of this specimen as *P. araucarinum*, the first recorded occurrence from the Oxfordshire Jurassic, was possible only by use of this procedure. Otherwise the specimen could at best have been determined as *Pagiophyllum* sp.

Figure 9.4 illustrates a latex replica of a portion of synangium of *Qasimia*, a *Marattia*-like fern from the Permian of Saudi Arabia. The locule in the centre of the photo. has an *in situ* spore-mass. Such spore-masses were in fact first noticed in this fern during a routine examination of latex replicas using the 'Environmental Chamber' facility of an ISI 60A SEM. The Environmental Chamber is a recently developed device which comprises a Robinson back-scattered electron detector in conjunction with a so-called Charge-Free Anti-contamination System (CFAS) which functions at high vacuum. This facility permits examination of uncoated specimens, as in this instance, and allows large areas of material to be examined without the need for mounting on stubs. For these reasons it is extremely useful for examining Type and figured or otherwise scarce and irreplaceable material without damage. Owing to its operation in the backscatter mode, it also acts as a crude atomic number detector. This is particularly useful for discriminating organic material such as cuticle or latex from adherent rock matrix. In this figure (9.4) the bright regions represent siliceous matrix. Although cuticles/epidermises and spores/pollen can be examined in the Environmental Chamber without coating, which removes the need for de-coating prior to subsequent study e.g. by TEM, the results on palaeobotanical material to date have tended to be somewhat limited in quality by the effects of charging, with the best results still obtainable only with coated material. Robinson detectors are now available for a wide range of makes of SEM.

The rounding off of latex and of silicone rubber becomes limiting at higher magnifications, *c.* x1000 to x30000, for examination of minute surface features (a micron or smaller in size) of replicated impressions, particularly of pollen and spore walls. As shown in Figs. 9.5A-F, replication is then achieved more effectively using cellulose acetate or nitrocellulose. Figure 9.5A is a direct view of the impressions in the matrix of a group of compressed *in situ* spores of *Qasimia*. The original compressions have long since vanished and therefore replicas were prepared to re-create their surface features. Various materials for replicating were used, a selection of which is illustrated in Fig. 9.5B–F, all prepared by repeated replication of the impressions shown in Fig. 9.5A. The latex replica in Fig. 9.5B almost completely fails to reproduce the ornamentation and artefactual 'stringing' is visible at top left. Of the silicone rubbers, Dow Corning 9161 with catalyst 9162 (Fig. 9.5C) faithfully replicated the positions of the ornamentation elements but shows them as warts which are larger than the pits in the original impression (Fig. 9.5A), whilst Xantopren Blau Light Body (Fig. 9.5D) is almost useless even for replicating the spore outlines. Despite this fact, Xantopren Blau has been used widely for SEM replica work (cf. Pameijer 1978, p.832). Cellulose acetate sheet dissolved onto the spore impressions with acetone and left for one hour at *c.* 20°C before removal is illustrated in Fig. 9.5E. The ornamentation elements are replicated as small, more or less spiny granules, appearing as would be expected from the pits seen in the impressions in Fig. 9.5A, and more like them than the warty granules of the Dow Corning silicone rubber replica in Fig. 9.5C. Identical results were obtained with nitrocellulose ('Durofix' glue) in amyl acetate.

In order to obtain an independent comparison of these results, a Carbon replica was prepared (Fig. 9.5F). This shows an almost identical appearance of ornamentation elements to Fig. 9.5E, suggesting therefore that cellulose acetate or nitrocellulose are the most effective replicating materials.

What to do with the information?

Having just reviewed how characters of epidermises/cuticles and pollen or spores may be observed, the question needs to be asked of what to do with the information thus acquired. Such information for its own sake is scientifically of little value. This leads back to the context of character comparison discussed on p. 124 and, in turn, to the wider question of which method of comparison (phenetics, cladistics, or intuition) is most appropriate to the particular circumstances. Discussion of these is beyond the scope of this chapter but, for example, Alvin and Dalby (1986) give an example in a phenetic context and Hill and Crane (1982) and Hill and Camus (1986) make some suggestions in a phylogenetic (cladistic) context.

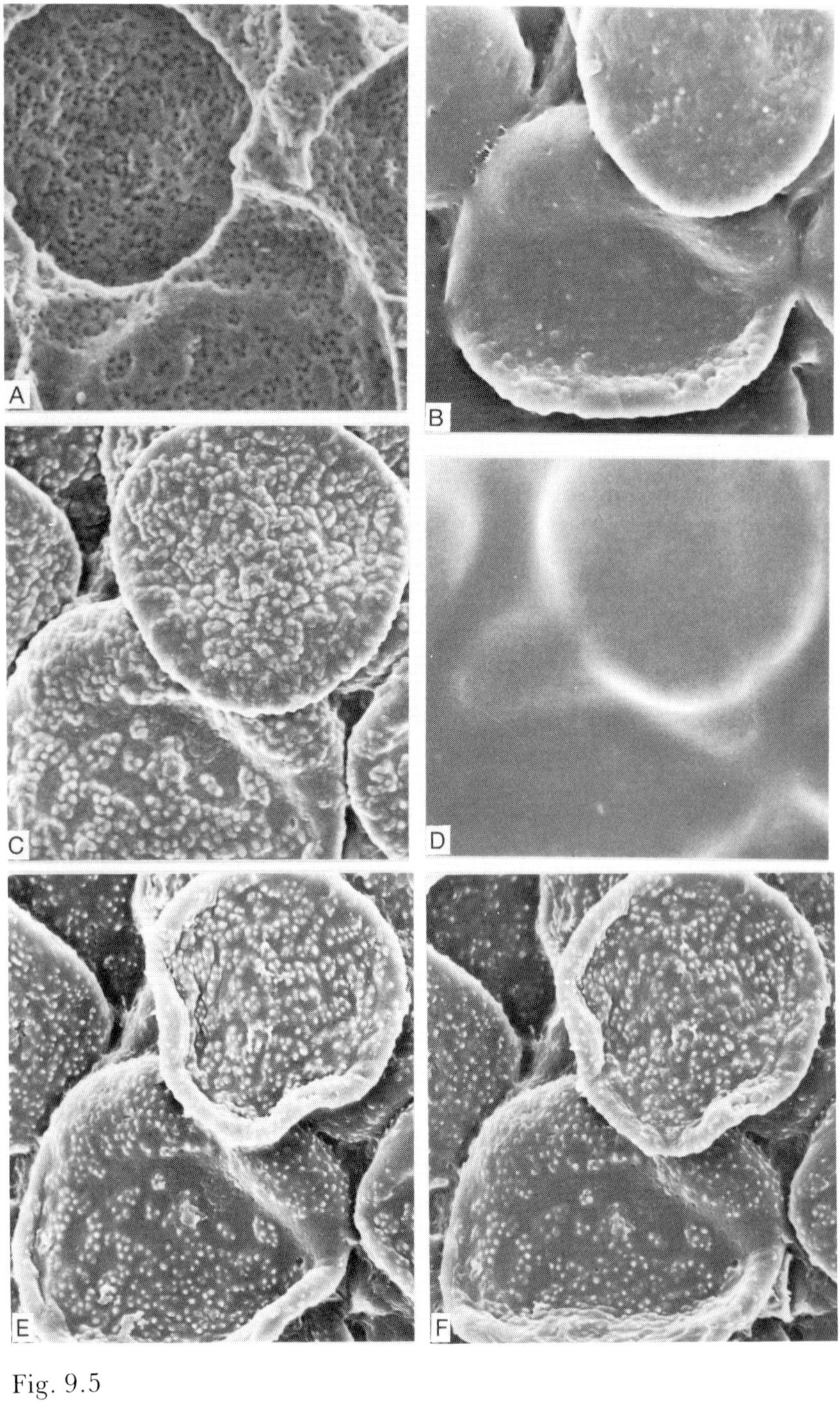

Fig. 9.5

Beyond systematics

Taxonomy and systematics play a vital role in enabling palaeobotanists to characterize and delimit species and higher taxonomic units. This brings order to what otherwise would be a chaotic jumble of information and provides units which are of use when pursuing broader questions. Phylogenetics has just been hinted at as being one of the broader issues, but there are also reproductive and vegetative palaeobiology, biostratigraphy, palaeoecology, and taphonomy, all of which today make up the lively subject of palaeobotany. *In situ* pollen and spores need to be studied in more depth if we are to gain an adequate picture of former vegetational composition and cover from the evidence provided by dispersed spores. Cuticles are of much value in reconstructing 'whole' plants from fragmentary fossils, as are pollen-grains within the micropyles of seeds (Harris 1940). Cuticles in the dispersed condition have been surprisingly little utilized, but using bulk maceration (Harris 1926) they have some promise both for determining vegetational composition and, as with dispersed miospores and megaspores, for biostratigraphy and palaeogeography (see Kovach and Dilcher (1984) for a useful review, and Collinson *et al.* (1985)). Cuticles and pollen in coprolites and gut or stomach contents (Hill 1976, Koenigswald and Schaarschmidt 1983) provide a means for determining the diets of ancient animals.

The range of possibilities for examining broader issues by using cuticles/epidermises and spores or pollen is therefore considerable. In this respect, particularly utilizing electron microscopy to examine the comparative anatomy of leaf and seed cuticles (Fig. 9.6), the subject, even at the level of systematics, is still very much in its infancy. Many exciting challenges lie ahead for those who strike out in new directions.

Acknowledgements

I am grateful to Mr. D. Claugher, British Museum (Natural History), for preparing the Carbon replica in Fig. 9.5F and to Dr. T.N. Taylor, Ohio State University, for the TEM photo in Fig. 9.2.

Fig. 9.5. Capacity of different materials for replicating fine surface detail at high magnifications. All photos. x2000, showing the same group of *in situ* spores of *Qasimia schyfsmae* (Lemoigne) Hill, Wagner and El-Khayal. A. direct view of impression of originally compressed spores in the rock matrix, i.e. the template from which the replicas shown in B–F were prepared: B. latex, C. Dow Corning silicone rubber 9161 with catalyst 9162, D. Xantopren Blau Light Body silicone rubber, E. Cellulose acetate, F. Carbon replica.

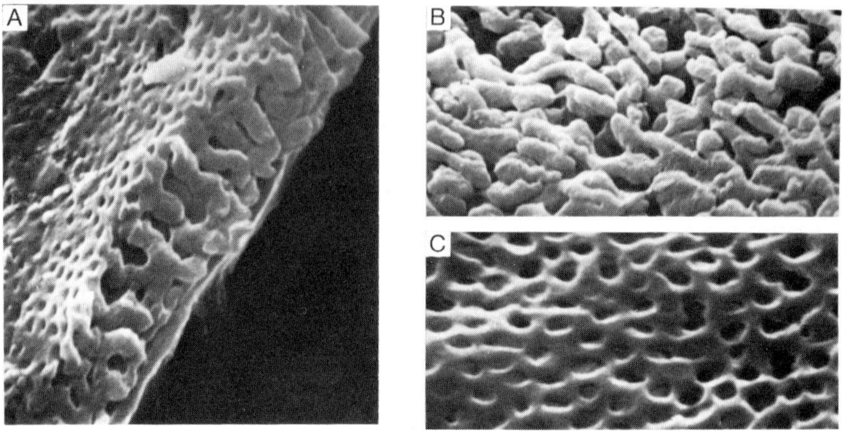

Fig. 9.6. Megaspore membrane of a fossil non-Angiospermous seed, viewed with SEM. *Allicospermum* sp., Middle Jurassic, Yorkshire, UK. A. broken edge showing three-dimensional architecture of branching rods of 'sexine', approx. x5175. B. inner surface of sexine showing branching rods, approx. x4500. C. outer surface of sexine showing pitting, approx. x9000. Little comparative work on megaspore membranes of seed plants has been published, despite the potential importance of such work in phylogenetic studies (see e.g. Zimmermann and Taylor 1970, Hill and Crane 1982, pp. 345–6).

References

Alvin, K.L. (1970) The study of fossil leaves by SEM. *Scanning Electron Microsc.* **3**, 121–8.

—— (1971) *Weichselia reticulata* (Stokes et Webb) Fontaine from the Wealden of Belgium. *Mem. Inst. r. Sci. nat. Belg.* no. 166.

—— and Dalby, D.H. (1986) Numerical analysis of some Yorkshire Jurassic conifers. *Rev. Palaeobot. Palynol* (in press).

Appert, O. (1973) Die Pteridophyten aus dem Oberen Jura Des Manamana in Sudwest-Madagaskar. *Schweiz. palaeont. Abh.* **94**, 1–62.

Archangelsky, S., Taylor, T.N., and Kurmann, M.H. (1986) Ultrastructural studies of fossil plant cuticles: *Ticoa harrisii* from the early Cretaceous of Argentina. *Bot. J. Linn. Soc.* (in press).

Barnes, S.H. (1985) Plant ultrastructure: freeze fracture and cytoplasmic maceration. *Proc. r. microsc. Soc.* **20**, 25–6.

—— and Blackmore, S. (1984) Scanning electron microscopy of chloroplast ultrastructure. *Micron Microsc. Acta* **15**, 187–94.

Blackmore, S. and Barnes, S.H. (1985) *Cosmos* pollen ontogeny: a scanning electron microscope study. *Protoplasma* **126**, 91–9.

Chaloner, W.G. and Gay, M.M. (1973) Scanning electron microscopy of latex casts of fossil plant impressions. *Palaeontology* **16**, 645–9.

Collinson, M.E., Batten, D.J., Scott, A.C. and Ayonghe, S. (1985) Palaeozoic, Mesozoic and contemporaneous megaspores from the Tertiary of southern England: indicators of sedimentary provenance and ancient vegetation. *J. geol. Soc.* **142**, 375–95.

Crankshaw, O.S. (1984) Instruction of replica techniques for scanning electron microscopy. *Scanning Electron Microsc.* (**4**), 1731–7.

Cutler, D.F., Alvin, K.L. and Price, C.E. (1982) The plant cuticle. *Linnean Society Symposium Series* no. 10. London, pp. 461.

Dilcher, D.L. (1973) A revision of the Eocene flora of southeastern North America. *Palaeobotanist* (1971) **20**, 7–18.

Doher, L.I. (1980) Palynomorph preparation procedures currently used in the Paleontology and Stratigraphy laboratories, U.S. Geological Survey. *Circ. U.S. geol. Surv.* No. 830.

Erdtman, G. (1971) *Pollen morphology and plant taxonomy. Angiosperms.* Hafner, New York, pp. xii, 553.

—— and Sorsa, P. (1971) *Pollen and spore morphology/plant taxonomy. 4. Pteridophyta.* Stockholm, pp. 302.

Gray, J. (1965) Palynological techniques. In *Handbook of paleontologic techniques* (eds B. Kummel and D. Raup), pp. 471–81, Freeman, San Francisco and London.

Hamilton-Waters, P. (1983) A review of the moulding and casting materials and techniques in use at the Palaeontology laboratory, British Museum (Natural History). *Conservator* **7**, 37–43.

Harris, T.M. (1926) Note on a new method for the investigation of fossil plants. *New Phytol.* **25**, 58–60.

—— (1940) On *Caytonia* Thomas. *Ann. Bot.* n.s. **4**, 713–34.

—— (1961) *The Yorkshire Jurassic Flora, I. Thallophyta-Pteridophyta.* British Museum (Natural History), London.

—— Millington, W. and Miller, J. (1974) *The Yorkshire Jurassic Flora, IV. Ginkgoales and Czekanowskiales.* British Museum (Natural History), London.

Hill, C.R. (1976) Coprolites of *Ptilophyllum* cuticles from the Middle Jurassic of North Yorkshire. *Bull. Br. Mus. nat. Hist.* (Geol.) **27**, 289–94.

—— and Crane, P.R. (1982) Evolutionary cladistics and the origin of Angiosperms. In *Problems of Phylogenetic reconstruction* (eds. K.A. Joysey and A.E. Friday), Systematics Association Special Vol. **21**, pp. 269–361.

—— and Camus, J.M. (1986) Evolutionary cladistics of marattialean ferns. *Bull. Br. Mus. nat. Hist.* (Botany) **14**, 219–300.

—— Wagner, R.H. and El-Khayal, A.A. (1986) *Qasimia* gen. nov., an early *Marattia*-like fern from the Permian of Saudi Arabia. *Scr. geol.* **79**, (in press).

Koenigswald, W.V., and Schaarschmidt, F. (1983) Ein Urpferd aus Messel, das Weinbeeren frass. *Natur Mus., Frankf.* **113**, 79–84.

Kovach, W.L. and Dilcher, D.L. (1984) Dispersed cuticles from the Eocene of North America. *Bot. J. Linn. Soc.* **88**, 63–104.

Krassilov, V.A. (1978) Bennettitalean stomata. *Palaeobotanist* (1976) **25**, 179–84.

Lacey, W.S. (1968) Palaeobotany. In *The preservation of natural history specimens, II* (eds R. Wagstaffe and J.H. Fidler), pp. 243–60, Witherby, London.

Martin, J.T. and Juniper, B.E. (1970) *The cuticles of plants.* Arnold, London, pp. xx, 347.

Pameijer, C.H. (1978) Replica techniques for scanning electron microscopy, a review. *Scanning Electron Microsc.* **2**, 831–6.

Pfefferkorn, G. and Boyde, A. (1974) Review of replica techniques for scanning electron microscopy. *Scanning Electron Microsc.* (1), 75–82.

Phipps, D. and Playford, G. (1984) Laboratory techniques for extraction of palynomorphs from sediments. *Pap. Dep. Geol. Univ. Qd.* **11**(1), 1–23.

Stace, C.A. (1965) Cuticular studies as an aid to plant taxonomy. *Bull. Br. Mus. nat. Hist.* (Botany) **4**, 1–78.

Taylor, T.N. and Alvin, K.L. (1984) Ultrastructure and development of Mesozoic pollen: *Classopollis. Am. J. Bot.* **71**, 575–87.

—— and Rothwell, G.W. (1982) Studies of seed fern pollen: development of the exine in *Monoletes* (Medullosales). *Am. J. Bot.* **69**, 570–8.

Van Cotthem, W. (1970) Comparative morphological study of the stomata in the Filicopsida. *Bull. Jard. bot. natn. Belg.* **40**, 81–239.

Watson, J. and Alvin, K.L. (1976) Silicone rubber casts of silicified plants from the Cretaceous of Sudan. *Palaeontology* **19**, 641–50.

Wesley, A. (1954) A short synopsis of some microscopical methods in palaeobotany. *Proc. Leeds philos. Soc. (sci. Sect.)* **6**, 168–79.

Wilkinson, H.P. (1979) 10. The plant surface (mainly leaf). In *Anatomy of the dicotyledons,* I. Systematic anatomy of the leaf and stem, with a brief history of the subject (eds. C.R. Metcalfe and L. Chalk; 2nd edn), pp. 97–165, Clarendon Press, Oxford.

Zimmermann, R.P. and Taylor, T.N. (1970) The ultrastructure of Paleozoic megaspore membrances. *Pollen Spores* **12**, 451–68.

10. Classifying the earliest gymnosperms

G. W. Rothwell

Abstract

Remains of the earliest gymnosperms occur in Upper Devonian and Lower Carboniferous strata, but they consist only of isolated organs and fragments of organs. Of these, ovules and cupules that contain ovules are unequivocal gymnosperms. Possible relationships of the remaining, putatively gymnospermous debris to the ovulate organs and to one another are largely suppositional. In an attempt to construct a phylogenetic classification with such limited data, features of the most ancient ovules and ovulate cupules have been evaluated. To determine which characters are of the greatest systematic significance, synapomorphies that define the earliest gymnosperm remains have been interpreted and primitive gymnospermous characters identified. Contrary to previous interpretations, ovule and cupule symmetry are of little value for interpreting major systematic relationships. Instead, features of the pollen chamber are stressed because they reveal complex structure, characteristic developmental patterns, and an elegant reproductive mechanism that could function in the absence of an integument. Reproduction of this type is distinctly different from that of all pteridophytes and extant gymnosperms. It characterized a single major group throughout the Late Devonian and Early Carboniferous, and is termed 'hydrasperman reproduction'. Gymnospermous taxa with more derived modes of reproduction are interpreted as having originated from this ancestral clade during the Late Carboniferous and, possibly, the basal Permian.

Introduction

At the present time we are witnessing a dramatic revival of interest in the origin and evolution of seed plants. Together with the recent popularisation of cladistic methodology in interpreting plant systematics, this interest has spawned a renaissance of phylogenetics in the formulation of gymnosperm classification schemes. Important additions to our knowledge include the discovery of new, and the recognition of previously described Devonian fossils as ovulate cupules (Matten, Lacey and Lucas 1980, Gillespie, Rothwell and Scheckler 1981, Scheckler 1985), and the stratigraphic correlation of Lower Carboniferous remains from Great Britain and Europe (Scott, Galtier and Clayton 1984) so we can access the chronology of appearance for both gymnospermous taxa and characteristic gymnospermous organography. The description of a diverse assemblage of gymnospermous remains at the Devonian/Lower Carboniferous boundary (Matten, Lacey and Edwards 1975, Matten, Lacey, May and Lucas 1980, May and Matten 1983, Matten, Fine, Tanner and Lacey 1984), and a recently initiated programme to reconstruct the first Lower Carboniferous gymnosperms (late Visean) based on attachments of organs (Rothwell and Scott 1985) also are adding significantly to the data base. As a result, we are developing a dramatically improved understanding of the most primitive gymnosperms.

The data base

When attempting to generalise about the most primitive seed plants, one is constrained by the knowledge that all Devonian and Lower Carboniferous fossils currently attributed to gymnosperms consist only of isolated fragments of disarticulated organisms. Therefore, we have an insufficient data base for all but the most tentative interpretations of systematics and phylogeny. Clearly, it is imperative that we reconstruct by the most rigorous and reliable methods possible (Rothwell 1985) representatives of the earliest gymnosperms to preclude the possibility that our classification schemes either are based on aggregations of organs that were not produced by the same type of plant, or that they reflect no more than 'organ phylogenies'. However, until such data are available, cautious classifications of the most primitive, putative gymnosperm organs can serve as tentative hypotheses to help focus ongoing research.

Currently known fossils include cupules, ovules, aggregated or fused sporangia with trilete grains, actinostelic stems with mesarch xylem maturation, additional protostelic stems with varying degrees of vitalization or dissection, eustelic stems, lateral organs with features both of stems and leaves, frond-like organs and woody roots. By their diagnostic nature, the ovules and ovulate cupules represent gymnosperms. One extraordinary specimen (Long 1977a) has sporangia attached in the same cupule with ovules (*Hydrasperma longii sensu* Matten,

Lacey and Lucas 1980). This specimen demonstrates that at least some of the most primitive seed plants produced sporangia with trilete prepollen, and it documents that bisporangiate fructifications characterized some of the earliest gymnosperms.

Other Devonian and Lower Carboniferous remains are assigned to gymnosperms only because they bear a structural resemblance either to comparable organs of more recent plants known to be gymnosperms, or because they conform to what we expect the organs of the most primitive gymnosperms to be. In several instances, the remains in question are more similar to organs of aneurophytalean progymnosperms than they are to those of more recent gymnosperms (e.g. *Tetrastichia* stems, Gordon 1938, *Buteoxylon* and *Triradioxylon* frond axes, Barnard and Long 1973, 1975), so our systematic judgments must be rendered with caution. Because anatomically-preserved remains typically display a larger suite of characters than do plant fossils preserved by other modes, the current analysis focuses on permineralized ovules and ovulate cupules.

The role of Symmetry

Both utilitarian classifications and systematically oriented schemes that utilize reproductive features, typically stress ovule and/or cupule symmetry as an important character (e.g. Brongniart 1881, Oliver 1904, Seward 1917, Meyen 1984, Crane 1985.) Ovules that are round or that have three or more angles in cross section are considered to be radial (= radiospermic of Oliver 1904) and ovules with two sides or that are flattened in cross section are considered to be bilateral (= platyspermic of Oliver 1904). Radial versus bilateral symmetry are interpreted as alternate character states, and often are given substantial weight in the formulation of classification schemes (e.g. Meyen 1984). Structures that impart the symmetry are either the integument of the ovule or the cupule, or both. This includes the nucellus in ovules where the latter structure is interpreted by some authors as having been derived (at least in part) from thc integument (Meyen 1984).

Unfortunately, there are several problems inherent in utilizing symmetry as a systematically diagnostic character for the earliest gymnosperms. These include (1) symmetry traditionally has been interpreted imprecisely, inconsistently and often inaccurately (2) the most ancient ovules typically do not exhibit either radial or bilateral symmetry, but have several intermediate types of symmetry (3) many ovules display more than one type of symmetry (4) relative flatness or roundness of ovules occurs for different reasons in different groups of plants and, (5) among extant gymnosperms ovule symmetry may be variable within a single taxon.

1. Interpreting Ovule Symmetry

Definitions of types of symmetry are highly variable, and typically each is devised for a particular usage. In palaeobotany, 'bilaterally symmetrical' and 'platyspermic' often are used interchangeably for flattened or bifacial ovules, while 'radially symmetrical' or 'radiospermic' are employed for the more rounded forms (e.g. Seward 1917). However, several dissimilar shapes of ovules may fall within each category, and many ovules grouped together actually may be less similar to one-another than each is to some ovules placed in the other category. Therefore, if we wish to use this feature as an indicator of systematic relationships, it is important that we attempt to define more precisely the symmetry displayed by different types of ovules.

Several two-dimensional shapes that relate closely to the cross-sectional configurations of ovules are presented in Fig. 10.

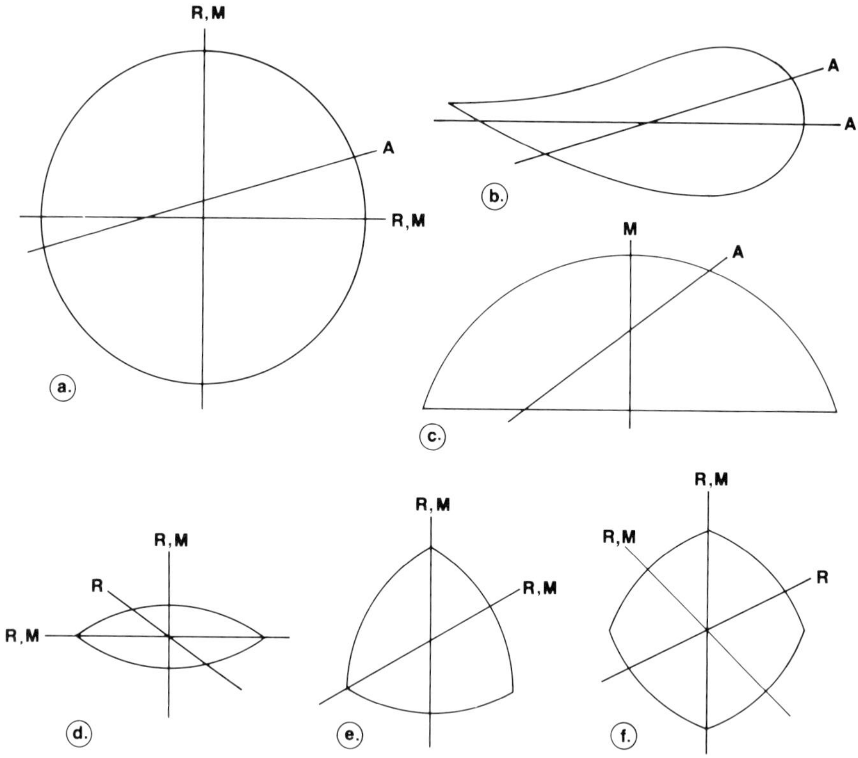

Fig. 10.1 Symmetry associated with gymnospermous ovules. Figure 1a is radially symmetrical. 1b is asymmetrical, 1c is bilaterally symmetrical, and 1d–f display additional types of symmetry. Lines through the shapes designate lines of asymmetry (at 'A'), mirror-image symmetry (at 'M') and rotational symmetry (at 'R'). See text for details.

1. Figures 10.1a and 10.1c–f may be interpreted as displaying symmetry because at least one line that passes through the centre of each dissects the shape into equal or congruent halves (i.e. lines labelled 'R', 'M' and 'R, M' are lines of symmetry). Line 'A' in Fig. 10.1a is not a line of symmetry because it does not pass through the centre of the figure.

Among symmetrical shapes, congruence can be visualised in two different ways. One half of the figure can be folded over on top of the other at the line of symmetry to reveal mirror-image symmetry (i.e., lines identified by 'M' in Fig. 10.1). One half of the figure also can be rotated around the centre of the object to another position where it will exactly cover another part of the figure to reveal rotational symmetry. Lines identified by 'R' in Fig. 10.1 are lines of rotational symmetry. Rotational symmetry may be subdivided further by the number of degrees that one half of a figure must be rotated until it exactly covers another part of the figure. Some lines define both mirror-image and rotational symmetry (i.e. lines 'R, M' in Fig. 10.1).

Figure 10.1b is asymmetrical because no line passing through the centre (e.g. lines 'A') dissects it into congruent halves. Unless distorted by abnormal growth or taphonomic processes, none of the most ancient ovules is asymmetrical.

Objects with only one line of symmetry are defined as bilaterally symmetrical, and usually display mirror-image symmetry (viz. line 'M' of Fig. 10.1c). Among Carboniferous gymnosperms, examples of this type of symmetry are *Camptosperma* Long (1961a) and the ovules of *Lebachia lockardii* Mapes and Rothwell (1984). One is more or less round (viz. *Camptosperma*) like what have been imprecisely termed radially symmetrical (Radiospermeae of Oliver 1904), while the other is flattened (viz. the ovule of *Lebachia,* Fig. 10.3) like the 'Platyspermeae' of Oliver (1904).

Round objects such as the circle in Fig. 10.1a display an infinite number of lines of both rotational and mirror-image symmetry (viz. line 'R, M' can be rotated to any position and the resulting halves of the object will be congruent), and are truly radially symmetrical. No Devonian or Lower Carboniferous ovules display true radial symmetry throughout.

Figures 10.1d–f conform to the cross-sectional configurations of two-, three- and four-angled ovules. All display more than one line of both mirror-image and rotational symmetry, so none is truly bilaterally symmetrical. Likewise, all display only a limited number of lines of mirror-image and/or rotational symmetry, so none is truly radially symmetrical. Figure 10.1d is similar to flattened, two-sided ovules such as *Eosperma oxroadense* Barnard (1960, Fig. 10.2), and to typical cardiocarpalean ovules of cordaites (e.g. *Mitrospermum*, Fig. 10.6) or

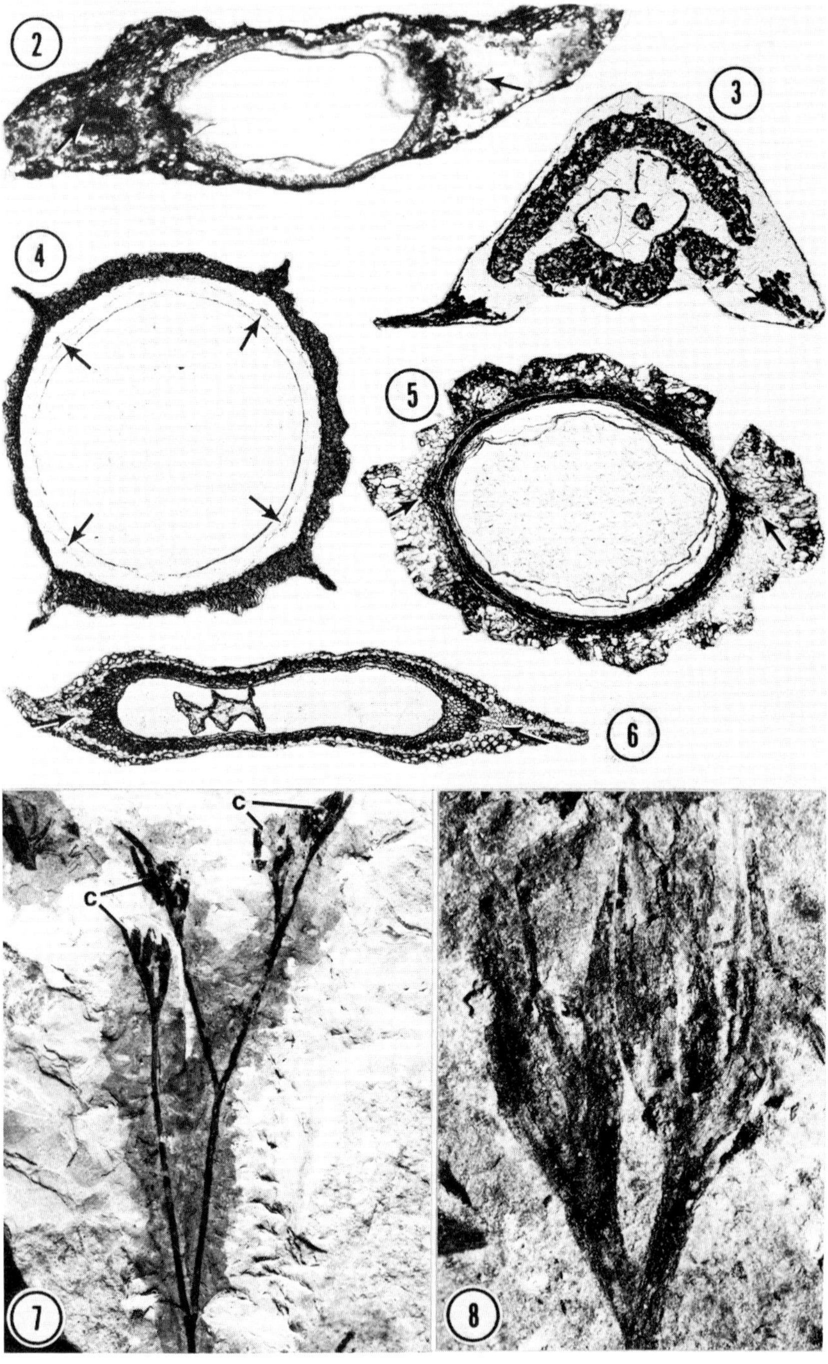
2
3
4
5
6
7
8
c
c
c

seed ferns (e.g.*Callospermarion*, Fig. 10.5). Note that none has bilateral symmetry like the ovules of *Lebachia* (Fig. 10.3). Rather, each has two lines of mirror-image symmetry and an infinite number of lines of 180 degree rotational symmetry (i.e. line 'R' can be placed on any radius and either side rotated 180 degrees to exactly cover the other).

Figure 10.1e is similar to the basically three-angled trigonocarpalean ovules, and has three lines each of rotational and mirror-image symmetry (each passes through an angle and the centre of the opposite side). Note that if the half of the figure above the line is rotated clockwise 120 degrees, it will fit over the right half of the figure. Figure 10.1f shows four angles like many lagenostomalean and some Lower Carboniferous ovules (e.g. *Conostoma anglo-germanicum,* Fig. 10.4 and *Eurystoma angulare* Long 1960a). Note that lines passing through opposite angles or through the centre of opposite sides define positions where one side either can be folded over or rotated 90 degrees to fit over another half of the figure. Therefore, there are four lines of both mirror-image and 90 degree rotational symmetry to this figure. Note also that a line passing through the centre of the figure on any radius (e.g. line 'R') separates halves that can be rotated 180 degree to fit over one another. Therefore, there also are an infinite number of lines of 180 degrees symmetry to this figure.

Symmetrical shapes (or ovules) with larger numbers of angles and sides each display distinct and different combinations of types of symmetry. However, none is truly radially symmetrical in the sense of the circle (Fig. 10.1a). It is clear that characterizations of ovules as radially or bilaterally symmetrical neither reflect the numerous differences in shape displayed by the fossils nor adequately portray the features of any.

2. Combinations of Types of Symmetry

Many ovules display the same type of symmetry throughout, while others show different types of symmetry depending upon the level at which they are cut, or the tissue that is considered. For example, as recently reinterpreted by Long (1977b), the Upper Carboniferous lagenostomalean ovule *Conostoma oblongum* is round toward the apex (Text fig. 7d-h of Long, 1977b), where it has a radially symmetrical exterior. However, if the disposition of integumentary vascular bundles

---

Figs. 10.2–8 Primitive ovules and cupules. Figs 10.2–6 illustrate cross-sectional configurations of ovules. Arrows indicate positions of integumentary bundles. Fig. 10.2. *Eosperma oxroadense.* Fig. 10.3. *Lebachia lockardii.* Fig. 10.4. *Conostoma anglo-germanicum.* Fig. 10.5. *Callospermarion undulatum.* Fig. 10.6. *Mitrospermum vinculum.* Figs. 10.7–8 illustrate the most ancient ovulate cupules from West Virginia. Fig. 10.7. Branching system with terminal cupules (at c). Fig. 10.8. Single cupule showing prominent hemispheres formed by basal fork.

is considered, this region displays the same type of symmetry as Fig. 10.1f. At the midregion the same ovule becomes ovoid in cross section and has symmetry comparable to *Eosperma* (Fig. 10.1d, 10.2) and to cardiocarpalean ovules (Figs. 10.5, 6). At the chalaza the vascular bundle of *C. oblongum* enters the ovule obliquely from one side (Text Fig. 7a-b of Long, 1977b) so that it is bilaterally symmetrical like the ovules of *Lebachia lockardii* (Fig. 10.3) and Fig. 10.1c. Similar but less pronounced differences in symmetry occur at different levels of the Lower carboniferous ovules *Deltasperma* Long (1961a) and *Eccroustosperma* Long (1961b). It is clear that ovule symmetry may vary from level to level and from tissue to tissue in the same specimen.

3. *Factors that determine Ovule Shape*

Among the most ancient ovules, those with a vascular strand that does not enter the centre of the chalaza (viz. those that may not have been orthotropous) display bilateral symmetry, at least near the chalaza (e.g. *Camptosperma. Deltasperma*). Other ovules have larger numbers of lines of symmetry and a shape that reflects the number of vascular bundles in the integument. For example, those with two bundles have two angles and are correspondingly flat (Table 10.1), those with three bundles have three corners and are triangular (e.g. *Eurystoma trigona* Long 1969), those with four bundles have four corners and are quadrangular (e.g. *Conostoma anglo-germanicum*, Fig. 10.4), those with five bundles have five sides (e.g. some specimens of *Tantallosperma* Barnard and Long 1975), and so on.

Within some genera different species have different numbers of vascular bundles in the integument (e.g. *Eurystoma* has 3–5, Long 1969, *Conostoma* has 4, 6 and 8-9, Stubblefield *et al.* 1984), and some species have specimens with different numbers of vascular bundles (Table 10.1). In the past, ovules with three or more integumentary bundles all have been lumped together as 'radially symmetrical' while those with two bundles have been described (often inaccurately) as bilaterally symmetrical. However, there is no reason to consider the difference between two and three integumentary bundles to be more phylogenetically significant than the difference between three and four or between four and five bundles when other features of the ovules are basically similar (Barnard 1960, Long 1961a).

Among extant gymnosperms, ovule shape often is less closely correlated with vascularization. Many conifers have flattened ovules and an unvascularized integument. These (e.g. *Pinus*) develop between two closely-spaced cone scales, and are flattened as a reflection of the space in which they have to grow. The flattened appearance also may be accentuated by the wing of cone–scale tissue that detaches with the ovule. In

Table 10.1. Variations in the structure of 23 currently known species of permineralized ovules from Devonian and Lower Carboniferous strata. Divergent features are those that vary from the Devonian representatives. All ovules either are borne in cupules (C) or the mode of attachment is unknown (U). Specimens found on a short stalk are recorded as attachment unknown.

Taxon	No. of Bundles	Attachment	Divergent Features
Calathospermum fimbriatum	4–6	C	Flattened, nucellar pad of tracheids.
Calathospermum scoticum	9–11	C	Flattened, bilaterally symmetrical.
Camptosperma berniciense	2	U	
Deltasperma fouldense	2	U	
Dolichosperma pentangonum	6–7	U	
Dolichosperma sexangulatum	5–7	U	Flattened, bilaterally symmetrical.
Eccroustosperma langtonense	2	U	Flattened
Eosperma oxroadense	2	U	
Eurystoma angulare	4–5	C	
Eurystoma burnense	3	C	
Geminitheca scotica	6–9	C	Bundles undulate from sarcotesta-endotesta-sarcotesta.
Genomosperma kidstoni	6–11	U	
Genomosperma latens	7–8	U	Nucellus and integument free.
Hydrasperma longii	8–10	C	Nucellus and integument free, nucellar pad of tracheids.
*Hydrasperma tenuis**	8–10	U (C)	
Lyrasperma scotica	2	U	Nucellus and integument free, nucellar pad of tracheids.
Mitrospermum bulbosum	?	U	
Salpingostoma dasu	5–6	U	
Sphaerostoma ovale	8	C	
Stamnostoma huttonense	5–6	C	
Stamnostoma bifrens	3–?	U	Flattened, bilaterally symmetrical.
Tantallosperma setigera	4–6	U	Pollen chamber floor and central column absent?
West Virginia specimens	4–5	C	

* *Hydrasperma tenuis* is a genus and species established for structurally distinct, isolated ovules from Berwickshire, and for which the mode of attachment is unknown. All of the Devonian ovules are of this basic type. Also, several of the Lower Carboniferous ovules are of this type, but they occur in cupules with differnt morphologies (e.g. *Hydrasperma tenuis* and *H. longii* of Matten, Lacey and Lucas 1980). If topotype specimens of *H. tenuis* subsequently are found to occur in cupules similar to *H. tenuis* from the Tournaisian Devonian of Ireland (Matten, Lacey and Lucas 1980), then there will be documentation that *H. tenuis* specimens from the two sources represent the same cupulate species.

contrast, the ovules of *Podocarpus* develop without the space constraints of *Pinus*, and the vascularized ovules have a round outer form. The ovules of *Ginkgo* also develop in the absence of space constraints, and have an essentially round exterior (i.e. sarcotesta). This is despite the fact that they usually have two integumentary vascular bundles and two ribs on the surface of the sclerotesta (Bierhorst 1971). However, some *Ginkgo* ovules have more than two integumentary bundles, and the number of ribs equals the number of bundles as in the most primitive ovules discussed above. There is no doubt that ovule shape is descriptively useful and taxonomically diagnostic for many gymnosperms, and that ovule symmetry may have some value for formulating classifications. However, until ovule shape and symmetry are carefully examined, precisely defined and fully understood for each group, correlations to systematic relationships are highly suspect.

4. Cupule Symmetry

Like ovule symmetry, presumed cupule symmetry has been employed by some authors as a systematically diagnostic character (e.g. Meyen 1984). However, symmetry has been even less accurately interpreted, and is less completely understood for cupules than for ovules. As more data become available, comparative evaluations of branching patterns and cupule organisation (e.g. Matten and Lacey 1981) are proving to be far more valuable for defining relationships than are simple interpretations of presumed symmetry.

Evidence and interpretation

1. The Most Primitive Ovulate Fructifications

There now are five ovulate fructifications from late Famennian (the most ancient ovules from West Virginia, *Archaeosperma arnoldii, Xenotheca devonica, Moresnetia zalesskya*), and Tournaisian (*Hydrasperma tenuis sensu* Matten, Lacey and Lucas, from Ireland) Devonian sediments (Rothwell, 1985). The four Famennian taxa are remarkably similar in size and complexity, in branching architecture, and in ovule morphology. The slightly younger *Hydrasperma tenuis* of Matten, Lacey and Lucas (1980) shows some variability in complexity and number of ovules, but otherwise is similar to the Famennian forms. Also where known, all are borne on a similar type of branching system, and have similar features of the ovule vascularization and nucellus. Cupules are terminal on systems of terete, forking axes (Fig. 10.7). The systems have all of the segments at one side slightly overtopping those on the other side (Fig. 10.7). In the West Virginia fossils, successive divisions are at right angles, forming a cruciately-forking system. In *Moresnetia* and *Xenotheca* it is not known if the ovulate branching system has cruciate forking, but in other features of basic structure they are strikingly similar to the cupulate systems from West Virginia (Scheckler, personal communication).

The cupules are constructed on a somewhat modified, cruciately forking pattern, with the basalmost fork producing two prominent hemispheres (Fig. 10.8). As originally described, each hemisphere of *Archaeosperma* was interpreted as one of a cupule pair (Pettitt and Beck 1968). However, the common enclosure formed by the two halves acts as a functional unit that more recently has been recognized as a single cupule (Matten, Lacey and Lucas 1980, Gillespie *et al.* 1981). Subsequent forks are closely spaced, forming quadrats that each bear one ovule toward the centre of the cupule and three of four telomes toward the periphery. The telomes of each quadrat arise from two forks in the same plane, and all of the Famennian cupules have four ovules (Scheckler 1985).

In the Irish fossils (viz. *H. tenuis sensu* Matten, Lacey and Lucas 1980) there may be two-six units that each correspond to a quadrant of the Famennian taxa. There also may be two-six ovules per cupule (with the ovule absent from some units), but otherwise they are similar to the Famennian taxa.

It should be emphasized that all features of the (1) fertile branching system (2) mode of cupule attachment (3) cupule architecture (4) ovule morphology and (5) ovule anatomy are similar (where known) for the earliest gymnosperms. Because of the profound structural similarities in numerous features of all of the Famennian taxa and only minor variations in the Tournaisian species, all may be interpreted as closely related systematically. If true, there is strong evidence for a monophyletic origin of gymnosperms (Rothwell 1980, 1982, 1985).

2. *The Most Primitive Ovules*

Ovules of all five genera are morphologically comparable to isolated specimens of *Hydrasperma*. They are more or less orthotropous and roughly round in cross section, with four to ten integumentary lobes that are fused at the base and free distally. Permineralized specimens have a single vascular bundle in each lobe, and an unvascularized nucellus (Matten, Lacey and Lucas 1980, Gillespie *et al.* 1981). Also, the nucellus and integument are more or less united below the level of the pollen chamber. Where known, the pollen chamber is roughly hemispherical and displays a lagenostome (= salpinx) and a membranous floor with a prominent central column. The latter feature is equivocal in the West Virginia ovules, but is well documented for the Irish material (Matten, Lacey and Lucas 1980). It also is crucial to interpreting both reproductive biology and systematics of the most primitive gymnosperms.

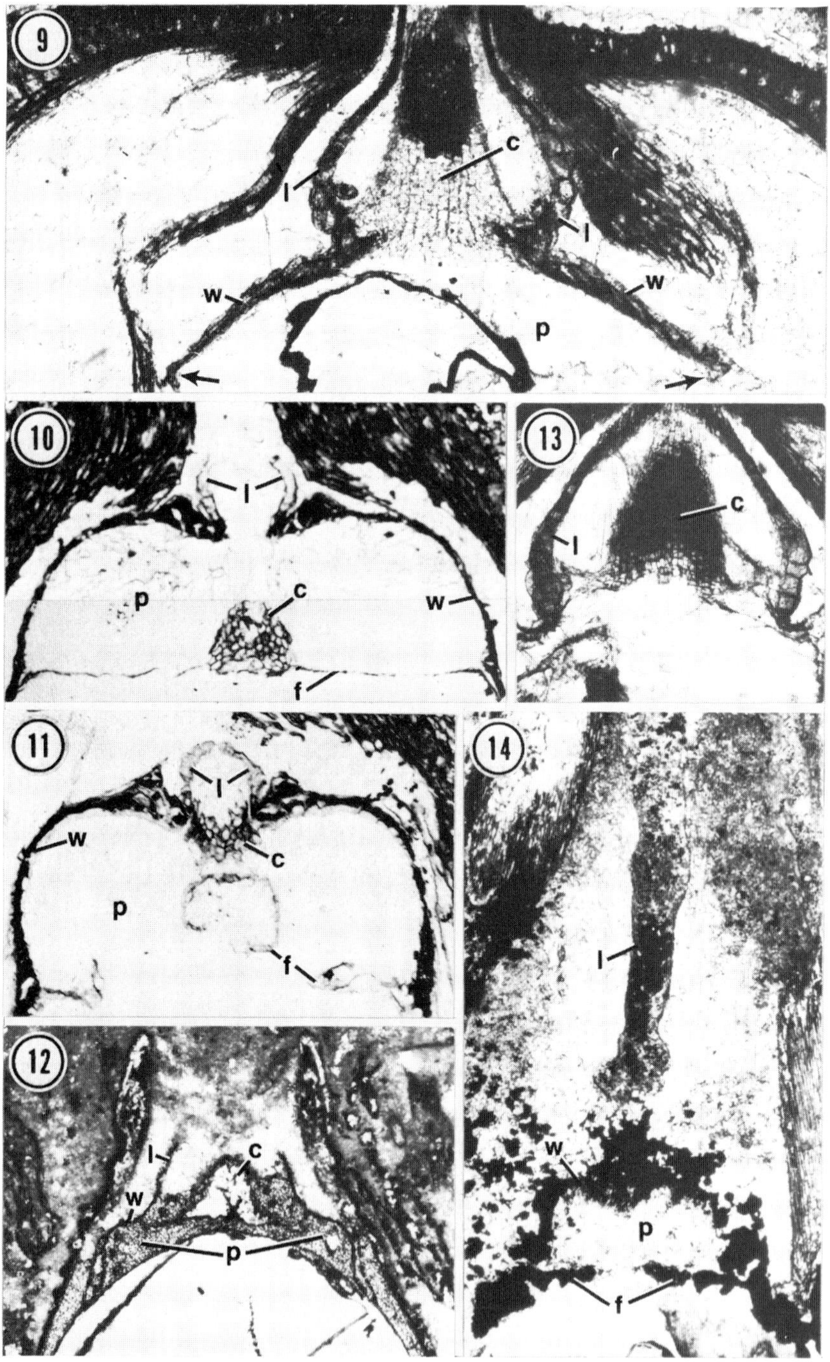
9
c
l
l
w
w
p
10
l
p
c
w
f
13
c
l
11
l
w
c
p
f
12
l
c
w
p
14
l
w
p
f

3. *Structure Of The Primitive Gymnospermous Pollen Chamber*

Features of the pollen chamber in the most primitive gymnosperm ovules are basically similar for all known species (Rothwell 1983, 1985). Each consists of two regions; the lower pollen chamber proper, and the upper lagenostome or salpinx (Rothwell 1985, Figs. 9-12, 10.14). There also is a membranous floor with a central column (Fig. 10.10). However, this basic similarity has not been recognized widely because of extremely confusing terminology used for the pollen-receiving mechanism, and because of the incomplete preservation of the pollen chamber in many specimens (Rothwell 1983). Structural features of the most primitive pollen chambers are best clarified by a historical review of the pertinent literature.

Ovules of this type first were studied by Williamson (1877), and include specimens now known as *Lagenostoma ovoides* (Figs. 54–75 of Williamson 1877), *Physostoma elegans* (Figs. 77–79 of Williamson 1877), *Conostoma oblongum* (Figs. 80–81 of Williamson 1877) and *Sphaerostoma ovale* (Figs. 82–84 of Williamson 1877). *Lagenostoma ovoides* was described first, and it is to this description that the current concepts of primitive pollen chambers may be traced. Williamson (1877) referred to the entire pollen chamber of *Lagenostoma* as the lagenostome (Williamson 1877, p. 234) and identified the central column, but his description led to confusion about which structures are integument and which represent nucellus. Also, Williamson did not observe the central column in his specimens of *Physostoma (Lagenostoma physoides* of Williamson 1877) so the pollen chamber of this species appeared to be quite different from that of *L. ovoides*.

Somewhat later, Oliver and Scott (1904) described *L. lomaxii*, and misinterpreted the obconical upper chamber to be the entire pollen chamber. Hence, the pollen chamber of *Lagenostoma* was thought to consist of only one region (the lagenostome of Oliver and Scott 1904). Most subsequent authors have followed Oliver and Scott's interpretation, such that the pollen chamber proper is usually thought to be obconical (e.g. Emberger 1944, Stewart 1983).

Among recent reconstructions, only that of Taylor and Millay (1981) accurately includes a lower, herispherical region in the pollen chamber

---

Figs. 10.9–14. Pollen chambers of ovules with hydrasperman reproduction. All have a basal chamber (p, the pollen chamber proper) surrounded by a uniseriate wall (w), and a more distal lagenostome or salpinx (1). There also is a membranous pollen chamber floor (f) with a prominent central column (c). All are at the post-pollination stage of development except 10.10, which is at the pollination stage. Fig. 10.9. *Lagenostoma lomaxii.* Fig. 10.10. *Conostoma williamsonii.* Fig. 10.11. *Conostoma williamsonii.* Fig. 10.12. *Stamnostoma cf. huttonense.* Fig. 10.13. *Lagenostoma lomaxii.* Fig. 10.14. *Salpingostoma dasu.*

of *Lagenostoma* (their Fig. 5A). In this genus structural features are difficult to interpret because the lower part of the pollen chamber is relatively small, and in oblique sections the upper region of the wall (viz. lagenostome) often appears to delimit the entire pollen chamber (e.g. Fig. 10.13). This appearance is accentuated because the central column typically is in contact with the base of the lagenostome as if it were in organic connection (Fig. 10.13). However, the specimen in Fig. 10.9 demonstrates that the subtending tissue delimits a lower, hemispherical region of the pollen chamber. The membranous floor, to which the central column was attached at an earlier stage of development, has not been found intact in *Lagenostoma*, but the former positions of attachment are marked by arrows in Fig. 10.9.

In 1911, Oliver and Salisbury monographed the genus *Conostoma*, adding the species *C. anglo-germanicum*. They interpreted the pollen chamber to be a relatively small lagenostome with a pad of soft tissue forming the floor. These authors did not realize that the 'pad of soft tissue' was a central column homologous to that of *Lagenostoma*, and they reported that a central column ('central cone of soft tissue' of Oliver and Salisbury 1911) like that of *Lagenostoma* is absent from *Conostoma*. In addition to the lagenostome, Oliver and Salisbury interpreted there to be a large, lower chamber in *Conostoma* that they termed the plinth. Many current authors refer to the plinth of Oliver and Salisbury as the pollen chamber, while others (e.g. Long 1977b) refer to the region immediately below the pollen chamber floor as a plinth. However, except for differences in the sizes and shapes of the various parts, we now know that the pollen chambers of *Lagenostoma* and *Conostoma* are structurally comparable (compare Fig. 10.9 and Fig. 10.11).

In 1941, Gordon described *Salpingostoma*, and recognized that the pollen chamber of this genus consisted of two parts; a hemispherical region surmounted by a tubular region (Fig. 10.14). Following established terminology, he referred to the pollen chamber (viz. lower region) as the lagenostome and interpreted the upper region to be new and different from anything that had been described before. The apical region he termed the salpinx.

By this time, it was believed that the pollen chambers of *Lagenostoma, Conostoma, Physostoma, Sphaerostoma* and *Salpingostoma* were all quite complex and different from one another (e.g. Fig. 2 of Gordon 1941). The confusion has been compounded even further by more recent studies, where the disparate terminologies used by earlier workers for homologous structures (and the same terms used for different structures) are perpetuated (e.g. Rothwell 1971, Long 1977b).

It should be stressed that all Devonian and Lower Carboniferous ovules (with the possible exception of *Mitrospermum bulbosum*, see Rothwell, 1985) and all lagenostomalean ovules have pollen chambers that in basic features are identical (Rothwell 1983, 1985). It also is of note that none of the ovules of post basal Permian gymnosperms have a pollen chamber of this type. Because the most ancient unequivocal occurrence of this pollen-receiving apparatus is in ovules of the *Hydrasperma*-type, it may be termed a 'hydrasperman pollen chamber'.

Development of the Primitive Gymnospermous Pollen Chamber. Ontogeny of the hydrasperman pollen chamber has been interpreted by several authors, and we now understand how it could function in the absence of a fully formed integument (Rothwell 1971, Stubblefield *et al.* 1984). At the time of pollination, the membranous pollen chamber floor was relatively flat and intact, and the lagenostome or salpinx was open so that pollen grains could enter the pollen chamber (Fig. 10.10). After pollination, growth of the megagametophyte pushed the floor upward to press the central column into the base of the lagenostome (or salpinx, Figs. 10.9, 11-13). It also ruptured the membranous floor so that the megagametophyte and microgametophyte(s) no longer were isolated from one another (Rothwell 1971, Figs. 11–12). More than 99 per cent of known specimens with a hydrasperman pollen chamber are preserved in the post-pollination stage of development. This demonstrates that ovule ontogeny was rapid, and that pollen-chamber development continued even if pollination did not occur.

4. *Hydrasperman Reproduction*

The ovules of extant gymnosperms all have a well formed micropyle that reflects the crucial role in reproductive biology played by the integument. Integumentary functions include several types of specializations for pollination, and postpollination closing of the micropyle to seal the integument (Singh 1978) to produce an environment conducive to continued gametophyte development, fertilization and embryogeny. The absence of a fully enclosing integument with a well-formed micropyle from all Devonian ovules and many Lower Carboniferous ovules dictates that these functions were performed by other tissues (viz. the nucellus) in the most primitive gymnosperms.

If one compares the hydrasperman pollen chamber to the pollen chambers in more recent ovules, features that could have performed these functions, and that also are unique to the earliest gymnosperms become obvious. These are the membranous pollen-chamber floor and the prominent central column (Fig. 10.14). Additionally, the differentiated (tubular or ellipsoidal) distal part of the pollen chamber wall (viz. salpinx = lagenostome) is distinctive (Rothwell 1985). Because the earliest unequivocal evidence for such a mechanism occurs in Devonian ovules of the *Hydrasperma*-type (viz. *H. tenuis* of Matten, Lacey and

Lucas, 1980) it may be termed 'Hydrasperman Reproduction'. The 'nucellar beak' of many more recent ovules is homologous to the lagenostome (or salpinx), but a membranous floor with a central column is absent from these. Therefore, it is the pollen chamber floor with a central column that is unique to, and that reflects reproductive biology in the most primitive gymnosperms.

Hydrasperman reproduction can be interpreted as a specialization of pteridophytic reproduction that employs three interdependent modifications. These are:

(1) Cells at the tip of the megasporangium (= nucellus) break down to form the pollen chamber, rather than sporangial dehiscence occurring. The outer wall of the megasporangial tip forms the lagenostome (or salpinx) and the lower region of the pollen chamber wall. A membranous pollen-chamber floor with a central column remains after lysigenous dissolution of the cells in the megasporangial tip. The latter also may have formed a droplet to aid in pollination (Gordon 1941), but this remains suppositional and the mechanism could have operated effectively without a pollination droplet.

(2) The microgametophytes enter the pollen chamber, rather than fall near a dehisced magasporangium or isolated megaspore. As the megagametophyte grows in size, it presses against the pollen-chamber floor and forces the central column into the base of the lagenostome. This seals the pollen chamber in a fashion analogous to sealing the micropyle in more recent gymnospermous ovules.

(3) The ovule or ovule-bearing structure (viz. cupule) or both abscise to facilitate dissemination of the ovule/seed. Replacing sporangial dehiscence with megasporangial abscission in the most primitive gymnosperms significantly altered the nature of the disseminule, and the additional enclosing structures may have had a profound effect on the ability of such propagules to survive, germinate and grow in dry habitats. If true, this reproductive mechanism would have enhanced dramatically the ability of the earliest gymnosperms to exploit new or previously little-occupied ecological space (viz. dry habitats).

Classification

It must be emphasized that all Devonian and Lower Carboniferous ovules (for which the anatomical features are well known) display hydrasperman reproduction. If the structure, development and reproductive biology of such organisms are correctly interpreted, they form an extremely elaborate complex of interdependent characters that is most unlikely to have evolved independently in two or more major groups (Knoll and Rothwell 1981). As such, hydrasperman reproduction may be regarded as a synapomorphy that characterizes the most primitive gymnosperms. This interpretation is supported by the common features of Devonian cupules and ovules (elaborated above) that are not known for more ancient plants. In this regard, these features may be viewed as an additional complex of synapomorphies. If true, then the Devonian and Lower Carboniferous gymnosperms (and their descendants) form a monophyletic group.

Evolutionary radiation of primitive gymnosperms

If the Famennian fossils discussed above represent the archetype of gymnosperm ovulate cupules, a comparison of such structures with remains from Tournaisian and Visean sediments allows one to assess the early evolutionary radiation of gymnospermous fructifications. The earliest evidence of this radiation may be the variations in cupule complexity and ovule numbers displayed by the Tournaisian Devonian *Hydrasperma tenuis*.

Cupules like those from Devonian sediments (e.g. *Stamnostoma huttonense* Long 1960b) also are preserved in Lower Carboniferous deposits. In addition, larger and more complex cupules (such as *Calathospermum*), and cupules with more variable branching patterns (i.e. *Hydrasperma longii* of Matten, Lacey and Lucas 1980) appear in Tournaisian and/or Visean strata. Differences among the various forms primarily result from (1) variations in the total number of ovules and sterile segments produced (2) variations in the planes of sequential divisions of the system (3) variations in branching from forking to pinnate and (4) variations in the degree to which the segments are fused laterally (Matten and Lacey 1981). All of these are relatively minor changes in branching pattern of telomic systems, and are similar in magnitude to structural variations among relatively closely-related extant plants (perhaps genera of a common family).

There also are several types of variations in structure among the cupulate and isolated permineralized ovules from Lower Carboniferous strata (Table 10.1). Most commonly, variations are in the numbers of integumentary lobes (= numbers of integumentary vascular bundles), in the degree of fusion of the integumentary lobes, and in the amount of fusion of the nucellus and integument. Like the differences among the Devonian and Lower Carboniferous ovulate cupules, variations of these types are consistent with what one would expect to find among the ovules of potentially closely-related plants. Also, like *Hydrasperma*, many ovules with these types of variations could be assigned to the form-order Lagenostomales (Seward 1917). Possible exceptions are the flattened ovules with only two integumentary vascular bundles (for example,

Eosperma), but as stressed above, this feature probably has little significance to phylogenetic systematics.

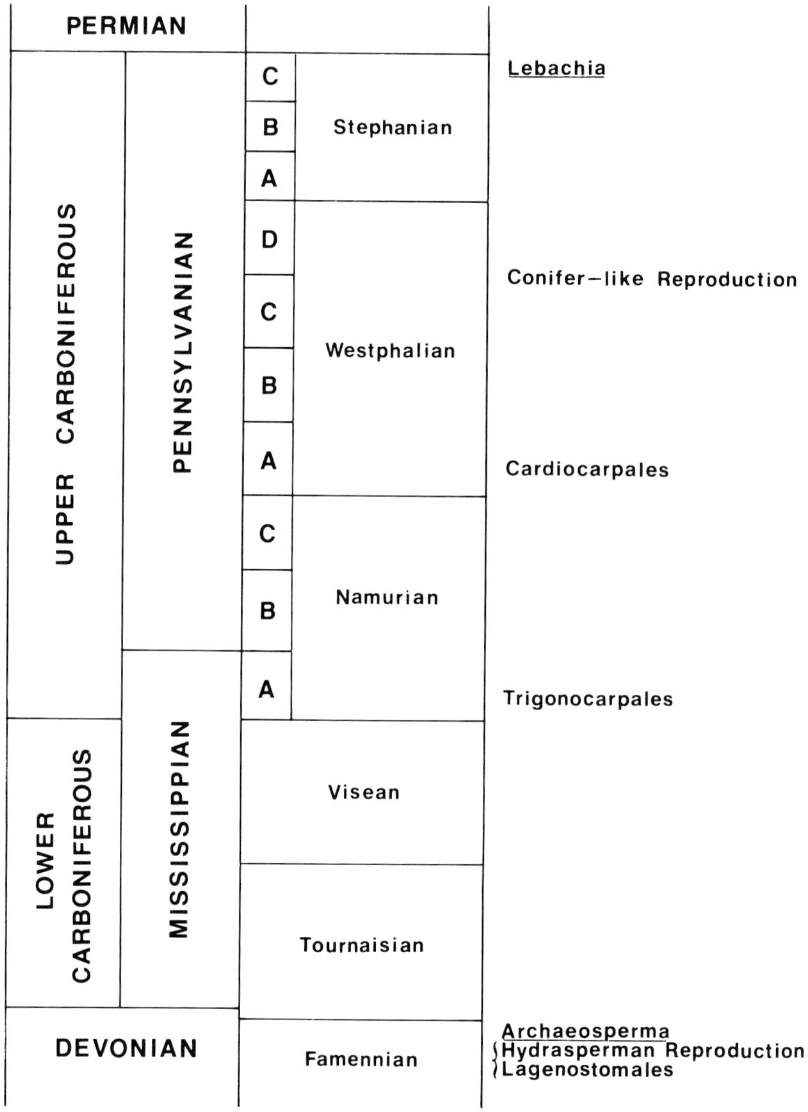

Fig. 10.15. The evolution of gymnospermous reproduction. The earliest record each of several types of gymnospermous reproduction suggest that all Devonian and Lower Carboniferous gymnosperms had hydrasperman reproduction, and that additional types of reproductive biology originated successively through the Upper Carboniferous. Also, the Famennian Devonian taxa had cupulate systems similar to that of *Archaeosperma arnoldii*. See text for details.

Additional variations occur in the basic vascularization of some ovules. A vascularized nucellus (viz. pad of tracheids at the base) occurs in *Camptosperma* and *Genomosperma*, and the integumentary bundles enter the sarcotesta in *Eurystoma angulare* and *Lyrasperma* (Table 10.1). Features of these types are reminiscent of ovules produced by some more recent gymnosperms, but by displaying hydrasperman reproduction such ovules compare most favourably with the other Devonian and Lower Carboniferous ovules.

At the present time there is no evidence for a transition from hydrasperman reproduction to the more derived reproductive biology of more recent gymnosperms, nor for the origin of additional gymnospermous taxa (clades?). However, currently available data do provide a basis for interpreting the chronology of grade changes that produced reproductive mechanisms associated with several other major groups of seed plants (Fig. 10.15).

If additional evidence confirms the somewhat equivocal features of the pollen chamber in most ancient ovules from West Virginia (Gillespie, *et al.* 1981), then the origin of hydrasperman reproduction (and thus gymnosperms) was not later than Famennian Devonian. Because the earliest ovules have the appropriate features of shape, vascularization and integument/nucellus adnation, they also mark the earliest occurrence of Lagenostomales. All other form-orders of Palaeozoic ovules (Seward 1917) and all other types of Carboniferous gymnospermous reproduction require that the ovules have a fully formed integument with a micropyle, and probably did not evolve earlier than the Upper Carboniferous.

Present evidence suggests that the chronology of evolutionary radiation was as follows (Fig. 10.15):

(1) The origin of Trigonocarpales occurred by Namurian A as is evidenced by the presence of *Rhynchosperma quinnii* Taylor and Eggert (1967b) in Mississippian sediments of Arkansas. If ovules of this type are characteristic of Medullosales, as popularly interpreted and supported by the occurrence of *Medullosa* and *Quaestora* from the same sediments (Taylor and Eggert 1967a, Mapes and Rothwell 1980), then this group evolved no later than the earliest Upper Carboniferous.

(2) The origin of Cardiocarpales occurred by Westphalian A as is evidenced by the presence of *Mitrospermum compressum* Arber (1910) in coal balls from England and comparable strata of Europe (Phillips 1980). These ovules probably were produced by cordaites as popularly interpreted and supported by associational evidence (e.g. Daghlian and Taylor 1979). However, credible evidence for compressed cordaitean remains has been described from as low as Namurian B sediments so that it would not be surprising if

permineralized remains from comparable sediments also contain cordaitean cardiocarps.

(3) The origin of modern conifer-like reproduction (viz. microgametophytes produced by true pollen, pollination via a pollination drop mechanism and siphonogamous fetilization) occurred by Westphalian D as is evidenced by the pteridosperm *Callistophyton* in Desmoinesian strata of North America (Rothwell 1981). Reproduction of this type is more like extant conifers than any other known Palaeozoic plant, and suggests a phylogenetic relationship between conifers and some seed ferns (Rothwell 1982).

(4) The origin of primitive conifer biology occurred by the Stephanian as is evidenced by *Lebachia lockardii* Mapes and Rothwell (1984) in Upper Pennsylvanian sediments of Kansas. Plants of this type had prepollen, rather than the true pollen that characterizes extant conifers and the more ancient *Callistophyton boyssetii*, and may have originated by Westphalian time as is indicated by compressed remains (Mapes and McComas 1984).

Alternative hypotheses

Some other current students of seed-plant evolution have suggested a biphyletic origin for gymnosperms (e.g. Beck 1976, 1981, Chaloner, Hill and Lacey 1977, Stewart 1981). Evidence for this view consist of (1) two lines of putative gymnosperm ancestors in Late Devonian sediments (viz. aneurophytalean and archaeopteridalean progymnosperms; Beck 1976, Stewart 1981) and (2) the occurrence of what has been interpreted to be a second type of noncupulate, ovulate structure in uppermost Devonian sediments (e.g. *Spermolithus devonicus*). The latter specimens consist of isolated, flattened bodies (thought to be ovules; Chaloner, Hill and Lacey 1977, Meyen 1984). However, there is no evidence for the mode of attachment of such structures (they equally could have been borne in cupules), nor that they actually do represent ovules. It should be emphasized that such structures also could represent magasporangia of unknown free-sporing pteridophytes with extreme heterospory (namely one functional megaspore per megasporangium) or some other organ. Even if they are ovules, they may differ from those of the other Devonian taxa only by the presence of two rather than four-ten integumentary lobes. The correlative evidence presented above supports the latter interpretation. When viewed in this light the flattened bodies assume far less systematic significance, and do not compromise interpretations of a monophyletic origin of gymnosperms.

An additional hypothesis that recognizes three major clades of gymnosperms in Lower Carboniferous strata recently has been elaborated by

Meyen (1984). This hypothesis is based largely on the recognition of taxa from presumed associations of isolated organs, and upon supposed parallel or convergent evolution of ovules with similar structural features. Meyen did not emphasize the structural, developmental and reproductive biological data regarded by this author as being most diagnostic of the earliest gymnosperms (hydrasperman reproduction), and he had access to less data on the Famennian Devonian ovulate cupules than was utilized in this study. More detailed evaluations of the Meyen hypothesis recently have been presented in critiques by Beck (1985), Miller (1985) and Rothwell (1985).

Summary

(1) Remains of the earliest gymnosperms occur in uppermost Devonian sediments, with those from Famennian, Tournaisian and Visean strata being represented only by fragments of disarticulated organisms. As a result, the ovules and ovule-bearing structures currently are the only unequivocally gymnospermous organs, and they provide the only reliable data for formulating classifications.

(2) Ovule and cupule symmetry have been inaccurately and inconsistently interpreted in the past. They also are unreliable characters for interpreting the phylogenetic systematics of primitive gymnosperms.

(3) The most primitive ovulate cupules from Famennian strata have basically similar structure, and this suggests a monophyletic origin of gymnosperms.

(4) The most ancient ovules and lagenostomalean ovules all have a common type of pollen chamber (the 'hydrasperman pollen chamber') that is not found in the ovules of any of the more recent gymnosperms.

(5) Ontogenetic features of this pollen chamber reveal that it could function in the absence of an ovular integument. This demonstrates the presence of a distinctive mode of reproductive biology, 'hydrasperman reproduction', among all of the most ancient gymnosperms.

(6) Hydrasperman reproduction may be regarded as a synapomorphy that defines a clade of the most primitive gymnosperms.

(7) Additional groups of gymnosperms diverged from the ancestral clade throughout the Upper Carboniferous and possibly the basal Permian.

Acknowledgements

The author is indebted to several individuals for assistance in completion of this work. Stephen E. Scheckler, Virginia Polytechnic Institute and State University provided data and informative discussions about the most ancient cupules from Belgium and Great Britain. Gary L. Floyd, The Ohio State University was instrumental in initiating a critical reevaluation of ovule symmetry. Andrew C. Scott made available specimens, laboratory space and facilities at Chelsea College, University of London, where Lower Carboniferous gymnospermous remains were studied. Cedric Shute, Department of Palaeontology, British Museum (Natural History) provided the negatives for Figs. 10.9, 10.13 and 10.14, and invaluable field assistance during collection of Lower Carboniferous remains was provided by Richard Bateman, Stephen Brindley and Keith Stephens, Chelsea College, University of London. Figures 10.1 and 10.15 were drafted by Catherine Steiner, Ohio University Cartographic Center.

References

Arber, A. (1910) On the structure of the Palaeozoic seed *Mitrospermum compressum* (Williamson). *Annals of Botany* **24**, 491–509.

Barnard, P.D.W. (1960) On *Eosperma oxroadense* gen. et sp. nov.: A new Lower Carboniferous seed from East Lothian. *Annals of Botany* New Series. **23**, 285–96.

—— and Long, A.G. (1973) On the structure of a petrified stem and some associated seeds from the Lower Carboniferous rocks of East Lothian, Scotland. *Transactions of the Royal Society of Edinburgh* **69**, 91–108.

—— and —— (1975) *Triradioxylon*—a new genus of Lower Carboniferous petrified stems and petioles together with a review of the classification of early Pterophytina. *Transactions of the Royal Society of Edinburgh* **69**, 231–49.

Beck, C.B. (1976) Current status of the Progymnospermopsida. *Review of Palaeobotany and Palynology* **21**, 5–23.

—— (1981) *Archaeopteris* and its role in vascular plant evolution. In *Paleobotany, Paleoecology and Evolution*. Vol. I (ed. K.J. Niklas, Praeger Publishers, New York, pp. 193–230.

—— (1985) Gymnosperm Phylogeny—A Commentary on the views of S.V. Meyen. *The Botanical Review*. **51**, 273–94.

Bierhorst, D.W. (1971) *Morphology of Vascular Plants*. The Macmillan Co., New York. 560 pp.

Brongniart, A. (1881) *Researches sur les graines fossiles silicifiees*. G. Masson, Paris. 93pp., 21 pl.

Chaloner, W.G., Hill, A.J. and Lacey, W.S. (1977) First Devonian platyspermic seed and its implications in gymnosperm evolution. *Nature* **265**, 233–5.

Crane, P.R. (1985) The morphology and relationships of the Bennettitales. (this volume).

Daghlian, C.P. and Taylor, T.N. (1979) A new structurally preserved Pennsylvanian cordaitean pollen organ. *American Journal of Botany* **66**, 290-300.

Emberger, L. (1944) *Les plantes fossiles dans leurs rapports avec les vegetaux vivants*. Masson et Cie, Paris.

Gillespie, W.H., Rothwell, G.W. and Scheckler, S.E. (1981) The earliest seeds. *Nature* **293**, 462-4.

Gordon, W.T. (1938) On *Tetrastichia bupatides*: A Carboniferous pteridosperm from East Lothian. *Transactions of the Royal Society of Edinburgh* **59**, 351-70.

—— (1941) On *Salpingostoma dasu*: A new Carboniferous seed from East Lothian. *Transactions of the Royal Society of Edinburgh* **60**, 427-64.

Knoll, A.H. and Rothwell, G.W. (1981) Paleobotany: Perspectives in 1980. *Paleobiology* **7**, 7-35.

Long, A.G. (1960a) On the structure of '*Samaropsis scotica*' Calder (emended) and '*Eurystoma angulare*' gen. et sp. nov., petrified seeds from the Calciferous Sandstone Series of Berwickshire. *Transactions of the Royal Society of Edinburgh* **64**, 261-80.

—— (1960b) '*Stamnostoma huttonense*' gen. et sp. nov.—A pteridosperm seed and cupule from the Calciferous Sandstone Series of Berwickshire. *Transactions of the Royal Society of Edinburgh* **64**, 201-15.

—— (1961a) On the structure of '*Deltasperma fouldenense*' gen. et sp. nov., and '*Camptosperma berniciense*' gen. et sp. nov., petrified seeds from the Calciferous Sandstone Series of Berwickshire. *Transactions of the Royal Society of Edinburgh* **64**, 282-95.

—— (1961b) Some pteridosperm seeds from the Calciferous Sandstone Series of Berwickshire. *Transactions of the Royal Society of Edinburgh* **64**, 401-19.

—— (1969) *Eurystoma trigona* sp. nov., a pteridosperm ovule borne on a frond of *Alcicornopteris* Kidston. *Transactions of the Royal Society of Edinburgh* **68**, 171-82.

—— (1977a) Some Lower Carboniferous pteridosperm cupules bearing ovules and microsporangia. *Transactions of the Royal Society of Edinburgh* **70**, 1-11.

—— (1977b) Observations on Carboniferous seeds of *Mitrospermum, Conostoma* and *Lagenostoma*. *Transactions of the Royal Society of Edinburgh* **70**, 37-61.

Mapes, G. and McComas, M. (1984) Upper Pennsylvanian Conifers from the Appalachian Basin. *American Journal of Botany* **71** (No. 5, Pt. 2), 115-6.

—— and Rothwell, G.W. (1980) *Quaestora amplecta* gen. et sp. nov., a structurally simple medullosan stem from the Upper Mississippian of Arkansas. *American Journal of Botany* **67**, 636-47.

—— and —— (1984) Permineralized ovulate cones of *Lebachia* from Late Palaeozoic limestones of Kansas. *Palaeontology* **27**, 69-94.

—— Fine, T.I., Tanner, W.R. and Lacey, W.S. (1984) The megagametophyte of *Hydrasperma tenuis* Long from the uppermost Devonian of Ireland. *American Journal of Botany* **71**, 1461-4.

Matten, L.C. and Lacey, W.S. (1981) Cupule organization in early seed plants. In *Geobotany II* (ed. R.C. Romans), Plenum Press, New York, pp. 221-34.

—— —— and Edwards, D. (1975) Discovery of one of the oldest gymnosperm floras containing cupulate seeds. *Phytologia* **32**, 299-303.

—— —— and Lucas, R.C. (1980) Studies on the cupulate seed genus *Hydrasperma* Long from Berwickshire and East Lothian in Scotland and County Kerry in Ireland. *Botanical Journal of the Linnean Society* **81**, 249–73.

—— —— May, B.I. and Lucas, R.C. (1980) A megafossil flora from the Uppermost Devonian near Ballyheigue, Co. Kerry, Ireland. *Review of Palaeobotany and Palynology* **29**, 241–51.

May, B.I. and Matten, L.C. (1983) A probable pteridosperm from the uppermost Devonian near Ballyheigue, County Kerry, Ireland. *Botanical Journal of the Linnean Society* **86**, 103–23.

Meyen, S.V. (1984) Basic features of gymnosperm systematics and phylogeny as evidenced by the fossil record. *The Botanical Review* **50**, 1–111.

Miller, C.N. Jr., (1985) A critical review of S.V. Meyen's 'Basic features of gymnosperm systematics and phylogeny as evidenced by the fossil record'. *The Botanical Review* **51**, 295–318.

Oliver, F.W. (1904) Notes on *Trigonocarpus*, Brongn. and *Polylophospermum* Brongn., two genera of Palaeozoic seeds. *New Phytologist*. **3**, 96–104.

—— and Salisbury, E.J. (1911) On the structure and affinities of the Palaeozoic seeds of the *Conostoma* group. *Annals of Botany* **25**, 1–50.

—— and Scott, D.H. (1904) On the structure of the Palaeozoic seed *Lagenostoma lomaxii* with a statement of the evidence upon which it is referred to *Lyginodendron*. *Philosophical Transactions of the Royal Society of London* 197B, 193–247.

Pettitt, J.M. and Beck, C.B. (1968) *Archaeosperma arnoldii*—A cupulate seed fromthe Upper Devonian of North America. *Contributions from the Museum of Palaeontology. The University of Michigan* **22**, 139–154.

Phillips, T.L. (1980) Stratigraphic and geographic occurrences of permineralized coal-swamp plants—Upper Carboniferous of North America and Europe. In *Biostratigraphy of fossil plants*. (ed. D.L. Dilcher and T.N. Taylor) Dowden, Hutchinson and Ross, Inc., Stroudsburg. PA. pp. 25–92.

Rothwell, G.W. (1971) Additional observations on *Conostoma anglo-germanicum* and *C. oblongum* from the Lower Pennsylvanian of North America. *Palaeontographica* **131B**, 167–78.

—— (1975) The Callistophytaceae (Pteridospermopsida): I. Vegetative Structures. *Palaeontographica* **151B**, 171–96.

—— (1980) Permineralized parts of putative primitive gymnosperms. *Botanical Society of America Miscellaneous Series Publication* **158**, 98.

—— (1981) The Callistophytales (Pteridospermopsida): Reproductively sophisticated Palaeozoic gymnosperms. *Review of Palaeobotany and Palynology* **32**, 103–21.

—— (1982) New interpretations of the earliest conifers. *Review of Palaeobotany and Palynology* **37**, 7–28.

—— (1983) Functional morphology in ovule development, and the evolution of gymnospermous reproduction. *American Journal of Botany* **70** (No. 5, Pt. 2), 77.

—— (1985) The role of comparative morphology and anatomy in interpreting the systematics of fossil gymnosperms. *The Botanical Review* **51**, 319–27.

—— and Scott, A.C. (1985) Ecology of the Lower Carboniferous plant remains from Oxroad Bay, East Lothian, Scotland. *American Journal of Botany*, 72, 899.

Scheckler, S.E. (1985) Seed plant diversity in the Late Devonian. *American Journal of Botany*, **72**, 900.

Scott, A.C., Galtier, J. and Clayton, G. (1984) Distribution of anatomically-preserved floras in the Lower Carboniferous of Western Europe. *Transactions of the Royal Society of Edinburgh: Earth Sciences* **75**, 311–40.

Seward, A.C. (1917) Vol. 3. *Fossil Plants*. Cambridge University Press, Cambridge.

Stewart, W.N. (1981) The Progymnospermopsida: The construction of a concept. *Canadian Journal of Botany* **59**, 1539–42.

—— (1983) *Paleobotany and the evolution of plants*. Cambridge University Press, Cambridge.

Singh, H. (1978) *Embryology of gymnosperms*. Gebruder Borntraeger, Berlin.

Stubblefield, S.P., Rothwell, G.W. and Taylor, T.N. (1984) *Conostoma williamsonii* n. sp., a lagenostomalean ovule from the Pennsylvanian of the Illinois Basin. *Canadian Journal of Botany* **62**, 96–101.

Taylor, T.N. and Eggert, D.A. (1967a) Petrified plants from the Upper Mississippian (Chester Series) of Arkansas. *Transactions of the American Microscopal Society* **86**, 412–6.

—— and —— (1967b) Petrified plants from the Upper Mississippian of North America. I: The seed *Rhynchosperma* gen. n. *American Journal of Botany* **54**, 984–92.

—— and Millay, M.A. (1981) Morphologic variability of Pennsylvanian lyginopterid seed ferns. *Review of Palaeobotany and Palynology* **32**, 27–62.

Williamson, W.C. (1877) On the organization of the fossil plants of the Coal-measures—Part VIII. Ferns (continued) and gymnospermous stems and seeds. *Philosophical Transactions of the Royal Society of London* **167**, 213–70.

11. The morphology and relationships of Bennettitales

P.R. Crane

Abstract

Recent cladistic analyses suggest that the Bennettitales are among the gymnosperms most closely related to the flowering plants. Studies of some Upper Triassic bennettitalean 'gynoecia' suggest that interseminal scales are homologous to ovules and that the bennettitalean ovule is primitively bitegimic. As has been hypothesized for flowering plants, the outer integument could be derived from the 'cupule' of a *Caytonia* or corystosperm megasporophyll. Upper Triassic material also suggests that the primitive bennettitalean microsporophyll is a leaf-like organ bearing sporangia singly, rather than in the synangia typical of Jurassic and Cretaceous members of the group. Further investigations of Triassic Bennettitales have the potential to considerably clarify relationships within the clade and improve our understanding of how the Bennettitales may be related to other seed plants.

Introduction

In a recent paper (Crane 1985) cladistic techniques were applied to evaluating the relationships between major groups of seed plants. The results suggested that the Bennettitales, Gnetales and *Pentoxylon* are the gymnosperms most closely related to flowering plants. Similar conclusions have been reached by subsequent cladistic analyses on an expanded set of characters (Doyle and Donoghue 1986, 1987). Although all of these analyses incorporate difficulties of various kinds and have failed to provide a definitive solution to the persistent problem of the origin of angiosperms, they appear to have significantly narrowed the spectrum of potentially relevant gymnosperm groups, and suggest that increased understanding of primitive Bennettitales and Gnetales is likely to be of particular value in clarifying the relationships of the flowering plant clade (Crane 1985). In this paper I briefly discuss the possible homologies of bennettitalean reproductive structures and consider how they may relate to those of angiosperms and other seed plants. Ovulate and pollen organs are both reviewed, focusing in particular on five unusual microsporophylls from the Upper Triassic of Europe and east Greenland.

Bennettitalean Ovulate Reproductive Structures

All bennettitalean ovulate reproductive structures so far described consist of a crowded mass of interseminal scales and ovules (see Crane 1985 for review). The interseminal scales are always extremely numerous, but the number of ovules ranges from two to 20 in *Vardekloeftia* (Harris 1932) to over a hundred in most species of *Williamsonia*. Typically the ovules are embedded in the 'gynoecium' with only the micropyles projecting above the surface formed by the tightly packed heads of the interseminal scales. Interseminal scales are unique to the Bennettitales and are most probably homologous with ovules (Seward 1913, 1917). Early developmental stages of interseminal scales and ovules in silicified Jurassic material from the Rajmahal Hills, India, are identical (Sharma 1974, see also Delevoryas, 1968), and the cuticle of the ovule in *Bennetticarpus crossospermus* (Harris 1932) and some specimens of *Vardekloeftia* (1932) is very similar to the cuticle of the interseminal scale heads in the same gynoecia. Taken together this evidence strongly suggests that both phylogenetically and ontogenetically the interseminal scales are modified sterile ovules.

Bennettitalean ovules are usually considered to be unitegmic, consisting of a single integument surrounding a basally attached nucellus (Seward 1917, Taylor 1981, Stewart 1983). However, there is evidence to suggest that the bennettitalean ovule is fundamentally bitegmic (Crane 1985). In *Vardekloeftia* and *Bennetticarpus crossospermus* (Upper Triassic, east Greenland; Harris 1932) the integument (inner integument of Crane 1985) is surrounded by a further cutinized layer. In *B. crossospermus* this outer layer forms a thickly cutinized 'micropylar plate' through the centre of which the elongated micropylar tube projects. In *Vardekloeftia* the integument is surrounded by an outer cutinized 'cupule' (Harris 1932). In small ovules the cuticle in the distal part of the 'cupule' resembles the 'micropylar plate' of *B. crossospermus* (Harris 1932). I interpret both the 'cupule' and the 'micropylar plate' (Harris 1932, Fig. 50) as clear evidence for a second integument in Bennettitales (Crane 1985). Further evidence is provided by the 'cup-shaped basal supporting husk' in *Cycadeoidea* ovules (Wieland 1906, p. 120), the 'cupule or aril of elongated tubular cells' in *Bennettites albianus* (Stopes 1918, p. 410), the outer layer of *Bennetticarpus wettsteinii* seeds (Kräusel

1949) and the 'tubular layer' of *Cycadeoidea* (*Bennettities*) *morieri* (Seward 1917, p. 398). The most straightforward interpretation of these fossils is that the Bennettitales are fundamentally bitegmic and that the failure to recognize two integuments in most bennettitalean ovules that are preserved as compressions (e.g. Harris 1969) is attributable to sparse cutinization of integumentary tissues in ovules that are almost completely enclosed in interseminal scales.

If the ovules of Bennettitales are bitegmic, then the origin of the outer integument must be accounted for in any interpretation of the origin or relationships of the group. In angiosperms several authors (Gaussen 1946, Stebbins 1974, Doyle 1978, Stewart 1983) have suggested that the outer integument of bitegmic ovules is homologous to the 'cupule' of Mesozoic 'seed ferns' such as *Caytonia* or corystosperms. This view receives some support from the 'anatropous' arrangement of the 'cupule' in both Mesozoic groups (Thomas 1925, 1933, Harris 1964), developmental evidence from extant angiosperms suggesting that the

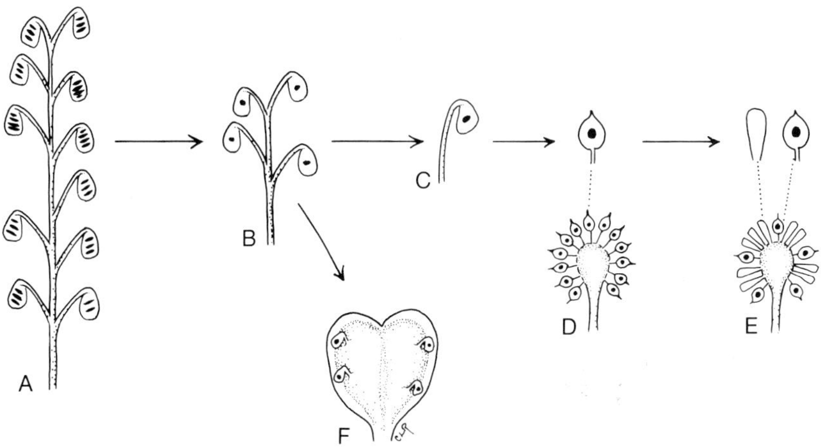

Fig. 11.1. Possible homologies between the 'cupulate megasporophylls' of *Caytonia*, and corystosperms and the bitegmic ovules of angiosperms, *Pentoxylon* and Bennettitales. A. *Caytonia*-like megasporophyll with numerous 'cupules', each containing numerous unitegmic ovules. B. Corystosperm-like megasporophyll with four 'cupules', each containing a single unitegmic ovule. C. Hypothetical 'unicupulate megasporophyll' consisting of a single recurved 'cupule' containing a single unitegmic ovule. D. *Pentoxylon*-like 'unicupulate megasporophyll' consisting of a single erect 'cupule' containing a single unitegmic ovule: the 'cupule' and ovule together are equivalent to a bitegmic ovule; several bitegmic ovules are clustered to form a single head. E. Bennettitalean 'unicupulate megasporophylls': sterile 'unicupulate megasporophyll' equivalent to an interseminal scale; fertile 'unicupulate megasporophyll' equivalent to a bitegmic ovule; several bitegmic ovules and interseminal scales are clustered to form a single head. F. Angiosperm 'cupulate megasporophyll' (carpel), containing four bitegmic ovules, opened and viewed adaxially.

inner inegument and nucellus are part of a single ontogenetic unit, while the outer integument, funicle and chalaza are part of another (Bor 1978) and the frequently pronounced anatomical differences between inner and outer integuments (Stebbins 1974). The developmental independence of the inner and outer integument is further suggested by the 'zig-zag' micropyles of many bitegmic angiosperm ovules (Stebbins 1974, Table 10.1).

If the outer integument of bennettitalean ovules could be accounted for in a similar way to the outer integument of angiosperm ovules (Crane 1985), then it would reconcile ideas that the Bennettitales are closely related to flowering plants ('Anthostrobilus Theory', Arber and Parkin 1907, 1908), with previously competing hypotheses that suggest an important role in the origin of angiosperms for certain Mesozoic 'seed ferns' (Stebbins 1974, Doyle 1978, Stewart 1983). The character transitions involved in interpreting the outer integument of both Bennettitales and angiosperms as homologous to the 'cupule' of Mesozoic seed ferns are illustrated in Fig. 11.1. Under this interpretation each ovule in the Bennettitales and *Pentoxylon* is homologous to a highly reduced 'cupulate megasporophyll' and thus Seward's (1917) megasporophyll interpretation is modified to account for the presence of a second integument. The Bennettitales are still further advanced with respect to *Pentoxylon* in having some of the ovules sterilized and transformed into interseminal scales. The relationships embodied in this hypothesis place *Pentoxylon* as the sister group to the Bennettitales. Consideration of additional characters (see Crane, 1985 for details)

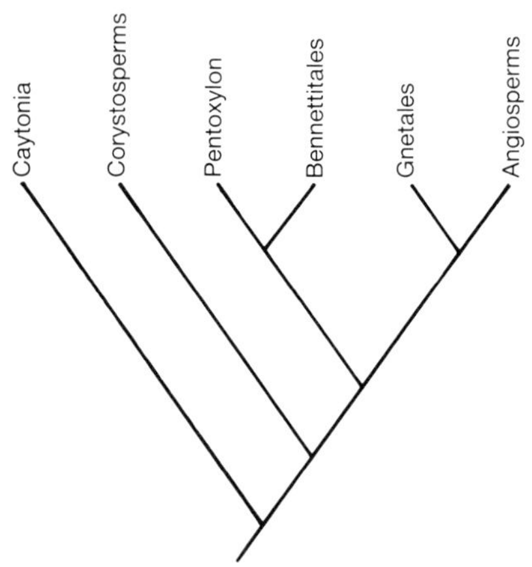

Fig. 11.2. Cladistic relationships between *Caytonia*, corystosperms, Bennettitales, *Pentoxylon*, Gnetales and angiosperms, based on Fig. 11.1 and additional characters discussed in Crane (1985).

marginally favours the view that the Gnetales are the sister group to flowering plants, as suggested by Arber and Parkin (1908). Bennettitales plus *Pentoxylon* form the sister group to angiosperms plus Gnetales (Fig. 11.2). Corystosperms then form the sister group to the Bennettitales plus *Pentoxylon* plus Gnetales plus angiosperm clade, and *Caytonia* is placed as the sister taxon to all other taxa considered (Fig. 11.2).

Systematically, if this pattern of relationships is accepted, it strongly suggests that two of the most widely used higher taxa within seed plants, the Cycadopsida and Pteridospermales (*sensu* Bierhorst 1971) have little phylogenetic utility. Bennettitales, for example, are more closely related to angiosperms than to *Caytonia* or cycads (Crane 1985), and the corystosperms are more closely related to the Bennettitales than to *Caytonia* or medullosans (Crane 1985). Just as progress in evaluating the phylogenetic relationships of early land plants involved dismantling the Psilophyta (Banks 1975), progress in understanding phylogenetic relationships within seed plants may also necessitate dismantling the Cycadopsida and Pteridospermales.

The major outstanding problem with the interpretation of bennettitalean ovules given above ('cupulate megasporophyll' interpretation) is the morphology of *Westersheimia pramelreuthensis* Krasser from the Upper Triassic of Lunz, Austria (Fig. 11.3; Kräusel, 1949). *W. pramelreuthensis* is a laterally borne ovulate reproductive structure apparently bearing

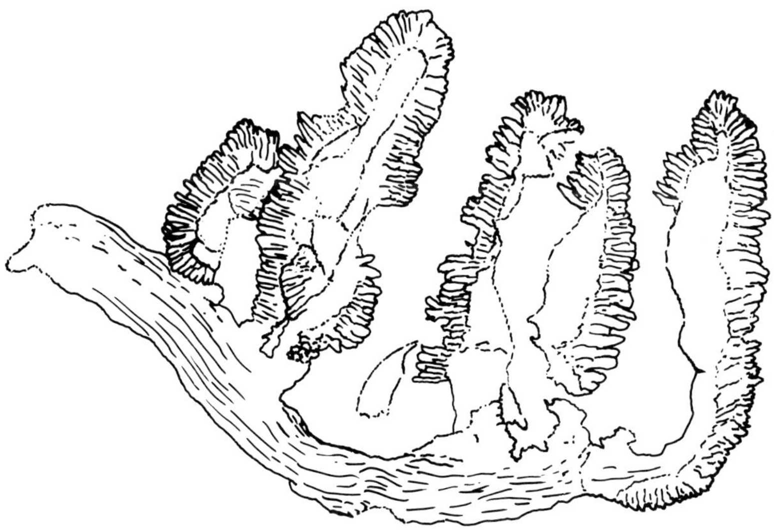

Fig. 11.3. *Westersheimia pramelreuthensis* Krasser; Upper Triassic; Lunz, Austria. Line drawing showing six 'gynoecia' based on Kräusel (1949, pl. 16, Fig. 4), x7.

several bennettitalean 'gynoecia', each composed of interseminal scales and ovules. Several specimens are known, and one is attached laterally to a main axis (Kräusel, 1949, text-fig. 14). Morphologically it is unclear whether this organ is a branch, with further lateral branches on which the ovules and interseminal scales are borne, or a single megasporophyll bearing bennettitalean 'gynoecia' on lateral pinnae. The second interpretation seems more likely from the gross morphology of *Westersheimia* but this clearly conflicts with the apparently terminal position of most bennettitalean 'flowers'. In *Westersheimia* none of the gynoecia appear to have been subtended by 'perianth' parts, and as in other Bennettitales the ovules and interseminal scales are not known to exhibit a clear phyllotactic arrangement. Clearly the morphology of *Westersheimia* requires careful reinvestigation, and if it is shown to be a single megasporophyll then the interpretation of individual ovules as reduced megasporophylls must be rejected.

Bennettitalean pollen organs

The pollen organs of most Bennettitales are aggregated into 'flower-like' reproductive structures that are either unisexual or form part of single hermaphrodite 'flower' (Crane 1985). Each pollen producing 'flower' or partial 'flower' is composed of several whorled microsporophylls, which may be either fused into a basal cup (*Weltrichia*, *Cycadeoidea*; Wieland 1906, Harris 1969), or borne individually around a central gynoecium (*Williamsoniella*, Harris 1944, 1969, 1974). Each microsporophyll is flattened, more or less bilaterally symmetrical, and frequently pinnately organized. Pollen sacs are typically arranged in characteristic bivalved synangia, which may be sessile on the adaxial surface of the microsporophyll lamina (*Weltrichia pecten*, *Weltrichia whitbiensis*, Harris 1969), or borne on adaxially oriented fertile lateral pinnae (*Cycadeoidea*, *Weltrichia sol*, *Williamsoniella*, Wieland 1906, Harris 1944, 1969, 1974). However, just as the ovulate reproductive structures of Upper Triassic Bennettitales (e.g. *Bennetticarpus crossospermus*, *B. wettsteinii*, *Vardekloeftia*) differ significantly from the relatively stereotyped morphologies of Jurassic and Cretaceous forms, bennettitalean microsporophylls from the Triassic of Europe and east Greenland also exhibit unusual morphologies.

Bennettistemon amblum Harris.

B. amblum (Harris 1932) from the Upper Triassic of east Greenland is an ovate microsporophyll about 20 mm long and 10 mm wide (Fig. 11.4A). The apex is rounded, the base is unknown but possibly contracted into a short stalk (Harris 1932), and the margin is entire. The cuticle of the

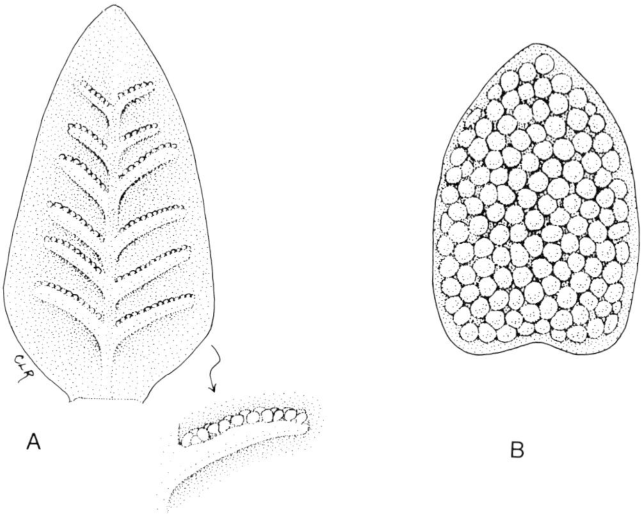

Fig. 11.4. Reconstructions of bennettitalean pollen organs from the Upper Triassic of east Greenland; based on descriptions and illustrations in Harris (1932). A. *Bennettistemon amblum* Harris, x2.5. B. *Bennettistemon ovatum* Harris, x7.

lower surface has numerous bennettitalean stomata. The upper surface bears about seven flanges at angles of 70°–90° on either side of the midrib. Each flange ('semicircular plate' Harris 1932) bears about 20–30 sporangia, each about 1 mm in diameter. The sporangia are borne singly and not in bivalved synangia. The pollen is ellipsoidal, approximately 35 μm long, monosulcate and psilate.

Harris (1932) compared *B. amblum* with *Haitingeria krasseri* (Schuster) Krasser (see below), which was then thought to be a megasporophyll (Krasser 1919). The similarities are still more striking now that *H. krasseri* is known to be a microsporophyll (Kräusel 1949).

Bennettistemon ovatum Harris.
B. ovatum (Harris 1932) from the Upper Triassic of east Greenland is a broadly ovate microsporophyll about 7 mm long and 6 mm wide (Fig. 11.4B). The apex is rounded, the base cordate and the margin entire. The cuticle of the lower surface has occasional bennettitalean stomata. The upper surface is completely covered with numerous sporangia, except close to the margin. The sporangia were 0.6 mm in diameter and borne singly, not in bivalved synangia. The pollen is ellipsoidal, approximately 30 μm long, monosulcate and psilate.

B. ovatum is morphologically similar to *Androlepis ambigua* Nathorst (1902) and *A. hartzi* Harris (1932), both of which are laminar structures with sporangia borne over one surface (Harris 1932). The cuticular information necessary to pursue these comparisons further is unavailable.

Leuthardtia ovalis Kräusel and Schaarschmidt.
L. ovalis from the Upper Triassic of Neuewelt near Basel, Switzerland (Kräusel and Schaarschmidt 1966), is probably a microsporophyll (Fig. 11.5A). It is broadly elliptical, 10–13 mm long and 6–9 mm wide. The apex is rounded, the base contracted into a short broad stalk, and the margin entire. The cuticle has numerous bennettitalean stomata. The

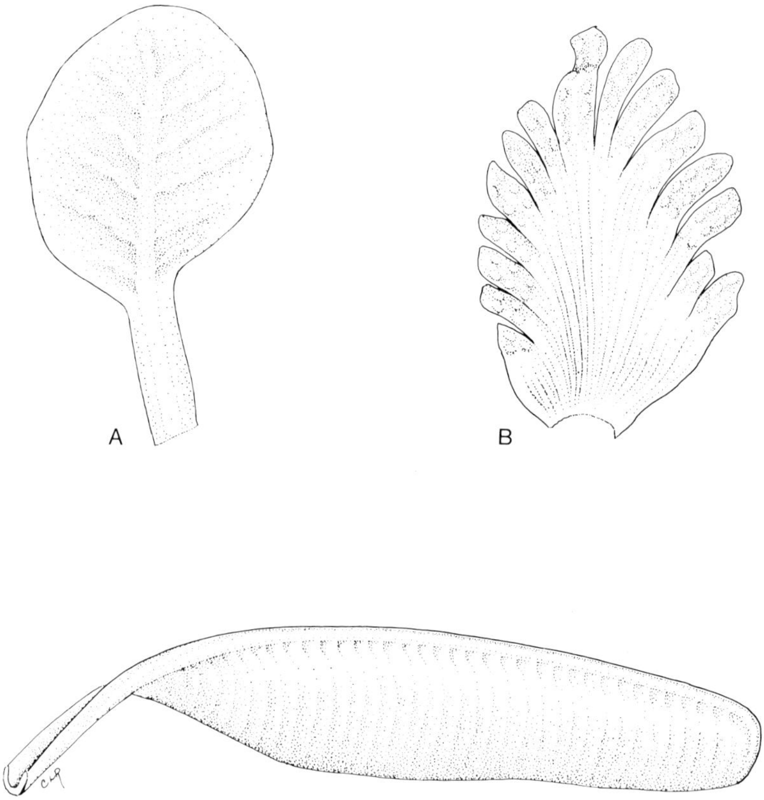

Fig. 11.5. Reconstruction of bennettitalean pollen organs from the Upper Triassic of Europe. A. *Leuthardtia ovalis* Kräusel and Schaarschmidt, Neuewelt; Switzerland; based on Kräusel and Schaarschmidt (1966, text-fig. 7), x3. B. *Haitingeria krasseri* (Schuster) Krasser; Lunz, Austria; based on Kräusel (1949, text-figs. 5, 6), x1. C. *Leguminanthus siliquosus* (Leuthardt) Kräusel and Schaarschmidt; Lunz, Austria; based on Kräusel and Schaarschmidt (1966, pl. 14, 15) and Naturhistorisches Museum Wien 1886I.16, x1.

(? upper) surface bears about five ribs at angles of 50°–90° on either side of the midrib. Each rib is comprised of three or four raised areas interpreted as sporangia. The pollen of *L. ovalis* is unknown.

Haitingeria krasseri (Schuster) Krasser.
H. krasseri from the Upper Triassic of Lunz, Austria (Kräusel 1949), is a broadly elliptical microsporophyll 30–70 mm long and 20–40 mm wide (Fig. 11.5B). The base is truncate and the upper two-thirds of the margin is divided into about seven pinnately arranged lobes on either side of the midrib. The cuticle has numerous bennettitalean stomata. Each lobe bears up to six pairs of pollen bearing structures approximately 1 mm in diameter. It is uncertain whether these are sporangia or synangia (Kräusel 1949). The pollen apparently extracted from the pollen bearing areas is ellipsoidal, large, 120–160 μm long, and psilate.

Leguminanthus siliquosus (Leuthardt) Kräusel and Schaarschmidt.
L. siliquosus (Kräusel and Schaarschmidt 1966) from the Upper Triassic of Neuewelt, Switzerland, and Lunz, Austria, is a leaf-like microsporophyll, resembling a legume pod (Fig 11.5C). It has a short broad stalk, which was probably clasping at the base, and continues distally into a broad midrib running the length of the organ. The two halves of the lamina are folded together *adaxially*, and the midrib is generally adaxially curved in its distalmost part. At maturity, however, the whole organ is apparently twisted at the level of the stalk so that the adaxial surface and open margin of the lamina is oriented *abaxially*. The microsporophylls vary considerably in size ranging from 20–210 mm long and 5–33 mm wide. The lamina has pinnately arranged secondary veins arising at approximately 90° from the midvein and extending to the margin. The cuticle of the lamina has numerous bennettitalean stomata. The adaxial (inner) surface of the lamina bears numerous rows of overlapping ellipsoidal pollen sacs approximately 0.9 mm long and 0.25 mm broad. Some of the pollen sacs appear to be arranged in loose groups of four, but they were not borne in bivalved synangia. The pollen is ellipsoidal, monosulcate and 14–20 μm long.

Several *Leguminanthus* microsporophylls often occur together on the same hand specimen (Fig. 11.6), perhaps suggesting that these organs were borne clustered on the plant. However, the extreme variation in size (Fig. 11.6) and the broad basal attachment make it unlikely that they were borne in a single whorl. It is therefore possible that the *Leguminanthus* microsporophylls (perhaps also *Bennettistemon amblum*, *B. ovatum*, *Leuthardtia* and *Haitingeria*) were borne clustered, possibly in helices, rather than in the whorls characteristic of most other Bennettitales.

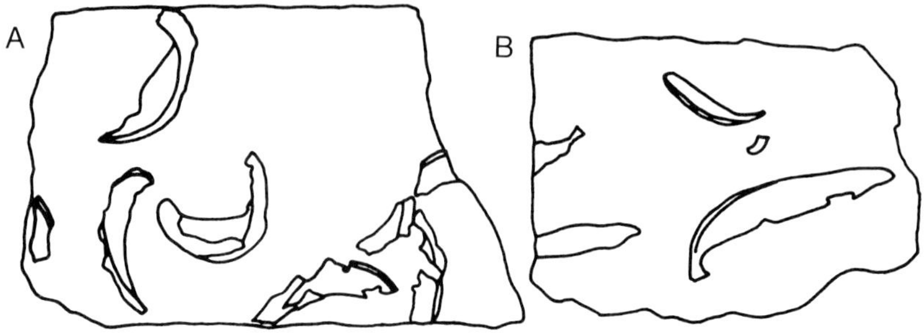

Fig. 11.6. *Leguminanthus siliquosus* (Leuthardt) Kräusel and Schaarschmidt; Upper Triassic; Lunz, Austria. Line drawing of two hand specimens showing distribution of microsporophylls and their variation in size. A. Redrawn from Kräusel and Schaarschmidt (1966, Fig. 10), x0.35. B. Naturhistorisches Museum Wien 1886I.16, x0.35.

L. siliquosus is most similar to the probable microsporophyll *Brightonia arota* Haris (1932) from the Upper Triassic of east Greenland, but apparently differs in its stomatal structure and other features. Nevertheless the type specimen of *B. arota* in which the two halves of the lamina are pressed together (Harris 1932, pl. 18, Figs. 10, 11) is strikingly like the specimens of *L. siliquosus* from Europe.

Bennettitalean microsporophylls.

The five pollen organs described above are among the few bennettitalean microsporophylls that may have been borne helically rather than in whorls, and in which the pollen sacs are not clearly borne in bivalved synangia. Although the precise phylogenetic relationships of these fossils remain to be determined, comparisons with other groups of seed plants suggest that these taxa with probable helical phyllotaxy and non-synangiate sporangia are probably primitive within the Bennettitales as a whole.

Discussion

If the interpretations presented above are accepted, then the ovules and pollen organs of the Bennettitales are morphologically homologous. Pending reinvestigation of *Westersheimia* the ovules can be regarded as highly reduced, simple megasporophylls consisting of a single bitegmic (integument + 'cupule') ovule. Each bennettitalean ovule is homologous to a 'cupulate megasporophyll' under this interpretation. The pollen organs can be regarded as much less reduced, pinnately organized microsporophylls with numerous pollen sacs that at least primitively were not borne in bivalved synangia. Bivalved synangia could be formed by inrolling of fertile pinnae bearing rows of sporangia as in *Haitingeria* (see also Harris 1932). This situation in the Bennettitales with respect to megasporophylls and microsporophylls could be interpreted as the converse of the situation in flowering plants. In angiosperms the megasporophylls (carpels) are much less reduced and enclose several bitegmic ovules, while the microsporophylls are more reduced, simple, non-pinnate and bear only a few pollen sacs (see also Arber and Parkin 1907). The structure of Upper Triassic Bennettitales is clearly critical to these interpretations. Genera such as *Bennetticarpus*, *Bennettistemon*, *Haitingeria*, *Leguminanthus*, *Leuthardtia*, *Vardekloeftia* and *Westersheimia* significantly expand the morphological diversity known in the group. They provide important information that must be accounted for in resolving homologies both within the Bennettitales and in relating the Bennettitales to other groups of seed plants. The adaxially folded, almost conduplicate, microsporophyll of *Leguminanthus*, combined with the recognition that bennettitalean ovules are bitegmic, also lends some support to Meyen's gamoheterotopy hypothesis (Meyen, personal communication 1984). The idea that a bennettitalean microsporophyll could be transformed into a carpel by the developmental replacement of pollen sacs by ovules also seems much less problematic in the light of plausible hypotheses in other areas of botany that invoke equally dramatic positional changes (Iltis 1983). Both the gamoheterotopy hypothesis and the morphological issues raised in this paper deserve further consideration and testing as our knowledge of early Bennettitales during the Trassic increases.

Acknowledgements

I wish to thank R.A. Spicer and B.A. Thomas for the invitation to participate in this symposium, M.N. Bose, W.A. DiMichele J.A. Doyle and K. Raungaard Pedersen for helpful discussions, and J. Kovar and the staff of the Naturhistorisches Musuem, Wien, for facilitating study of material from the Triassic of Austria. C. Richardson drew the illustrations and E. Zeiger assisted in preparing the typescript. This research was supported in part by National Science Foundation Grant BSR-8314592.

References

Arber, E.A.N. and Parkin J. (1907) On the origin of angiosperms. *J. Linn. Soc. Bot.* **22**, 29–80.

—— and —— (1908) Studies on the evolution of angiosperms. The relationship of the angiosperms to the Gnetales. *Ann. Bot.* (*Lon.*) **22**, 489–515.

Banks, H.P. (1975) Reclassification of the Psilophyta. *Taxon* **24**, 401–13.

Bierhorst, D.W. (1971) *Morphology of vascular plants*. MacMillan, New York.

Bor, J. (1978) A note on anatropy versus orthotropy. *Phytomorphology* **28**, 219–24.

Crane, P.R. (1985) Phylogenetic analysis of seed plants and the origin of angiosperms. *Ann. Miss. Bot. Gdn.* **72**, 716–793.

Delevoryas, T. (1968) Some aspects of cycadeoid evolution. *J. Linn. Soc. Bot.* **61**, 137–64.

Doyle, J.A. (1978) Origin of angiosperms. *Ann. Rev. Ecol. Syst.* **9**, 365–92.

—— and Donoghue M.J. (1986) Relationships of angiosperms and Gnetales: a numerical cladistic analysis. Oxford University Press, (this volume).

—— and —— (1987) The origin of angiosperms: a cladistic approach. In *The origin of angiosperms and biological consequences* (eds. E.M. Friis, W.G. Chaloner and P.R. Crane), Cambridge University Press, Cambridge (in press).

Gaussen, H. (1946) Les gymnosperms, actuelles et fossiles. *Trav. Lab. Forest. Toulouse.* Tome 2. *Ethol. Dendrol.* Sect. 1, (3, 5), 1–26.

Harris, T.M. (1932) The fossil flora of Scoresby Sound East Greenland. Part 3: Caytoniales and Bennettitales. *Meddel Grøland* **85**(5), 1–133.

—— (1944) A revision of *Williamsoniella. Philos. Trans. Roy. Soc.* **231**, 313–28.

—— (1964) *The Yorkshire Jurassic flora.* II. *Caytoniales, Cycadales and Pteridosperms.* British Museum (Natural History), London.

—— (1969) *The Yorkshire Jurassic flora.* III. *Bennettitales.* British Museum (Natural History), London.

—— (1974) *Williamsoniella lignieri*: its pollen and the compression of spherical pollen grains. *Palaeontology* **17**, 125–48.

Iltis, H.H. (1983) From Teosinte to Maize: the catastrophic sexual transmutation. *Science* **222**, 886–94.

Krasser, F. (1919) Studien uber die fertile Région der Cycadophyten aus den Lunzer Schichten: makrosporophylle. *Denkschr. Akad. Wiss. Wien* **97** 1–32.

Kräusel, R. (1949) Koniferen und andere gymnospermen aus der Trias von Lunz, Nieder-Osterreich. *Palaeontographica Abt. B.* **84** 35–78.

—— and Schaarschmidt F., (1966) Die Keuperflora von Neuewelt bei Basel I.V. Pterophyllen und Taeniopteriden. *Schweiz Paläontol. Abh.* **84**, 1–64.

Meyen, S.V. (1983) Systematics of peltaspermaceous pteridosperms and their position in gymnosperm phylogeny. *Bull. Mosk.* Obshch. Inspt. Pirody, Otd. Biol. **88**, 3–14. (In Russian).

Nathorst, A.G. (1902) Beitrage zur kenntnis einiger Mesozoischen Cycadophyten. *Kongl. Svenska Vetenskaps-Akademiens Handlingar* **36**(4), 1–28.

Seward, A.C. (1913) A petrified *Williamsonia* from Scotland. *Philos. Trans. Roy. Soc.* **203**, 101–26.

—— (1917) *Fossil plants. III. Pteridospermeae, Cycadofilices, Cordaitales, Cycadophyta.* Cambridge University Press, Cambridge.

Sharma, B.D. (1974) Ovule ontogeny in *Williamsonia* Carr. *Palaeontographica Abt. B* **148**, 137–43.

Stebbins, G.L. (1974) *Flowering plants: evolution above the species level.* Harvard Univ. Press, Cambridge.

Stewart, W.N. (1983) *Paleobotany and the evolution of plants*. Cambridge Univ. Press, New York.

Stopes, M.C. (1918) New bennettitalean cones from the British Cretaceous. *Phil. Trans. Roy. Soc.* **208**, 389–440.

Taylor, T.N. (1981) *Paleobotany. An introduction to fossil plant biology*. McGraw Hill, New York.

Thomas, H.H. (1925) The Caytoniales, a new group of angiospermous plants from the Jurassic rocks of Yorkshire. *Phil. Trans. Roy. Soc.* **213**, 299–363.

—— (1933) On some pteridospermous plants from the Mesozoic rocks of South Africa. *Phil. Trans. Roy. Soc.* **222**, 193–265.

Wieland, G.R. (1906) American fossil cycads. Structure. *Publ. Carnegie Inst. Wash.* **34**, 1–277.

12. Relationships of angiosperms and Gnetales: a numerical cladistic analysis

J.A. DOYLE AND M.J. DONOGHUE

Abstract

The significance of the numerous similarities between Gnetales and angiosperms has long been controversial, some considering them evidence of relationship, others as examples of convergence. A numerical cladistic (parsimony) analysis of seed plants plus progymnosperms supports the concept of Arber and Parkin, and Crane that angiosperms are related to Gnetales but not derived from them. Angiosperms are the sister group of Bennettitales and Gnetales, and gnetalian reproductive structures are best interpreted as highly reduced bennettitalian flowers. Although some angiosperm-gnetalian similarities (flowers, siphonogamy, tunica, lignin chemistry, reduced megaspore wall, granular exine) are probably homologous, others (vessels, dicot-like leaves, envelope around the seed, gametophyte reduction) are convergent. Computer experiments on modified data sets support the unity of the angiosperm–bennettitalian–gnetalian clade and the conclusion that coniferopsid characters of Gnetales are convergent. However, exact relationships with seed fern groups are more equivocal. Megafossils of Mesozoic Gnetales and better information on anatomy, chemistry, and embryology of Bennettitales and Caytoniales would provide further tests of relationships.

Introduction

The Gnetales (*Ephedra, Welwitschia, Gnetum*) are unique among gymnosperms in having so many features otherwise restricted to angiosperms (Martens 1971). All three genera have vessels in the wood and reproductive structures (compound strobili) made up of minute flower-like units, with either a perianth and a whorl of more or less fused microsporophylls or a terminal ovule surrounded by an additional envelope. *Gnetum* is remarkably dicot-like in habit and leaf architecture, and *Welwitschia* and *Gnetum* show almost as much reduction of the male gametophyte as in angiosperms, a partially free-nuclear female gametophyte, and cellular embryogeny. The significance of these similarities has long been a subject of controversy, some considering them evidence that Gnetales constitute a link between other gymnosperms and angiosperms, others considering them prime examples of convergence. Recent paleobotanical studies have added an ecological dimension to the parallels between angiosperms and Gnetales. Striate 'ephedroid' pollen comparable to *Ephedra* and *Welwitschia* in both general morphology and ultrastructure (Trevisan 1980) underwent a major Early Cretaceous radiation centered in the Northern Gondwana semiarid tropical belt, the same area where angiosperm pollen first appears and undergoes its most vigorous early diversification (Brenner 1976, Doyle *et al.* 1977, Doyle 1984); the two groups even tend to be abundant in the same local sedimentary facies (fluvial deposits: Doyle *et al.* 1982).

Previous ideas on relationships

Gnetales played a central role in Wettstein's (1907) theory of the origin of angiosperms, often associated with the Englerian school of systematics. Wettstein homologized the inflorescences of the wind-pollinated Amentiferae, made up of apetalous, unisexual flowers, with the compound strobili of Gnetales, and he interpreted the showy, bisexual flowers of magnoliids and other groups as pseudanthia derived by aggregation of unisexual units.

Arber and Parkin (1907, 1908) also believed that angiosperms and Gnetales are related, but they rejected Englerian concepts of angiosperm evolution, arguing that angiosperms, Gnetales, and the Mesozoic Bennettitales were derived from a hypothetical common ancestor with a bisexual pro-anthostrobilus. Thus they interpreted the 'flowers' of Gnetales as reduced from bisexual strobili rather than primitive (as they interpreted Amentiferae within angiosperms) citing among other evidence the presence of an abortive terminal ovule in the staminate flower of *Welwitschia*.

Views similar to those of Arber and Parkin have been expressed by Takhtajan (1969) and Ehrendorfer (1976), but in general the notion of a close relationship between angiosperms and Gnetales has fallen into disfavour. This is partly due to more thorough examination of characters and changing concepts of angiosperm phylogeny. Recognition of vesselless angiosperms suggested that angiosperms were derived from some group without vessels, not a group with them. Furthermore, vessel members appear to be derived from tracheids with different kinds of pitting in the two groups (scalariform in angiosperms, circular bordered in Gnetales: Thompson 1918, Bailey 1944). Increasing recognition of putatively ancestral characters in magnoliid dicots (vesselless wood, gymnosperm-like monosulcate pollen, leaflike carpels) and derived characters in Amentiferae (advanced vessels, triporate pollen) cast further doubt on Englerian views. More recently, palynological studies of Cretaceous rocks have shown that the first recognizable angiosperm pollen is monosulcate, while pollen of the amentiferous type does not appear until later, apparently as the culmination of a long series of modifications (reviewed in Doyle 1978).

Several lines of evidence have led to the idea that Gnetales are related to the so-called coniferopsid gymnosperms (conifers, ginkgos, cordaites) while angiosperms are related to cycadopsids (cycads, seed ferns, Bennettitales). For example, Gnetales have circular bordered pits in the metaxylem and even the protoxylem, like conifers and *Ginkgo*, but metaxylem tracheids in angiosperms have scalariform pitting, as in cycadopsids (Bailey 1944). Eames (1952) homologized the flowers of *Ephedra* with the axillary fertile short shoots of *Cordaites*, with the perianth of the staminate flower and the envelope around the seed both corresponding to the sterile scales below the stalk-like sporophylls. Eames rejected relationships of *Ephedra* with *Welwitschia* and *Gnetum*, largely on ontogenetic and anatomical arguments (not confirmed by more recent studies: Martens 1971) that the ovule is appendicular in *Ephedra* but truly terminal in *Gnetum* and *Welwitschia*, and that a shift from one condition to the other is impossible. He favoured relationships of *Welwitschia* and *Gnetum* with Bennettitales, partly because all three groups have syndetocheilic stomata (Florin 1931). However, others have noted that *Welwitchia* and *Gnetum* have the same coniferopsid wood features as *Ephedra* and that their reproductive structures can be interpreted in terms of the same basic plan (Bailey 1944, Bierhorst 1971, Doyle 1978). At the same time, closer connections have been proposed between angiosperms and specific cycadopsid groups. Thus angiosperm carpels have been homologized with the megasporophylls of Mesozoic seed ferns (*Caytonia*, Corystospermaceae), which bore reflexed cupules (equated with anatropous, bitegmic ovules) in two rows along the rachis, or with fertile structures of Permian glossopterids, which consisted of a leaf bearing one or more cupules on its adaxial surface (Gaussen 1946, Stebbins 1974, Doyle 1978, Retallack and Dilcher 1981). Since the recognition of the progymnosperms in the 1960's, it has been widely suggested that cycadopsids and coniferopsids were independently derived from Devonian progymnosperms (cycadopsids, some of which are protostelic, from the more primitive Aneurophytales; coniferopsids from more advanced types such as *Archaeopteris*: Beck 1981, Meeuse 1963). Taken together, these views would imply that angiosperms and Gnetales belong to evolutionary lines that have been separate since the Devonian.

Major doubts concerning aspects of this consensus have been expressed recently. Muhammad and Sattler (1982) found scalariform perforations in vessel elements of *Gnetum* and therefore questioned anatomical arguments against relationships between angiosperms and Gnetales, while Young (1981) argued on grounds of parsimony that the first angiosperms had vessels and these were lost in several early lines. Others have challenged the division of gymnosperms into cycadopsids and coniferopsids and the idea that the two groups were independently derived from progymnosperms. Thus Rothwell (1982) has documented remarkably coniferopsid-like characters in the Late Carboniferous seed fern *Callistophyton* (e.g. platyspermic ovules, saccate pollen) and has proposed that coniferopsids were derived from seed ferns by heterochronic substitution of scale leaves (cataphylls) for compound fronds, with concomitant changes in branching habit and stem anatomy. Meyen (1984) has proposed that cordaites and conifers are derived from one group of seed ferns (Lagenostomales) and ginkgos from another (Permian peltasperms, ultimately from Callistophytaceae).

Other challenges have come from cladistic studies of seed plant relationships. Cladistic analysis attempts to reconstruct the branching pattern of phylogeny (i.e. to recognize clades, or monophyletic groups in the strict sense of Hennig 1966) from the distribution of shared derived characters (synapomorphies), usually resolving character conflicts caused by convergence and reversal (homoplasy) on the principle of parsimony (i.e. preferring the hypothesis of relationships that requires the fewest character state changes, or steps). In an analysis that considered only extant groups, Hill and Crane (1982) favoured a scheme in which angiosperms are the sister group of conifers plus Gnetales, while cycads plus *Ginkgo* are the sister group of the angiosperm–conifer–gnetalian clade. This contrasts with the conventional interpretation, which, rephrased in cladistic terms, would regard cycads as the sister group of angiosperms, and cycads plus angiosperms (cycadopsids in a broadened sense) as the sister group of coniferopsids (including *Ginkgo* and Gnetales). This analysis has subsequently been completely revised by Crane (1985), taking into account fossil as well as living groups. In agreement with Rothwell, he concludes that coniferopsids were derived from platyspermic seed ferns, but he separates Gnetales from the coniferopsids and places them together with angiosperms, Bennettitales, and *Pentoxylon* in a clade related to Mesozoic seed ferns.

The only way to resolve the question of angiosperm–gnetalian relationships and thus to determine whether their shared features are homologous or convergent is in the context of a general understanding of seed plant relationships, and we agree that the best approach to this goal is through cladistic analysis, as the most explicit and logically sound method of phylogeny reconstruction. However, cladistic studies on seed plants have so far been inconclusive. There is reason to suspect this is due in part to problems with the data and/or methods of analysis. For example, over half the characters used by Hill and Crane (1982) are apomorphies of all seed plants, of angiosperms or other terminal groups, or of Gnetales, and are therefore not informative in determining relationships among seed plant groups. Furthermore, many of the remaining characters are questionable, including the four used to support the proposed angiosperm–conifer–gnetalian clade: pycnoxylic anatomy is considered derived from manoxylic, but outgroup comparison with progymnosperms implies that it is actually ancestral (i.e. polarity was assessed incorrectly); siphonogamy and non-motile sperm are aspects of the same character (i.e. they are redundant); presence of strobili refers to structures made up of very different-looking parts in different groups. In addition, scoring of groups is sometimes questionable; e.g. angiosperms are scored as pycnoxylic, but primitive dicots have multiseriate rays and xylem parenchyma. (It should be noted that most of these problems were corrected in Crane's study.) Ideally, it should be possible to gain a synoptic picture of the total morphology of any terminal group by tracing the changes from the base of a cladogram to the top, but characters of leaf architecture, positional relationships of reproductive structures, and vegetative anatomy are poorly represented in Hill and Crane's data set. In addition, they did not take advantage of computer-assisted (numerical cladistic) methods, which are all but essential when dealing with large groups where character conflicts are common and where it is easy to overlook equally or even more parsimonious alternatives because of the burden of mental inertia.

In an attempt to understand why ideas on seed plant relationships differ so much from each other and to evaluate their relative merits, we undertook our own phylogenetic analysis, using numerical cladistic methods. Since our goal was to test current hypotheses, we made a great effort to code characters in ways that avoid bias for or against competing theories of character evolution. In order to assess the robustness of the results and the relative merits of alternatives, we also used the computer for a series of experiments, with variously modified data sets. More extensive documentation of our data matrix, justification of character coding and scoring of groups (which may sometimes seem to conflict with widely held assumptions), and discussion of broader implications will be presented elsewhere; here, we stress aspects of our study related to the angiosperm–Gnetales problem.

Methods of analysis

For cladogram construction we used the PHYSYS package of programs for phylogenetic analysis (Mickevich and Farris 1982, Smith *et al.* 1982), specifically the revised Wagner algorithm, which is improved over earlier versions in accepting missing data and in other respects. This algorithm begins with a matrix of binary characters, where 'zero' (0) normally represents the ancestral state and 'one' (1) the derived state, and attempts to find the branching diagram that minimizes the total number of character state transitions; it treats forward changes and reversals equally. In order to improve the chances of obtaining the most parsimonious tree(s), we used routines for global branch-swapping (WAG.S) and shuffling the order of entry of taxa (PIM). In a few cases we were able to obtain additional equally parsimonious trees by adding 'dummy' synapomorphies to force particular clades together and then subtracting the extra steps after analysis. This program provides diagnoses for terminal taxa and nodes that can be used to place characters on the tree, but often other arrangements of characters are equally parsimonious; in many cases, we preferred alternatives (e.g. origin and subsequent loss of a feature vs. two origins).

In our analysis, we attempted to represent all seed plant groups, but subdivision and inclusion of groups were necessarily a compromise based on a combination of factors, including a desire to recognize only monophyletic groups, the variable quantity and quality of information available, and relevance to competing theories of seed plant (especially angiosperm) relationships. Thus we treated the three modern genera of Gnetales separately because of recurrent controversy over whether they are related, and we included progymnosperms because of the possibility that seed plants are diphyletic. We split up several conventional taxa such as seed ferns because they are presumably paraphyletic (primitive grade taxa, some members of which are more closely related to various higher groups than to each other). Since many members of these groups are incompletely known, we represented them in the analysis with a few of the better-reconstructed forms (e.g. *Aneurophyton, Peltaspermum*). This procedure should not lead to misconceptions as long as it is realized, for example, that other 'Aneurophytales' may be cladistically more closely related to *Archaeopteris* or seed plants than to *Aneurophyton*. On the other hand, we left several large, diverse groups such as conifers and angiosperms undivided because they have enough apomorphies that they can be safely assumed monophyletic. In order to score such groups, it was necessary to determine basic conditions within them. In most cases these are clear from comparison with any plausible outgroup (e.g. monosulcate pollen in angiosperms), or from concepts of relationships within the group; often the results are corroborated by the stratigraphic record. Thus we used magnoliid dicots and pre-Albian fossils as guides in coding angiosperms, and Lebachiaceae in coding conifers.

Terminal taxa were scored for 62 binary characters (Tables 12.1,

Table 12.1: *Characters.* In the case of binary characters, only the derived state is given, or the ancestral state is also indicated in parentheses. In the case of multistate characters, the meaning of each coding is given.

1: axillary branching
2: accessory axillary buds
3: leaves on homologues of progymnosperm penultimate order axes
4: opposite phyllotaxy
5–7: 000 = all leaves dichotomous; 100 = pinnately compound leaves + cataphylls; 110 = simple or dissected leaves with pinnate venation + cataphylls; XX1 = linear or dichotomous leaves + cataphylls, or scale leaves only
8: rachis simple (vs. bifurcate)
9: finest vein order reticulate
10: several vein orders reticulate
11: guard cell poles level with aperture (vs. raised)
12: syndetocheilic stomata
13: tunica-corpus
14–15: 00 = protostele; 10 = eustele with internal secondary xylem; X1 = eustele with external secondary xylem only
16: all stem bundles endarch
17: each leaf supplied by >2 stem bundles
18: metaxylem without scalariform pitting
19: scalariform pitting in secondary xylem
20: vessels
21: multiseriate rays
22–23: 00 = no secretory cavities or canals; 10 = secretory cavities; X1 = secretory canals
24: Mäule reaction
25–27: 000 = megasporangia/ovules on dichotomous structures on radial axis; 100 = on pinnately compound sporophyll; 110 = on once-pinnate sporophyll (or simple but with lateral ovules); XX1 = on one-veined stalk or sessile
28–30: microsporangiate structures, same coding as 25–27
31: ovule terminal
32: fertile appendages on homologues of progymnosperm last order axes
33–35: 000 = ovule(s) in radial cupule; 100 = directly on ± planate sporophyll; X10 = in circinate cupule, or ovule bitegmic and anatropous; 101 = in second integument derived from two appendages
36: 1 ovule per radial cupule
37: 1 ovule per circinate cupule/ovule
38: microsporangia abaxial
39: microsporangia ± fused
40: microsporophylls whorled
41: compound male and female strobili
42–43: 00 = no seeds; 10 = radiospermic; X1 = platyspermic
44: lagenostome
45: micropylar tube
46: vascularized nucellus
47: thick nucellar cuticle
48: heterospory
49–50: 00 = tetrad scar, no sulcus; 10 = sulcus, pollen tube; 11 = pollen inaperturate
51: pollen bilaterally symmetrical
52: pollen saccate
53: exine granular (vs. alveolar)
54: striate exine sculpture
55: megaspore tetrad linear (vs. tetrahedral)
56: megaspore wall thin or lacking sporopollenin
57: microgametophyte with prothallial but no sterile cell
58: siphonogamy, non-motile sperm
59: tetrasporic megagametophyte
60: egg a free nucleus
61: early embryogenesis cellular
62: embryo with feeder

Table 12.2. Data Matrix

	1 2 3 4 5 6
Aneurophyton	0X00000X00---0000000000-000000X0XXXXXX0X0000XXX0000000XX-0XX--
Archaeopteris	0X10000X00---X1000000---000000X1XXXXX0000000XXX100000000-00X--
Pitus	--0010000000-0000000100-000000000000X010X10100-10000X000-000--
Lyginopterids	100010000000-0000000100-100100000001X010X10100-10000X000-000--
Medullosa	100010000000-10010001X1-10010000100XX010X10001-1001000-0-0-0--
Callistophyton	1000100100-0-X100000110-10010000100X0110XX1000-1101100-00--0--
Glossopterids	10X0110X10-0-X1-00000XX-XX0XX000100X0000XX100011101101-0---0--
Peltaspermum	--0-100100-0------------11010000100X0100XX10X-011010-0-0------
Corystosperms	--0010000000-1001000010-10011000X10X1100XX10X-011011-0-1------
Caytonia	--00100X1010------------11010000X10X0110XX100011101100-1------
Cycads	-XX0110X00000X1110001X101101100X100XX100X100010110100010000000
Bennettitales	10X0110X0001-X1100101X1-XX11100X0001X011X100100110101011--0--0
Pentoxylon	10X0110X0000-1000010100-XX10100X100XX0X1XX100--11010-0---------
Cordaites	1010XX1X0000-X100000010-X01X0101100XX0001X1001-1000100-000-0--
Ginkgos	1010XX1X00000X1101000100X01X0101100XX1000X10000110100010000000
Conifers	1010XX1X00000X1101000X10X01X0101100XX1000X10000100110010010000
Ephedra	11X1XX1X00001X1101011001XX1XX11X101XX0111X10100111101111010000
Welwitschia	11X1XX1X10010X11-1011X11XX1XX11X101XX0111X101001101011X1111111
Gnetum	11X1110X11011X1111111101XX1XX11X101XX0X111001001110010X1111111
Angiosperms	10X0110X11111X11101X10011101100XX10X1X10X100001110101011X10010

0: ancestral state; 1: derived state; -: information not available or questionable; X: not applicable (character not present), or precursor state unknown (in multistate characters).

12.2), of which 20 represent additive binary codings of eight multistate characters (cf. Brooks 1984). We made a great effort to amass potentially informative characters from all parts of the plant body, so as to give as complete a picture as possible of the morphology of terminal groups; 24 characters are vegetative, 38 reproductive. We did not include autapomorphies (derived characters unique to a single terminal taxon)—although autapomorphies are important in determining that the groups used are monophyletic, they contribute nothing to understanding relationships among groups, decrease the efficiency of computation, and give a false sense of the amount of information present in the matrix. We also attempted to eliminate redundant characters; sometimes these were recognized because they always changed simultaneously on cladograms obtained in preliminary analyses and could be plausibly attributed to the same morphogenetic factors (for example, loss of scalariform pitting in the metaxylem and presence of circular bordered pits in the protoxylem).

Polarities were determined by outgroup analysis (Maddison *et al.* 1984). Comparison of progymnosperms plus seed plants with Cladoxylales, sphenopsids, and ferns (the first outgroup) and trimerophytes (the second outgroup) indicated that *Aneurophyton* is primitive in all characters considered; hence it can be used as a functional outgroup for assessment of polarities in the remaining taxa (the functional ingroup: Watrous and Wheeler 1981). On this basis, deciding on polarity and the resulting coding was often relatively easy (e.g. homospory to heterospory, free-sporing reproduction to seeds, proximal to distal pollen germination, mesarch to endarch primary xylem). The greatest difficulties arose in coding multistate characters, where the ancestral state is clear but there are very different alternative interpretations on relationships of the several derived states to the ancestral state and to each other. In such cases, it is easy to bias the results by coding in a way that favours one pathway. For example, under the hypothesis that cycadopsids and coniferopsids were separately derived from progymnosperms (Meeuse 1963, Beck 1981), the radiospermic seeds of cycadopsids and the platyspermic seeds of coniferopsids both originated from no seeds; under the hypothesis that coniferopsids were derived from platyspermic seed ferns (Rothwell 1982), all platyspermic seeds are derived from radiospermic. Thus binary coding no seeds '00', radiospermic '10', and platyspermic '11', which might be appropriate within seed ferns under either hypothesis, biases against the Beck–Meeuse hypothesis by making the transition from no seeds to platyspermic take two steps, while coding them '00', '10', and '01' adds an artificial step to a direct transition from radiospermic to platyspermic. In order to avoid such biases, we devised a new method of coding partially ordered characters which could take advantage of the algorithm's ability to deal with missing data ('X's' in the data matrix) by choosing whatever value for 'X' gives the most parsimonious result in combination with other characters. For example, we coded no seeds '00', radiospermic '10', and platyspermic 'X1' (i.e. either '01' or '11'): platyspermic can hence be derived from either no seeds or radiospermic by one step.

Similarly, we coded the coniferopsid leaf condition (scale leaves only, or scale leaves plus linear or dichotomous leaves) as 'XX1', so that it can be derived by one step from the condition in progymnosperms (dichotomous leaves only, '000'), in seed ferns (pinnately compound fronds and cataphylls, '100'), or in cycads (simply pinnate leaves and

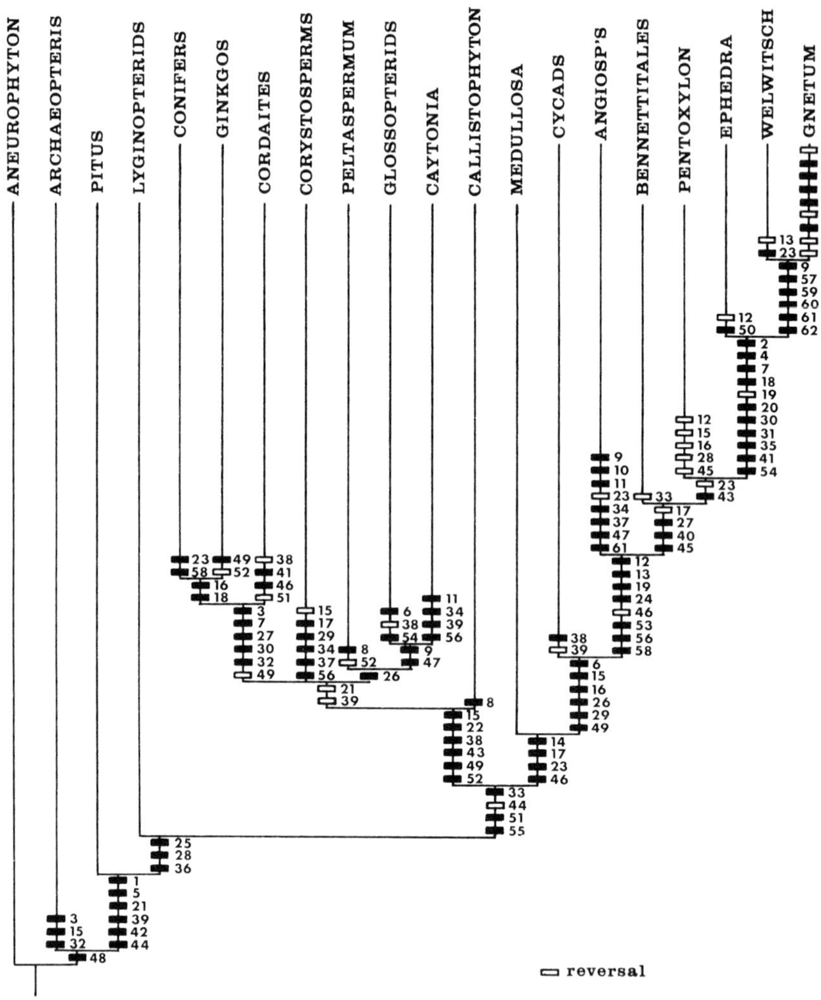

Fig. 12.1. One of four 130-step most parsimonious cladograms derived from the data matrix in Table 12.2.

cataphylls, '110'). As we will discuss in detail elsewhere, this technique does introduce a subtle, counterintuitive bias in some cases, by adding an artificial step when two or more groups with the X-coded state are positioned between groups with other codings in a cladogram, but this is not likely to pose a problem in the present case.

Results

Analysis of the data matrix in Table 12.2 resulted in four very similar most parsimonious cladograms requiring 130 character transitions. One of these is shown in Fig.12.1; how the others differ is discussed below. Lengths of the vertical internodes between branch points are proportional to the number of presumed synapomorphies, giving some idea of the relative support for particular connections. Since we excluded autapomorphies, characters on the lines leading to terminal taxa all change elsewhere on the tree, so they suggest the amount of homoplasy associated with the evolution of each terminal group.

At the base of the cladogram are a series of progymnosperms and Carboniferous seed ferns. *Aneurophyton* (including *Triloboxylon*) is primitive in all characters considered, while *Archaeopteris* is united with seed plants on heterospory (and possibly phloem structure: Wight and Beck 1984). The basal taxon within seed plants is the protostelic Early Carboniferous seed fern *Pitus*, reconstructed by Long (1979) as having pinnately compound leaves but multiovulate cupules on branch-like parts of a fertile frond, radiospermic seeds, microsynangia, and multiseriate rays. (A.C. Scott and G.W. Rothwell (personal communication) question Long's reconstruction of *Pitus*; however, because the characters of its supposed component parts are known in isolated organs of other early seed ferns and appear to be more primitive than those of later forms, we have retained Long's concept of *Pitus* in the analysis, as a plausible picture of the morphology of primitive seed ferns.) Lyginopterids in a strict sense (*Heterangium, Lyginopteris*), with planated fertile fronds and only one ovule per cupule, are the sister group of the remaining taxa, including all extant seed plants. These taxa form two major clades, united by lack of a cupule (possibly due to its transformation into a new integument) and lagenostome, bilaterally symmetrical pollen, and possibly a linear megaspore tetrad (the medullosan condition is controversial).

The basal taxon in one of the two clades is the late Carboniferous seed fern *Callistophyton*. Shared advances of this group (some lost in higher members) are a typical eustele, platyspermic ovules, saccate pollen, a sulcus and pollen tube, abaxial microsporangia, and secretory cavities;

since *Callistophyton* differs from the inferred ancestor in only one character (non-bifurcate rachis), we will refer to the whole group as the callistophyte clade. Besides *Callistophyton*, it includes not only the Mesozoic seed ferns, which have been associated with *Callistophyton* by several authors, but also glossopterids and coniferopsids. This generally supports Rothwell's (1982) view that coniferopsids originated from *Callistophyton*-like seed ferns rather than *Archaeopteris*-like progymnosperms. However, relationships within the clade are rather uncertain, probably because of poor information on characters in many groups. In Fig. 12.1, glossopterids and Mesozoic seed ferns are interpolated between *Callistophyton* and coniferopsids, based on their uniseriate rays and secondarily free microsporangia; this implies that the terminal microsporangia and radial pollen of cordaites and the lack of a sulcus in cordaites and primitive conifers (Mapes and Rothwell 1984) are secondary reversals. In one of the other 130-step trees, coniferopsids are the basal offshoot, and *Callistophyton* is nested well within the clade. In another, *Caytonia* and corystosperms together are placed below *Callistophyton*; according to this scheme, corystosperms are more primitive than other callistophytes in having a bifurcate rachis and probably medullosan stem anatomy (*Rhexoxylon*: Archangelsky 1968).

The basal member of the other clade is another late Carboniferous seed fern, *Medullosa*. The ancestor of this group could have been identical to *Medullosa*, with such advances as a eustele with internal secondary xylem, leaves supplied by traces from several primary xylem strands, secretory canals, and a vascularized nucellus. Excluding later modifications, the remaining members of this 'medullosan' clade are united by leaves with simply pinnate venation (either dissected, as in living cycads, or entire, as in *Taeniopteris*), a normal endarch eustele, once-pinnate sporophylls, and a sulcus. This group includes the cycads, which have often been considered medullosan derivatives, and a clade consisting of angiosperms, Bennettitales, *Pentoxylon*, and Gnetales. The fourth 130-step cladogram differs in reversing the order of Bennettitales and *Pentoxylon*, which does not significantly affect the results. Crane (1985) has proposed that the last four groups are related, but he links them with platyspermic seed ferns rather than cycads and medullosans. They are united by syndetocheilic stomata (lost in *Pentoxylon* and *Ephedra*), at least occasional scalariform pitting in the secondary xylem, granular exine structure, a reduced megaspore wall, and possibly several characters that are not preserved in Bennettitales or *Pentoxylon* but occur in both angiosperms and Gnetales: a tunica layer in the apical meristem, siphonogamy (which arises independently in conifers), and the Mäule reaction, an indication of lignin chemistry (Gibbs 1957).

Our results thus imply that Gnetales are the closest living relatives of angiosperms, but not that angiosperms were derived from (in cladistic terms, nested within) Gnetales, as proposed by Wettstein (1907). Instead, interrelationships of groups within the clade are more consistent with the scenario of Arber and Parkin (1908); that is to say, Gnetales are relatives of angiosperms and Bennettitales that underwent drastic floral reduction and aggregation in response to wind pollination, as did Amentiferae within angiosperms. Bennettitales and Gnetales share several advances over angiosperms—stalked or sessile ovules, whorled microsporophylls, and a micropylar tube—while the three genera of Gnetales are in turn united by opposite leaves (probably originally linear), multiple axillary buds, vessels, loss of scalariform pitting, simple microsporophylls, a single terminal ovule, a second integument apparently derived from the perianth of the ovulate flower, and striate pollen (lost in *Gnetum*). Within Gnetales, *Welwitschia* and *Gnetum* share additional advances over *Ephedra*: vein anastomoses, reduction of the male gametophyte, a tetrasporic megagametophyte with free nuclei serving as eggs, and a feeder in the embryo. Since the whole clade is related to groups with pinnately organized sporophylls, the angiosperm carpel can still be interpreted as a leaflike organ, while the solitary ovules of Bennettitales and Gnetales and the simple stamens of angiosperms and Gnetales are presumably reduced structures, as argued by Arber and Parkin. The fact that we obtained these results without assuming that the clade originally had bisexual strobili makes the agreements with Arber and Parkin all the more significant.

There are several reasons why these results should be treated with caution, and these should be considered before discussing additional implications. First, even our most parsimonious trees include a large amount of homoplasy—roughly one convergence or reversal per character. When homoplasy is common, many almost equally parsimonious arrangements can be expected, and the results may be unstable (sensitive to minor changes in interpretation of characters). Another reason for caution is that some of the strongest evidence for angiosperm–gnetalian relationships comes from chemical, embryological, and anatomical characters rarely preserved in fossils; one or more fossil groups may have had these (or other!) angiosperm apomorphies. Furthermore, although in general our results are gratifyingly consistent with the stratigraphic record, all the 130-step trees conflict at a few points with the known ranges of groups (e.g. cordaites extend down to the mid-Carboniferous, but none of their immediate sister groups in Fig. 12.1 are known before the Permian or Triassic, while the tree with coniferopsids at the base places *Callistophyton* among the younger groups). They also require morphological changes most would consider implausible (e.g. secondary loss of a sulcus and reappearance of a tetrad scar in coniferopsids, or reversion to synangia and multiseriate rays in

Callistophyton). Fortunately for our purposes, most such anomalies occur in the callistophyte clade and do not affect the question of angiosperm-gnetalian relationships.

To evaluate the nature and seriousness of these problems in more detail, we performed a series of experiments with modified data sets. We were able to determine lengths of many alternative trees by adding 'dummy' synapomorphies to the matrix to force particular relationships among taxa and then subtracting them from the resulting tree. Thus a tree in which coniferopsids are related to *Archaeopteris* (which would eliminate the reversals and stratigraphic anomalies mentioned above) is only two steps longer (132 steps) than the shortest trees; this suggests it would be premature to rule out the Beck–Meeuse hypothesis. Considering alternatives more directly relevant to the angiosperm–gnetalian question, trees in which Gnetales are forced into the coniferopsids are considerably longer (136 steps). However, when *Caytonia* and corystosperms are moved into the medullosan clade (which would fit the presumed medullosan anatomy of corystosperms), several trees of only 133 steps are obtained, some with cycads interpolated between *Caytonia* and corystosperms and angiosperms, others with angiosperms grouped with *Caytonia*. The best alternative found (132 steps) is obtained when angiosperms, Bennettitales, *Pentoxylon*, and Gnetales are moved into the callistophytes, as proposed by Crane, in which case they form part of a basal clade also including *Caytonia* and corystosperms. The last two alternatives are both consistent with the idea that the anatropous bitegmic ovules of angiosperms are homologous with *Caytonia* or corystosperm cupules. Trees in which the four groups are related to glossopterids are a little less parsimonious (134 steps); trees in which angiosperms alone are moved into the callistophytes are much worse (137 steps). We conclude that the exact relationships of angiosperms with seed fern groups are equivocal, and that several alternative theories on origin of the angiosperm carpel and bitegmic ovules (Stebbins 1974, Doyle 1978, Retallack and Dilcher 1981, Crane 1985) are still viable—but whatever the ultimate ancestry of angiosperms, their closest relatives are the Bennettitales and Gnetales.

As another test of the robustness of the results, we analysed subsets of the data matrix. Analysis of vegetative characters alone (1–24) separates Gnetales from Bennettitales and moves them into the coniferopsids; not surprisingly, angiosperms are associated with *Gnetum*. Analysis of macro-reproductive characters (25–43) also moves Gnetales into coniferopsids, but *Pentoxylon* and Bennettitales go with them, while angiosperms remain with cycads. When only micro-reproductive characters are analysed (44–62), the coniferopsids are broken up, while angiosperms, Bennettitales, and Gnetales form a clade related to Mesozoic seed ferns, as proposed by Crane, whose analysis emphasizes such characters. Removing nine characters which are rarely or never preserved in Mesozoic groups (13, 24, 55, 57–62) has relatively few effects on the results, perhaps the most interesting being that *Gnetum* is the basal taxon in Gnetales rather than *Ephedra*, underlining the importance of embryological characters in uniting *Welwitschia* and *Gnetum*. Significantly, none of these experiments broke up the Gnetales, as suggested by Eames (1952) and others. On the whole, these exercises support relationships among angiosperms, Bennettitales, and Gnetales, since either all three groups or pairs of them are linked in all cases.

Assuming that the results of our analysis of the whole data set are substantially correct, the links between Gnetales and coniferopsids in vegetative features (e.g. leaf morphology, anatomy) and between both Gnetales and Bennettitales and coniferopsids in macro-reproductive features (e.g. solitary ovules) are due to convergence. The fact that angiosperms too are nested within coniferopsids on vegetative data is largely attributable to convergence with Gnetales, especially *Gnetum*. Micro-reproductive features seem to give the most accurate indications of relationship, at least for the groups in question. It is easy to accept the macro-morphological similarities with coniferopsids as convergences, since most of them involve simple structures that may look similar simply because they are reduced. On the other hand, it is harder to imagine that many of the cycadopsid features of angiosperms (e.g. pinnately organized carpels) or Bennettitales (e.g. pinnate leaves) arose by convergence from a coniferopsid prototype: this would require elaboration of more complex structures, and it seems less likely that results similar to those in other groups would be obtained.

Critical re-examination of gnetalian structures often cited as indicating relationships with coniferopsids further supports our results. Thus Gnetales are like cordaites in having compound strobili consisting of bracts and axillary fertile shoots, but their microsporophylls are whorled and bear fused pollen sacs (except in *Gnetum*, which has only two sacs per flower), and their ovules have a micropylar tube and a reduced megaspore membrane, features anomalous in coniferopsids but typical of Bennettitales. Similarly, although Gnetales resemble conifers and *Ginkgo* in their pycnoxylic secondary xylem and lack of scalariform pitting in the primary xylem, they are like cycadopsids in having multiseriate rays.

Furthermore it appears, with hindsight, that several features which we excluded from the original data set because of uncertainty concerning polarity or homologies of parts may be additional synapomorphies of angiosperms, Bennettitales, and Gnetales. The three groups also show several striking parallel trends; these cannot be used as synapomorphies, but they may reflect shared genetic advances. For example, Ehrendorfer (1976) notes that Gnetales differ from other gymnosperms and resemble angiosperms in having relatively small chromosomes, less repetitive DNA, and extensive polyploidy. Although the ancestral condition in Bennettitales is controversial, our data are consistent with the idea that all three groups originally had flower-like bisexual strobili and a perianth, and hence that insect pollination was prevalent in the clade from its beginning. The three groups also show a strong tendency for adaptation to dry conditions: this is clearest from the Early Cretaceous record of angiosperms and Gnetales, but it is also true of Bennettitales, which were one of the dominant groups in southern Eurasia during the Late Jurassic, a period of widespread aridity (Vakhrameev 1970). These observations could greatly affect scenarios of early angiosperm evolution, since insect pollination, rapid reproduction, and ability to colonize arid or unstable habitats have been interpreted as key adaptations intimately associated with the origin of angiosperms and their occupation of a new adaptive zone (Stebbins 1974, Doyle and Hickey 1976, Doyle 1984). If these features arose earlier in evolution of the clade to which angiosperms belong, they cannot be the whole reason for the unique success of angiosperms.

Conclusions

The hypothesis that angiosperms are related to Bennettitales and Gnetales considerably simplifies some aspects of angiosperm evolution, by reducing the number of changes required in the origin of the group. For example, it implies that the angiosperm flower is a structure with direct homologues in related groups, and the same can be said of the essentially bennettitalian wood anatomy, stomatal structure, and granular monosulcate pollen of primitive angiosperms. However, it does leave unanswered some of the problems that have most vexed botanists, particularly the origin of the carpel, since ovulate structures of Gnetales and Bennettitales are so reduced that evidence on basic sporophyll morphology must be sought in more distant relatives. It also provides no obvious explanation for origin of the second integument in angiosperms, although a cupule of some sort is known in Triassic Bennettitales (Harris 1932, Crane this volume). Both problems might be alleviated if new evidence associates the angiosperm-bennettitalian clade with *Caytonia* and corystosperms, an alternative that is already nearly as parsimonious (see Addendum to this Chapter).

Our results may make the origin of angiosperms somewhat less mysterious in another way, since they imply that the angiosperms were a product of the radiation of one of the most vigorous Mesozoic groups, rather than the final offshoot of a declining seed fern line. On the other hand, derivation of angiosperms and Bennettitales from a common ancestor implies that the angiosperm line existed at least as far back as Bennettitales (i.e. Late Triassic) whereas convincing angiosperm remains are not known until the Cretaceous. One explanation is that angiosperms were geographically or ecologically restricted before the Cretaceous. However, it is also possible that many typical advances of angiosperms did not evolve in the angiosperm line until the Cretaceous. Our scheme predicts that leaves, wood, and pollen of early angiosperm precursors (in fact, all but floral structures) might be easy to mistake for remains of Bennettitales (cf. Arber and Parkin 1907).

Our analysis also clarifies why angiosperm–gnetalian relationships have been so controversial. Previous workers were quite right in interpreting many striking similarities between the two groups as convergences, but they were incorrect in assuming that this ruled out all relationships. Thus, even if Young (1981) is right that vessels are ancestral in angiosperms, vessels presumably arose independently in the two groups, since they are absent in Bennettitales and *Pentoxylon*. Positional relationships of floral parts imply that the envelope around the seed in Gnetales is homologous with the perianth of Bennettitales (and presumably angiosperms), not the carpel wall. Judging from conditions in *Ephedra*, Bennettitales, and *Pentoxylon*, the dicot-like leaves of *Gnetum* and the reduced gametophytes of *Gnetum* and *Welwitschia* arose within Gnetales. Most of the chemical similarities between Gnetales and angiosperms noted by Gottlieb and Kubitzki (1984) are presumably also convergent, since they occur in the 'rosiflorean block' which appears to be nested within higher dicots on such advances as triaperturate pollen. The real homologies are more cryptic (siphonogamy, tunica-corpus, lignin chemistry, reduced megaspore wall, granular exine) and their significance is difficult to recognize except in the context of broader relationships, particularly when Gnetales are viewed as highly derived relatives of Bennettitales.

In addition, many arguments concerning angiosperm–gnetalian relationships have been clouded by questionable phylogenetic reasoning. Perhaps most importantly, the question has often been phrased in terms of whether or not Gnetales could be ancestors of angiosperms (e.g. Bailey 1944, Doyle 1978, Muhammad and Sattler 1982) which overlooks the possibility that they were derived from a common ancestor that would not be placed in either group. Many workers, particularly Eames, were guilty of extreme character weighting, elevating minor morphological distinctions (some of them probably imaginary, like ovule position in different Gnetales) to insurmountable obstacles to relationship, and dismissing characters as systematically worthless because they arose independently elsewhere. Although we have found much independent support for the conclusions of Arber and Parkin, it is easy to dismiss their original arguments as excessively speculative.

One of the great strengths of cladistic analysis is that it makes so many predictions to be tested by future work. On the neontological front, molecular data might provide valuable tests. Under our hypothesis, angiosperms probably diverged from Gnetales as recently as the Triassic, while all other connections among major extant seed plant groups would date at least to the Carboniferous; such a difference in divergence time might be detected by studies of appropriate DNA sequences. On the paleobotanical front, critical evidence might come from information on cryptic characters of Bennettitales (meristem type, lignin chemistry, embryology), to determine whether they have the predicted gnetalian–angiosperm conditions, and on diversity within the group, as stressed by Crane (this volume). Data on missing characters of Mesozoic seed ferns and organs of Mesozoic Gnetales other than pollen could also be critical. There are a few incompletely known megafossils that may be gnetalian, such as the Late Triassic genus *Dechellyia* (Ash 1972), which was conifer-like in habit but had opposite leaves and is associated with cones containing ephedroid pollen. If future work on *Dechellyia* reveals other apomorphies of Gnetales but not the bennettitalian features of the modern genera, it might support the coniferopsid alternative. On the other hand, Early Cretaceous gnetalian megafossils under study by Upchurch and Crane (1985) have loosely organized reproductive structures, supporting the idea that the compound strobili of modern Gnetales are aggregations of bennettitalian flowers rather than structures inherited from cordaites.

Acknowledgements

We are grateful to San Diego State University for computer time, to Michael Russell and Susan Fansler for assistance with computer analysis, to Wayne Maddison for discussion of coding partially ordered multistate characters, and to Charles Beck, Peter Crane, Gar Rothwell, and Garland Upchurch for discussion of characters and use of unpublished results.

Addendum

Since completing this study, we reanalysed our data matrix with revised codings of certain characters. Our primary purpose was to take into account the evidence presented by Crane (this volume) that *Pentoxylon* had a radial cupule and that Bennettitales may have been primitively platyspermic. In the process, we also discovered a subtle bias against Crane's (1985) hypothesis that the clade consisting of angiosperms, Bennettitales, *Pentoxylon*, and Gnetales is most closely related to

Mesozoic seed ferns with circinate cupules. As mentioned in the text, the use of X-coding in multistate characters biases against placement of X-coded taxa between taxa with different codings. However, under the present coding of the cupule character (33–35), Crane's hypothesis requires just this sort of arrangement, since it postulates that circinate cupules ('X10') were intermediate between leaf-borne, non-cupulate seeds ('100') and radial cupules ('000'). The computer counts this transformation series as three steps (1→0 in binary character 33, 0→1→0 in 34), one more than the actual number of morphological changes involved. We removed this bias by recoding non-cupulate seeds '100', circinate cupules '110', orthotropous cupules of Bennettitales and *Pentoxylon* '010', and the gnetalian condition 'X01'. To express the uncertainty concerning basic seed symmetry in Bennettitales and angiosperms (where the seeds are too reduced to show the usual criteria of symmetry, such as bundles in the inner integument), we re-coded both groups 'XX' for this character (42–43).

Mega- and microsporophylls of cycads (25–27, 28–30) differ greatly from those of other groups coded as once-pinnate, in bearing ovules directly on the rachis and pollen sacs scattered on either side of the midrib, rather than on leaflet-like segments. Since these observations suggest that cycad sporophylls may have been reduced from pinnately compound structures by a unique pathway, we recoded both 25–27 and 28-30 as '1XO', which treats derivation from pinnately compound or via once-pinnate intermediates as equally likely.

With these changes, trees in which the angiosperm–bennettitalian–gnetalian clade is related to Mesozoic seed ferns (*Caytonia*, corystosperms, and/or glossopterids) become two or three steps *more* parsimonious than trees in which it is related to cycads and medullosans. As noted in the text, such relationships have the advantage of providing homologues for the second integument of angiosperms and the cupule of Bennettitales and *Pentoxylon*, which have no obvious precursor structures in the schemes reported above.

References

Arber, E.A.N. and Parkin, J. (1970) On the origin of angiosperms. *J. Linn. Soc. Bot.* **38**, 29–80.

—— and —— (1908) Studies on the evolution of the angiosperms. The relationship of the angiosperms to the Gnetales. *Ann. Bot.* **22**, 489–515.

Archangelsky, S. (1968) Studies on Triassic fossil plants from Argentina. IV. The leaf genus *Dicroidium* and its possible relation to *Rhexoxylon* stems. *Palaeontology* **11**, 500–15.

Ash, S.R. (1972) Late Triassic plants from the Chinle Formation in north-eastern Arizona. *Palaeontology* **15**, 598–618.

Bailey, I.W. (1944) The development of vessels in angiosperms and its significance in morphological research. *Am. J. Bot.* **31**, 421–28.

Beck, C.B. (1981) *Archaeopteris* and its role in vascular plant evolution. In *Paleobotany, Paleoecology, and Evolution*, vol. I. (ed. K.J. Niklas), pp. 193–230, Praeger, New York.

Bierhorst, D.W. (1971) *Morphology of Vascular Plants.* Macmillan Co., New York, 560 pp.

Brenner, G.J. (1976) Middle Cretaceous floral provinces and early migrations of angiosperms. In *Origin and Early Evolution of Angiosperms.* (ed. C.B. Beck), pp. 23–47, Columbia Univ. Press, New York.

Brooks, D.R. (1984) Quantitative parsimony. In *Cladistics: Perspectives on the Reconstruction of Evolutionary History.* (ed. T. Duncan and T.F. Stuessy), pp. 119–32, Columbia Univ. Press, New York.

Crane, P.R. (1985) Phylogenetic analysis of seed plants and the origin of angiosperms. *Ann Missouri Bot. Gard.* **72**, 716–93.

Doyle, J.A. (1978) Origin of angiosperms. *Ann. Rev. Ecol. Syst.* **9**, 365–92.

—— (1984) Evolutionary, geographic, and ecological aspects of the rise of angiosperms. *Proc. 27th Internat. Geol. Congr. (Moscow, 1984)*, **2**, pp. 22–33, VNU Science Press, Utrecht.

—— Biens, P., Doerenkamp, A. and Jardiné, S. (1977) Angiosperm pollen from the pre-Albian Cretaceous of Equatorial Africa. *Bull Centres Rech. Explor.-Prod. Elf-Aquitaine* **1**, 451–73.

—— and Hickey, L.J. (1976) Pollen and leaves from the mid-Cretaceous Potomac Group and their bearing on early angiosperm evolution. In *Origin and Early Evolution of Angiosperms.* (ed. C.B. Beck), pp. 139–206, Columbia Univ. Press, New York.

—— Jardiné, S. and Doerenkamp, A. (1982) *Afropollis*, a new genus of early angiosperm pollen, with notes on the Cretaceous palynostratigraphy and paleoenvironments of Northern Gondwana. *Bull. Centres Rech. Explor.-Prod. Elf-Aquitaine* **6**, 39–117.

Eames, A.J. (1952) Relationships of the Ephedrales. *Phytomorphology* **2**, 79–100.

Ehrendorfer, F. (1976) Evolutionary significance of chromosomal differentiation patterns in gymnosperms and primitive angiosperms. In *Origin and Early Evolution of Angiosperms.* (ed. C.B. Beck), pp. 220–40, Columbia Univ. Press, New York.

Florin, R. (1931) Untersuchungen zur Stammesgeschichte der Coniferales und Cordaitales. Erster Teil: Morphologie und Epidermisstruktur der Assimilationsorgane bei der rezenten Koniferen. *Kungl. Sven. Vetenskapsakad. Handl.* Ser 3. **10**, 1–588.

Gaussen, H. (1946) Les Gymnospermes, actuelles et fossiles. *Trav, Lab. For. Toulouse*, **II** *Étud. Dendrol.*, sect. l, vol. 1, Fasc. 3, ch. 5, 1–26.

Gibbs, R.D. (1957) The Mäule reaction, lignin, and the relationships between woody plants. In *The Physiology of Forest Trees.* (ed. K. V. Thimann), pp. 269–312, Ronald Press, New York.

Gottlieb, O.R. and Kubitzki, K. (1984) Chemosystematics of the Gnetatae and the chemical evolution of seed plants. *Planta Medica* 1984, 380–5.

Harris, T.M. (1931) The fossil flora of Scoresby Sound East Greenland. Part 3:

Caytoniales and Bennettitales. *Meddelelser om Grønland* **85**(5), 1–133.

Hennig, W. (1966) *Phylogenetic Systematics*. Univ. Illinois Press, Urbana, 263 pp.

Hill, C.R. and Crane, P.R. (1982) Evolutionary cladistics and the origin of angiosperms. In *Problems of Phylogenetic Reconstruction*. (ed. K.A. Joysey and A.E. Friday), pp. 269–361, Systematics Association Special Volume 21, Academic Press, London.

Long, A.G. (1979) Observations on the Lower Carboniferous genus *Pitus* Witham. *Trans. Roy. Soc. Edin.* **70**, 111–27.

Maddison, W.P., Donoghue, M.J. and Maddison, D.R. (1984) Outgroup analysis and parsimony. *Syst. Zool.* **33**, 83–103.

Mapes, G. and Rothwell, G.W. (1984) Permineralized ovulate cones of *Lebachia* from late Palaeozoic limestones of Kansas. *Palaeontology* **27**, 69–94.

Martens, P. (1971) Les Gnétophytes. *Encyclopedia of Plant Anatomy*, **12**(2), Borntraeger, Stuttgart, 295 pp.

Meeuse, A. D.J. (1963) From ovule to ovary: a contribution to the phylogeny of the megasporangium. *Acta Biotheoret.* **16**, 127–82.

Meyen, S.V. (1984) Basic features of gymnosperm systematics and phylogeny as evidenced by the fossil record. *Bot. Rev.* **50**, 1-112.

Mickevich, M.F. and Farris, J.S. (1982) Phylogenetic analysis system (PHYSYS) *(FORTRAN V software system of cladistic and phenetic algorithms)*.

Muhammad, A.F. and Sattler, R. (1982) Vessel structure of *Gnetum* and the origin of angiosperms. *Am. J. Bot.* **69**, 1004–21.

Retallack, G. and Dilcher, D.L. (1981) Arguments for a glossopterid ancestry of angiosperms. *Paleobiology* **7**, 54–67.

Rothwell, G.W. (1982) New interpretations of the earliest conifers. *Rev. Palaeobot. Palynol.* **37**, 7–28.

Smith, J.D., Farris, J.S. and Mickevich, M.F. (1982) Documentation for phylogenetic analysis system (*online PHYSYS documentation*).

Stebbins, G.L. (1974) *Flowering Plants: Evolution above the Species Level*. Harvard Univ. Press, Cambridge, Mass., 399 pp.

Takhtajan, A.L. (1969) *Flowering Plants: Origin and Dispersal*. Smithsonian Institution, Washington, D. C., 310 pp.

Thompson, W.P. (1918) Independent evolution of vessels in Gnetales and angiosperms. *Bot. Gaz.* **65**, 83–90.

Trevisan, L. (1980) Ultrastructural notes and considerations on *Ephedripites, Eucommiidites* and *Monosulcites* pollen grains from Lower Cretaceous sediments of southern Tuscany (Italy). *Pollen Spores* **22**, 85–132.

Upchurch, G.R. and Crane, P.R. (1985) Probable gnetalean megafossils from the Lower Cretaceous Potomac Group of Virginia. *Am. J. Bot.* **72**, 903.

Vakhrameev, V.A. (1970) Yurskie i rannemelovye flory. In *Paleozoyskie i Mezozoyskie Flory Yevrazii i Fitogeografiya Etogo Vremeni*. (ed. V.A. Vakhrameev, I.A. Dobruskina, Ye. D. Zaklinskaya and S.V. Meyen), pp. 213–81, Nauka, Moscow.

Watrous, L.E. and Wheeler, Q.D. (1981) The outgroup comparison method of character analysis. *Syst. Zool.* **30**, 1–11.

Wettstein, R.R. von (1907) *Handbuch der systematischen Botanik*, II. Band, Franz Deuticke, Leipzig, Wien, 577 pp.

Wight, D.C. and Beck, C.B. (1984) Sieve cells in phloem of a Middle Devonian progymnosperm. *Science* **225**, 1469–71.

Young, D.A. (1981) Are the angiosperms primitively vesselless? *Syst. Bot.* **6**, 313–30.

13. A reconsideration of the Rhyniophytina Banks

D. EDWARDS AND D.S. EDWARDS

Abstract

The problems encountered in the description and classification of early vascular plants are exemplified by reference to the Rhyniophytina. Genera normally included in the subdivision are briefly reviewed and their assignment evaluated in the light of new information, including that on the nature of the presumed water-conducting tissues of early land plants. A new broader concept of the Rhyniophytina is presented encompassing a number of levels of organization ranging from isotomously branching forms with simple terminal sporangia to those that show a complexity in both branching pattern and sporangial characteristics.

Introduction

Most of the problems confronting the investigators of Silurian and early Devonian plants are similar to those presented elsewhere in this volume. However as might be anticipated in the study of one of the most innovative periods in plant history, that which encompasses early stages in the colonization of the terrestrial environment, the appearance of completely novel forms of plant construction and organization with attendant changes in chemistry has presented further challenges.

Although it might be expected that the rapid diversification (Niklas, Tiffney, and Knoll 1983) associated with the initial adaptive radiation of the pteridophytes would produce a complex of intergrading forms, Banks (1968a, 1975) was able to distinguish three major combinations of anatomical and morphological characters in the many plants assigned to the Psilophytales since its erection by Kidston and Lang (1920a).

Space does not allow a detailed re-evaluation of the Zosterophyllophytina and Trimerophytina, instead we concentrate on the Rhyniophytina. In that it contains the earliest records of presumed pteridophytes it provides an excellent opportunity to develop our brief—the systematic approach in a dynamic evolutionary context. We propose to use its putative members to illustrate the problems associated with the recognition and classification of early vascular plants emphasizing the dearth of morphological characters available for diagnoses and the vicissitudes of preservation that range from the complete lack of anatomy in the majority of coalified compressions and impressions to the exquisite cellular detail of the silicified petrifactions of the Rhynie Chert. Indeed it was mainly on the latter that Banks based his diagnosis of the family Rhyniaceae and hence his concept of the Rhyniophytina as comprising 'naked (leafless) plants, whose aerial branches fork dichotomously, some terminated by sporangia; horizontal rhizomes with rhizoids, roots apparently lacking; slender, solid xylem strands, probably centrarch' (Banks 1975, p. 407).

General considerations

Apart from the Rhynie Chert representatives, investigations on rhyniophytes are hindered by poor preservation, fragmentary remains and a paucity of diagnostic characters. Occurrences are rare and numbers of fertile specimens at any one locality generally small, seldom permitting assessment of variation.

1. Modes of preservation

The silicification of an autochthonous deposit as seen in the Rhynie Chert provides a remarkable insight into both plants and vegetation in the Lower Devonian, but its constituent taxa have not yet been recorded as compression fossils. In certain cases this might simply be a problem of recognition for example, *Rhynia* sporangia would be recorded as *Salopella* and *Aglaophyton* axes with narrow central strands as *Taeniocrada*, but plants which are morphologically distinct, such as, *R. gwynne-vaughanii* with its hemispherical bulges, would surely be easily identified. Edwards (1980a) has suggested that this is a preservational problem of a different sort — the Rhynie Chert plants with predominantly parenchymatous construction would have had low fossilization potential in typical Old Red Sandstone environments.

An associated problem is the difficulty of elucidating the gross morphology of plants preserved as petrifactions: this is feasible only by tedious reconstruction based on large numbers of peel sections. A good example of the importance of such work is seen in the reconstruction of *Nothia* (El—Saadawy and Lacey 1979a) which has radically altered our understanding of this plant once considered to be the terminal fertile parts of *Asteroxylon*.

Some anatomical detail can be derived from coalified compressions: film pulls show fragments of cuticle or tracheids (Lang 1937) but these are seldom isolated from fertile rhyniophytes and never from the earliest representatives. Work in progress in Cardiff suggests that such axes have far greater potential in elucidating three dimensional structure than previously anticipated. A systematic examination of less compressed axes by scanning electron microscopy from a few Pridoli and Gedinnian localities in the Welsh Borderland has yielded details of both cortex and xylem although not yet in fertile material. In addition the recent discovery of a pyritised axis in a rhyniophyte assemblage from the basal Devonian augurs well for the detection of protoxylem position in older members of the group.

2. *Reproductive versus vegetative specimens*

Except that they demonstrate the existence of early pteridophytes at a particular horizon, vegetative remains of rhyniophytes when preserved as featureless compression fossils are of little value. Even when found on the same bedding plane as axes bearing sporangia, it is impossible to unite them unequivocally into one plant, and when more than one fertile taxon is present the problem is compounded. Such sterile axes when dichotomously(isotomously) branched are called *Hostinella*. Høeg (1967) reviewing this genus and its numerous species presented a broader concept encompassing sterile smooth axes with dichotomous or pseudomonopodial branching sometimes bearing axillary tubercles. We prefer Halle's treatment when he considered *Hostinella* as 'convenient designation for repeatedly bifurcating branch systems which are not sufficiently well characterized to permit of a more satisfactory classification. Specimens of this kind normally lack distinctive characters on which a determination could be based and in most cases cannot be grouped into provisional species' (Halle 1916, p. 23). Description of anatomy would provide additional specific characters (e.g. Banks 1968b, Edwards 1980b), although in these examples the presence of a centrarch xylem strand is not particularly helpful in the further elucidation of the affinities of the plants, be they rhyniophyte or trimerophyte.

Halle also proposed that *Aphyllopteris* Nathorst be used as a form genus for sterile plants with stout main axis and more slender lateral branching. Such a growth pattern characterizes certain Zosterophyllophytina and most Trimerophytina, but is also seen in the more advanced members of the Rhyniophytina (e.g. *Renalia*). Another convenient form genus for sterile axes is *Taeniocrada* White. The type species *T. lesqueureuxi* from the Upper Devonian of New York State (White 1903) was based on *sterile* smooth dichotomously branching axes in which the central strand of tracheids was narrow compared with axis diameter. it is now an excellent example of a Devonian 'dust-bin' genus because a number of plants, united in the possession of stems typical of the type species but with a variety of sporangial arrangements, have been included in it; one with clusters of terminal sporangia, *T. decheniana* is a possible rhyniophyte. However in our opinion *Taeniocrada* is best used as a form genus for sterile compression axes with very narrow central strands. Fertile specimens undoubtedly belong to a number of different plants of diverse affinity and should be transferred to appropriate, sometimes new, taxa when more is known of their anatomy and reproductive characteristics (see below, listed taxa 6,7,8).

Morphologically simple axes also present problems in reconstructing the habit of the whole plant. Most specimens of *Cooksonia*, for example, are preserved as small fragments with just one or two, usually isotomous, branching points. The most complete specimens from Pridoli (*C. bohemica* Schweitzer) and Gedinnian (*C. caledonica* Edwards) strata show repetition of this branching pattern and hence a compact tufted growth form, with teminal sporangia all at approximately the same level above the ground. Howeve, in 1969, Ananiev and Stepanov described two plants from the Lower Devonian of Siberia showing marked pseudomonopodial branching with dichotomously branching lateral systems terminating in sporangia: they assigned them to *C. hemisphaerica* and *C. pertoni*. A thorough search at the type localities has failed to produce any evidence that the plants Lang originally described were lateral branches of a more complex system although a few specimens deviate slightly from strictly isotomous branching. We therefore consider that plants with overtopping producing pseudo–monopodial branching should not be included in the *Cooksonia* and that the Russian specimens are better transferred to another genus (Banks 1975, Gensel (1976). At the other end of the spectrum we question whether or not single sporangia terminating unbranched axes should be assigned to *Cooksonia*(e.g. see Edwards 1979). Such fragments are by far the commonest fertile elements in late Silurian strata of the northern hemisphere. In some cases the sporangia alone are identifiable; *C. pertoni* is often found in this state because branching may occur some distance (<4.5mm) below the sporangia. As yet we have not found a distinctive sporangial organisation consistently terminating an unbranched axis and so cannot recognise the 'short and wide sporangium' equivalent of *Sporogonites* Halle or *Tortilicaulis* Edwards, both of which we would exclude from the Rhyniophytina on the grounds that the sporangium-bearing axes are unbranched and vegetative anatomy is unknown.

3. *Delimitation of genera*

This involves choice of characters and the handling of intergrading forms; examples will again be taken from work on the earliest rhyniophytes.

Edwards and Fanning (1985, Fig. 6) recently figured a single specimen from the Lower Devonian that appears intermediate in growth form between *Cooksonia* and *Renalia*. A lateral branch of a pseudomonopodial system branches isotomously twice and terminates in oval sporangia of *Cooksonia* type but is not referable to an existing species. Gensel (1976) is of the opinion that the difference between dichotomous and pseudomonopodial branching is not of major taxonomic importance, citing as examples the variability of branching pattern in some early Cooksonias (Obrhel 1962) and its use at the specific level in *Psilophyton*. We believe that such a difference constitutes a major change in the growth pattern of vegetative axes with important repercussions on the reproductive behaviour of early land plants. The new specimen will therefore not be included in the genus *Cooksonia*, although we admit that the lateral systems if found isolated would be placed in a new species of that genus (Edwards and Fanning 1985). Whether or not it should be assigned to *Renalia* is more equivocal, as the new specimen is so fragmentary and appears to lack the sporangial features of that genus. A further example involves the genus *Cooksonia* itself. In the diagnosis sporangial shape is described as short and wide: apart from two exceptions, subsequent authors have interpreted this to include sporangia that are circular, semicircular, elliptical and reniform in outline. However *C. downtonensis* described by Heard (1939) from Pridoli strata in S. Wales had vertically elongate sporangia. His specimens are missing, but it seems very likely that they were poorly preserved examples of *Steganotheca striata* later described by Edwards (1970) from the same locality. The second example is more thought-provoking. We are currently studying in Cardiff a Lower Devonian plant with smooth isotomously branching axes and terminal elliptical sporangia often twice as high as wide. Each sporangium has a narrow border. The plants are different from any others we have examined and seem to us to comprise a new genus. Croft and Lang (1942), however, called them *Cooksonia* sp. and hence they must have fitted Lang's concept of his own genus! Perhaps he would have modified this view had he the benefit of our accumulated data!

Finally, there are difficulties related to poor preservation particularly in impression fossils. Hueber (1983) has suggested that *Yarravia*, usually considered an advanced genus in that the fructification is synangiate, is a compacted poorly preserved form of *Hedeia*. A similar difficulty in distinguishing between sporangia that are adnate or fused has recently been encountered in the identification of clusters of small sporangia from the Gedinnian of the Welsh Borderland (Edwards and Fanning 1985). In these examples, three sporangia appear fused proximally, but with free tips. Although named cf. *Yarravia* they may simply represent forms of *Salopella*.

Table 13.1. Summary of morphological and anatomical characters in genera assigned to the Rhyniophytina.

	No. of species accepted in this paper[1]	Stratigraphic Range[2]	Preservation Compression — C / Petrifaction — P	Branching Isotomous — X / Anisotomous — O	Axes smooth — X / Axes with enations — O	Stomata present — X / Stomata absent — O	Central strand present — X / Central strand absent — O	Centrarch strand present — X / Centrarch strand absent — O	Conventional — X / Tracheid ornamentation / Anomalous — O	Sporangia terminal on main axes — X / Sporangia terminal on lateral axes — O	Sporangia wider than long — X / Sporangia longer than wide — O
Rhynia g-v	1	S/E	P	X/O	O	X	X	X	X	X	O
R. major	1	S/E	P	X	X	X	X	X	O	X	O
Horneophyton	1	S/E	P	X	X	X	X	X	X	X	O
Cooksonia	6	W/G	C	X	X	?	X	?	?	X	X
Steganotheca	1	Lu-P	C	X	X	?	X	?	?	X	O
Salopella	3	Lu-S	C	X	X	?	X	?	?	X	O
**Eorhynia*	1	Lu	C	X	X	?	O	?	?	X	O
Dutoitea	3	MD	C	X	X/O	?	O	?	?	X	X/O
Eogaspesiea	1	E	C	X	X	?	O	?	?	X	O
+ *Taeniocrada*	1	S-E	C	X/O	X	?	X	?	X	X	O
Yarravia	2	Lu-S	C	X	X	?	X?	?	?	X	O
Hedeia	1	Lu-S	C	X	X	?	?	?	?	X	O
Nothia	1	S/E	P	X/O	O	X	X	X	O	N/A[6]	X
**Renalia*	1	E	C	X/O	X	X	X	?	X	O	X
**Hsüa*	1	E	C	X/O	X	X	X	X	X	O	X
Hicklingia	1	MD	C	X	X	?	X	?	?	N/A	X

Notes: 1. See text, p. 208 et seq. for details. 2. Initials as follows: W-Wenlock, Lu-Ludlow, P-Pridoli, G-Gedinnian, S-Siegenian, E-Emsian, MD-Middle Devonian. 3. Dehiscence by an apical split. 4. *Cooksonia* may have been indehiscent. 5. Dehiscence was possibly by an apical operculum. 6. Sporangia lateral forming loose spikes. * Genera erected since Banks (1975) paper: + *T. dechaniana* only: ? indicates lack of infomation.

4. *Delimitations of species*

Our approach to species delimitation is a pragmatic and phenetic one—a species is the smallest taxonomic unit that we can consistently identify. We can add little to the numerous discussions on the species concept in palaeontology (see e.g. Niklas, Tiffney and Knoll 1980), except to note that none of the simplest Silurian plants (e.g. *Cooksonia*)

Dehiscence not predetermined — X Dehiscence predetermined — O	Dehiscence by longitudinal slit — X Dehiscence by two valves — O	Spore Type	Key references
X	X	c.f. *Perotriletes*	Edwards, D.S. 1980
X	X	c.f. *Geminospora*	Edwards, D.S. in press
O	NA[3]	c.f. *Apiculiretusispora; Retusotriletes;* ? *Emphanisporites*	Kidston and Lang 1920a
X	?[4]	*Ambitisporites; Apiculiretusispora*	Lang 1937
?	NA[5]	—	Edwards, D. 1970
?	?	?Not comparable with dispersed spore genus	Edwards & Richardson 1974
?	?	—	Ishchenko 1975
?	?	—	Anderson & Anderson 1985
?	?	—	Daber 1960
O	?	—	Schweitzer 1980a
?	?	—	Lang & Cookson 1935
?	?	—	Cookson 1935
O	O	?Poorly preserved	El-Saadawy & Lacey 1979
O	O	Retusotriletes	Gensel 1976
O	O	?Not compared with dispersed spore genus	Li 1982
O	O	—	Edwards 1976

when given a Linnaean binomial could ever even approximate to the concept of a biological species as an interbreeding monophyletic taxon!

Most of our problems and experiences centre on the Silurian-early Devonian plants, for example *Cooksonia* and *Salopella*, with extremely simple morphological organization and little preserved anatomy. Banks (1980) has pointed out that over-emphasis of minor characters which would be thought insignificant in classifying more complex plants, may lead to 'splitting'; an observation at odds with the more conventional view that the more morphologically complex a plant the more 'splitting' occurs. Alternatively, simplicity in morphology may mask a diversity in physiology and breeding behaviour.

Our approach is to collect as many specimens as possible and to derive maximum information from them, not only to increase numbers of characters for diagnosis, but to allow statistical analysis (Edwards and Fanning 1985: current work in Cardiff). Most progress has been made with the type species *C. pertoni*; in particular scanning electron microscopy of uncompressed heavily coalified specimens, has produced information on shape, wall characteristics, and spores. The large number of specimens allow an assessment of variation in size and appearances in different preservation states. We therefore now consider *C. pertoni* a 'good' species, but in older strata identification of *C. pertoni* rests on gross morphology of very compressed specimens and perhaps should be preceded by 'cf.', a prefix which is *intended* to convey the doubts of the author to the reader (see Hughes 1981).

Taxa Assigned to the Rhyniophytina

Table 13.1 summarizes the characteristics of taxa assigned to the subdivision by Banks (1968a, 1975) and later workers (Gensel and Andrews 1984). Before presenting our own assessment of their affinity, we propose to comment briefly on each. This is not intended as an exhaustive survey: in most cases only the more recent references are given.

(1) *Rhynia gwynne-vaughanii* Kidston and Lang

Of the two species originally erected by Kidston and Lang (1917, 1920a) we propose that only *Rhynia gwynne-vaughanii*, the type species, be retained within the genus. The gross morphology of the plant has been shown to be more complicated than was at first supposed and although the main axes branch dichotomously, lateral (adventitious) branching is of far greater importance. Sporangia were abscissed after spore release, with concomitant overtopping of the sporangia by adventitious branches (Edwards, D.S. 1980). *Rhynia* is one of the few rhyniophytes for which three dimensional anatomical detail is know—a narrow, terete, centrarch conducting strand of several tracheids with annular, rarely spiral, thickenings is surrounded by a parenchymatous cortex which lacks a hypodermal layer with thickened walls.

(2) *Rhynia major* (Kidston and Lang)

This plant consists of upright dichotomously branched axes with large terminal fusiform sporangia. Although at first confused by Kidston and Lang with *Rhynia gwynne-vaughanii*, the plant has recently been assigned to a new genus on the basis of its simpler morphology and the structure of its conducting strand (Edwards, D.S. in press).

The latter consists of two regions—an outer zone of thin-walled elongate cells, which was interpreted as phloem in the original description and a central region of thicker-walled elongate cells, originally interpreted as xylem. Support for the interpretation of the outer zone as true phloem tissue was provided by Satterthwait and Schopf (1972) who described areas on the lateral walls which they considered represented pit fields; however Lemoigne and Zdebska (1980) suggested that these may be preservational in origin. Hébant (1977) pointed out the similarity between the cells of this tissue and the leptoids found in certain moss gametophytes and sporophytes (although leptoids frequently have thickened lateral walls, unlike the cells of this zone in *R. major*).

Two areas can be recognized in the central region of the strand—a central part in which the cells are narrow with thin walls and a peripheral region in which the cells are wider with thick walls. The vertical walls of the cells in both areas sometimes show a reticulate pattern which is similar to that sometimes found on epidermal cell walls and was considered by Lemoigne and Zdebska (1980) to result from deterioration of the wall. Since similar patterns are found on uniformly thickened walls it is likely that the walls of these central conducting cells were also uniformly thickened and are thus comparable to the hydroids found in the setae of some polytrichaceous mosses (see Hébant 1977, p. 53).

(3) *Horneophyton* (Kidston and Lang) Barghoorn and Darrah

The structure of the sporangium and the basal corm-like rhizome of this plant separate it from other rhyniophytes. The ellipsoidal to globose sporangia have a branched columella within a common spore cavity, up to three successive dichotomies of the columella resulting in a five-lobed sporangium being recorded by El-Saadawy and Lacey (1979b). Dehiscence of the sporangia was probably by an apical slit at the apex of each sporangial lobe, possibly as a result of differential contraction (Bhutta 1973, Eggert 1974).

The unvascularized rhizome bore upright naked axes which included a large vascular strand. Thickenings on the walls of the tracheids appear to be of an annular type and are preserved only in the basal and apical portions of the plant.

Reconstructions of the plant (Kidston and Lang 1921, Eggert 1974) appear to be very speculative, since nothing is known of the height, frequency of branching and abundance of sporangia. The morphology of the rhizome is also incompletely known.

(4) *Cooksonia* Lang

Erected by Lang (1937) for plants with short and wide sporangia terminating smooth, dichotomously branching axes, this represents both the simplest and oldest level of organization yet attributed to the Rhyniophytina. Its importance to studies on the origins of vascular plants and the early colonization of the land has resulted in very rigorous scrutiny of often very unimpressive compression fossils. The accumulated anatomical data is disappointing. Although central lines are visible on some axes, *in situ* tracheids have not yet been convincingly demonstrated and poor preservation of the cuticularized epidermis precludes the detection of stomata. Smooth trilete spores have been isolated from a few species. At the specific level, the dearth of characters available for their delimitation has been summarized elsewhere (Edwards 1979, Edwards, Bassett and Rogerson 1979, Edwards and Fanning 1985, Edwards and Edwards this paper, above). Indeed where species are very clearly delimited, there is a strong possibility that they are better placed in new genera. *C. crassiparietilis* from the late L. Devonian of Kazachstan (Yurina 1969) has large, often fan-shaped sporangia up to 9mm wide and 7mm high. Each appears to be supplied by a number of (possibly) vascular strands, and has a pronounced border, composed of thick walled cells, thought to be involved in dehiscence. The specimens themselves are fragmentary, but there is some dichotomous branching in the subtending axes. There are also indications that these were connected to much larger axes. With the possible exception of *C. caledonica*, where sporangia are surrounded by a rim, perhaps a preservational feature or possibly structural, species of *Cooksonia* have no well defined dehiscence mechanisms. Hence for a number of reasons we consider that *C. crassiparietilis* should be transferred to another genus, perhaps even outside the Rhyniophytina.

(5) *Steganotheca* Edwards

Apart from an extension to its stratigraphic range, (it is now known from Ludlow as well as Pridoli strata at the single locality), nothing has been added to the original description of the genus (Edwards 1970, Edwards, Bassett and Rogerson 1979). Tracheids have not been isolated from central lines seen on some axes, and whether or not the apical thickening was involved in dehiscence remains unknown.

(6) *Salopella* Edwards and Richardson

Erected as a form genus for longitudinally elongate, fusiform sporangia terminating smooth, dichotomously branching axes, *Salopella* has never been proven vascular. Recently described Australian species (Tims and Chambers 1984) show a central line in certain axes. The type species *S. allenii* (Edwards and Richardson 1974: L. Devonian, Welsh Borderland) is based on a single fragmentary specimen with branching just below the sporangia.

Closely associated fertile axes in both *S. australis* (Ludlow-L. Devonian) and *S. caespitosa* (L. Devonian) suggest that the plants were either tufted or grew from a rhizome. Unlike the type species they have not yielded spores, although the sporogenous tissue is thought to be represented by a darker area. A number of fragments with much smaller sporangia have recently been discovered in Pridoli and Gedinnian strata in the Welsh Borderland (Edwards and Fanning 1985). New species will be delimited on basis of sporangial shape, branching angles and proximity of branching to the sporangia. The growth habits of the complete plants remain unknown. The fragments may be parts of more complex plants e.g. *Hedeia* (see below) or *Psilophyton*, but it is considered more likely that the parent plants were repeatedly dichotomously branched with organization similar to that seen in *Rhynia major*.

(7) *Eorhynia* Ishchenko

The genus was erected to include compression and impression fossils with little or no preserved anatomical detail originally placed in *Rhynia major* (Ishchenko 1969, 1975). The fusiform sporangia terminating smooth isotomously branching axes look remarkably similar to those in *Salopella* Edwards and Richardson: the two taxa are probably congeneric.

(8) *Eogaspesiea* Daber

The single species, *E. gracilis* is based on tufts of very long (<9cm) narrow (<0.5mm) branches which branch irregularly and terminate in elongate oval sporangia. Daber (1960) thought that the aerial branches arose from a rhizomatous system from which tracheids have been isolated, but the two systems were not connected. Anatomy is not known from the aerial system.

(9) *Dutoitia* Høeg

Three species are illustrated in Anderson and Anderson (1985) and are included in their Psilophyta. Ellipsoidal, club-shaped or cup-shaped sporangia terminate dichotomizing axes which are smooth except in *D. pulchra* where minute projections are present. This species also differs from the others in that the sporangia are more *Cooksonia*-like. In *D. maraisia*, the elongate ellipsoidal sporangia are borne on pendulously arching axes. The diagnoses are based on numerous, although fragmentary specimens, and there is no indication that these were parts of more complex plants as might be anticipated from their Middle Devonian age. Thus on present evidence, *Dutoitia* is perhaps better assigned to the Rhyniophytina, although the diversity of sporangial form in the three species would suggest that more than one genus is present.

(10) *Taeniocrada* White

The genus is normally included under 'Questionable Rhyniophyta' (Banks 1975, Gensel and Andrews 1984) because only one of its species has terminal sporangia. The holotype of *T. decheniana* is sterile, but Kräusel and Weyland (1930) described groups of terminal ellipsoidal sporangia, in which the subtending axis is fused to one side of the sporangium so that it has an anatropous appearance (Schweitzer 1980a), associated with astomatous axes of typical *Taeniocrada* aspect. They thought that the axes were cylindrical near the base of the plant and flattened distally. A recent reconstruction of the fertile region (Schweitzer 1983) shows terete sporangium-bearing axes. This highlights the difficulties of elucidating the original three dimensional shape of plant organs in highly compressed fossils. In this case, the decision has implications for the putative habitat of the plant, in that *Taeniocrada* has been considered semi-aquatic. Having opined that *Taeniocrada* is better kept as a form genus for sterile axes we obviously believe that *T. decheniana* should be transferred to a new taxon. A second German species, *T. longisporangiata* (Schweitzer 1980a) has a cluster of sporangia terminating a short unbranched lateral axis; the sporangia themselves being reminiscent of *Dawsonites*. Recent discoveries of Lower Devonian plants with lateral sporangia from Belgium that were originally called *T. langii* by Stockmans have resulted in the erection of a new genus (Fairon–Demaret 1985). Its affinities are considered obscure, if indeed with the classical tracheophytes, because of the nature and arrangement of conducting elements in a terete xylem strand that defies description as centrarch, exarch or mesarch. Some Soviet species (e.g. *T. latissima* Senkevich) also have lateral sporangia but lack anatomical information. Finally, current anatomical investigations (Hueber 1982) on another sterile species, *T. dubia* (but see Schweitzer 1980a for possible sporangia) indicate that the central presumed conducting elements have a unique type of thickening.

(11) *Hedeia* Cookson

The most familiar reconstructions of this genus (e.g. Andrews 1964) show a fertile corymb-like branch system with dichotomously branching axes terminating in fusiform sporangia (>20). The Siegenian species, *H. corymbosa*, has no sterile branches, but these characterize new Ludlow representatives (Dr J.D. Tims, personal communication). Indeed she considers the corymb effect to result from a combination of overtopping and decrease in sporangial length towards the periphery so that all the sporangial tips are the same level. Although nothing is known about the anatomy of the fertile region nor of the nature of the remainder of the plant, Tims concludes that its affinities are more probably with the rhyniophytes than the trimerophytes.

(12) *Yarravia* Lang and Cookson
As originally diagnosed, the genus is advanced in that the fructification is synangiate comprising five to six elongate sporangia, radially arranged so that normally only three are visible. Again there is no anatomical detail and the most complete specimens are only 5cm long, with synangia terminating unbranched, smooth axes. Recent examination of new and museum collections casts doubts on the synangium interpretation. Tims (personal communication) suggests that *Yarravia* may have been an immature form of *Hedeia*, while Hueber believes it to be poorly preserved *Hedeia* (see p. 205).

(13) *Nothia* Lyon ex Hoeg
When first recognized as a distinct plant, *Nothia* was considered to have a similar morphology to *Rhynia* (see Kidston and Lang 1920b, Lyon 1964, Høeg 1967) but more recent critical work has shown it to be more comparable to members of the Zosterophyllophytina (El-Saadawy and Lacey 1979a). Only the terminal parts of the plant are known, the fertile region consisting of irregular lax spikes of lateral kidney-shaped sporangia with transverse dehiscence. The anatomy is, however, more rhyniophyte-like, with an elliptical centrarch strand of elongate conducting cells but these lack any differential wall thickenings (and any indication of the reticulate patterns characteristic of *R. major*).

(14) *Renalia* Gensel
This late Lower Devonian genus is based on compression fossils: rounded to reniform sporangia terminate lateral dichotomously branching systems. Three species have been described, *R. hueberi* Gensel, *R. grabertii* Schweitzer and *R. major* Schweitzer. The latter is not considered here, as sporangia appear to be lateral rather than terminal (Schweitzer 1980b, pls. 13, 14). Most information comes from the type species. The sporangia split along the convex distal margin into two equal valves, the dehiscence line being marked by a row of thick-walled cells. The central strand of xylem comprises spiral to scalariformly thickened tracheids (Gensel 1976), but the position of the protoxylem remains unknown.

(15) *Hsüa* (Li and Cai) Li
Originally described as a *Cooksonia*, this Emsian monotypic genus comprises long (< 24 cm), pseudomonopodially branching main axes with dichotomously-branching, lateral axes terminating in round to reniform sporangia which dehisced into two valves (Li

1982). Axillary tubercles are present and the sterile axes have circinate tips. The centrarch strand of scalariform tracheids is described as centrarch.

(16) *Hicklingia* Kidston and Lang
Edwards (1976) demonstrated that the sporangia were not terminal, but borne laterally on short stalks and aggregated into lax spikes. Such an organization suggests affinity with the Zosterophyllophytina, but anatomical information is also required to substantiate such an assignation.

A Revaluation of the Rhyniophytina

1. Tracheids in early vascular plants

As a preamble to a consideration of the diagnosis and membership of the subdivision, it is necessary to deliberate on the nature of conducting tissues in early land plants. When Banks split the psilophytes between three subdivisions within the Tracheophyta, for most workers there existed very precise demarcations between that division, the Bryophyta and algae, and a clear-cut concept of the nature of xylem in vascular plants. However recent investigations on early land plants have prompted renewed debate on such issues.

The controversy concerning evidence for the existence of vascular plants in pre-Devonian strata has largely centred on interpretation of microfossil remains (e.g. Gray and Boucot 1977, 1978, Taylor 1982, Niklas and Smocovitis 1983); megafossils, particularly fertile specimens, have until recently, largely been accepted as vascular without question (as in Stewart 1983). However the realization that a variety of physiological strategies could have existed in the initial radiation of land plants (Raven 1977, Stebbins and Hill 1980) and the discovery of novel forms of presumed conducting cells (e.g. Hueber 1982, Fairon-Demaret 1985) suggests that uncritical acceptance of the nature of conducting cells of some early land plants may have been misplaced.

Tracheids are defined on the basis of their structure, function and biochemistry; only structure can be studied in most fossil plants. Morphologically, tracheids differ from the hydroids of mosses in the possession of a differentially deposited secondary wall which leads to the recognition of different types of thickenings. Biochemically, lignification of tracheids suggests that they would be resistant to compression and alteration during fossilization; indeed in many fossil plants tracheids are the only anatomical details preserved. On the other hand, as pointed out by Edwards and Fanning (1985), the detection of lignin residues in megafossil remains does not necessarily prove the existence of tracheids in the conducting strand, as outer cortical tissues rather than conducting tissues may have been lignified in some early land plants (Niklas 1976). Other plants, for example some marine algae and bryophytes, may synthesize phenolic substances which may degrade to produce lignin-like products and these further complicate conclusions based on biochemical studies of early land plants (Niklas and Smocovitis 1983, Thomas, this volume).

The existence of a non-tracheidal conducting strand in *R. major* suggests that morphologically similar plants may have had moss-like conducting strands and thus not fit into our present concept of bryophytes and pteridophytes. In addition, the existence of plants with a conducting system not composed of tracheids or hydroids is indicated by the finding of spirally-thickened tubes with unique structure in axes of *Taeniocrada dubia* Kräusel and Weyland by Hueber (1982). Most Upper

Table 13.2.

A. From Banks 1975	B. This Paper
Subdivision RYHYNIOPHYTINA	RHYNIOPHYTINA
Family RHYNIACEAE	RHYNIACEAE
(Type genus: *Rhynia*)	(Type genus: *Rhynia*)
Rhynia	*Rhynia (R. gwynne-vaughanii)*
Horneophyton	*'Taeniocrada' (T. decheniana)*
Cooksonia	*Renalia* (pro parte)
Steganotheca	
Salopella	*Hostinella* (pro parte)
Dutoitia	*Aphyllopteris* (pro parte)
Eogaspesiea	*Taeniocrada* (pro parte)
Questionable Rhyniophytina	'rhyniophytoid'
	Eogaspeseia
Taeniocrada	*Cooksonia*
Hicklingia	*Steganotheca*
Nothia	*Salopella/Eorhynia*
Yarravia	*Hedeia/Yarravia*
Hedeia	
	Questionable Rhyniophytina
	Dutoitia
	Hsüa
	Horneophyton
	Excluded genera
	Nothia
	Rhynia major
	Hicklingia

Silurian and Lower Devonian compression fossils show indications of the presence of a conducting strand (see Table 13.1) but very few, mainly sterile, axes can be demonstrated to contain tracheids (Edwards, Bassett and Rogerson 1979). The difficulty lies in deciding whether the absence of a tracheidal conducting strand is due to poor preservation or is a real feature of the plant. In certain assemblages, where tracheids have been demonstrated in some sterile axes, (e.g. the Gedinnian Targrove assemblage, Edwards and Fanning 1985) there may have been a mixture of pteridophytes and non-vascular pteridophyte-like plants and it may prove impossible to differentiate these morphologically.

2. *Composition of the Rhyniophytina*

The genera Banks (1975) included in the Rhyniophytina are noted in Table 13.2A. In his characterization of the subdivision he was forced to rely heavily on anatomical data from the two Rhynie Chert genera *Rhynia* and *Horneophyton* while his concept of the whole plant was based on the classical reconstructions of the two Rhynia species which appeared to parallel the organization seen in earlier forms e.g. *Cooksonia*. Thus the plants in the Rhyniophytina seemed to conform to the kind of organization expected in the earliest and presumed ancestral group of pteridophytes.

Following more recent investigations on the Rhynie Chert assemblage, *Rhynia major* and *Nothia* are no longer considered rhyniophytes. The lateral position of the sporangia alone would be sufficient to exclude *Nothia*, but in addition, although the conducting strand is centrarch, the nature of the conducting elements raises the very fundamental question as to whether or not the plant should be considered vascular. *R. major* is also removed from the rhyniophytes on these latter grounds. More information is needed on the gross morphology and anatomy of *Horneophyton*; in addition, its unique branching sporangium with predetermined apical dehiscence make it difficult to place in the rhyniophytes with confidence.

Considering *Rhynia gwynne-vaughanii*, the major advance has been the discovery of sporangial abscission and the overtopping that follows this process, two advanced characteristics in a plant with dichotomous and adventitious branching and simple vegetative anatomy.

There still remains a complex of megafossil compressions embodying the concept of rhyniophytes as simple plants with smooth mainly isotomously branching axes and terminal sporangia. These include *Cooksonia*, *Salopella*, *Eogaspesiea*, *Eorhynia*, *Steganotheca*, *Taeniocrada decheniana* and possibly *Dutoitia*. Many show central strands, but with the exception of *T. decheniana*, none have been proven vascular. Although it is tempting to blame the lack of tracheids on the limitations of preservation, recent experience demands caution. We therefore propose to describe such fossils as rhyniophytoid, a designation intended as an informal grouping within the Rhyniophytina. As far as we are aware, the term was introduced by Pratt *et al.* (1978) who used it in two ways. Thus they wrote (p. 137), 'Two general types of non-vascular plants present in the Ludlovian and/or Downtonian are the nematophytes and rhyniophytoid plants e.g. *Steganotheca'* and later on the same page (137) that 'Nematophytes, rhyniophytoid plants with vascular tissue . . . extend into the Devonian'. Edwards *et al.* (1983) further confused the issue by adding the phylogenetic connotation, that such plants could have been intermediates between earliest vascular plants and algal ancestors. Here we propose a practical compromise, that rhyniophytoid be used for plants that look like rhyniophytes, but cannot be assigned unequivocally to that group because of inadequate anatomical preservation.

Of the compression fossils Banks originally included under questionable Rhyniophytina *Taeniocrada (decheniana)* is the most completely known (Schweitzer 1980a), its most noteworthy feature being the mode of attachment of sporangium to subtending axis. More needs to be learned of *Hedeia*, but repeated overtopping in the fertile branch system suggests that its organization is not that far removed from that in *R. gwynne-vaughanii* and so we propose, with some reservations to retain it within the rhyniophytes as an advanced member of the group.

Two new genera described since 1975, *Renalia* (Gensel 1976) and *Hsüa* (Li 1982) are united by the development of marked overtopping in sterile axes producing a pseudomonopodial system with terminal sporangia on dichotomous lateral branches. The Siberian specimens described as *C. hemisphaerica* (Ananiev and Stepanov 1969) also have this organization. *Renalia*, frequently included as a questionable rhyniophyte (Gensel and Andrews 1984), has often been cited as an intermediate between rhyniophytes and zosterophylls, mainly because of the mode of dehiscence of its sporangia. For the moment, in the absence of three dimensional anatomy, we prefer to consider *Renalia* as an advanced rhyniophyte showing specialization in both sporangia and growth pattern, and have already found examples of plants showing intermediate levels of organization (see above) earlier in the Devonian. It is not unlikely that splitting into two equal valves in globose sporangia is an example of convergence in the two groups. In addition, there is some evidence that the structural modifications associated with dehiscence are different in *Renalia* and *Zosterophyllum* (Edwards 1969). *Hsüa* is more problematic: the reconstruction shows considerable lateral branching producing trusses of sporangia, more typical (except in sporangial shape) of the trimerophytes although the illustrations of the fossils appear to show fewer sporangia.

Thus we hold a wider concept of the Rhyniophytina than previous authors (Table 13.2B). We include in its plants with smooth axes, lacking well-defined spines or leaves, showing a variety of branching patterns that may be isotomous, anisotomous, pseudomonopodial or adventitious. Elongate to globose sporangia were terminal on main axes or on lateral systems showing limited branching. It seems probable that the xylem, comprising a solid strand of tracheids, was centrarch. We decided to exclude plants which did not possess conventional tracheidal thickenings, and are aware that this may appear to constitute an extreme example of overweighting. However we did so because we wish to emphasize the probability that early in the colonization of the land there were a number of 'evolutionary' experiments involving both the structure and functioning of water conducting tubes. Whether or not such plants should be considered vascular is discussed elsewhere (Edwards, D.S., in manuscript).

We see the Rhyniophytina emerging as a group encompassing a number of levels of organization ranging from the dichotomously branching simple or more generalized members to those that show a variety of derived characters both in sporangial characteristics and branching pattern. This is consistent with the adoption of a variety of strategies in growth and reproductive behaviour in the initial radiation of the vascular plants. The majority of problems encountered in the description and classification of the members reflect this unique evolutionary position.

Schweitzer (1983), on the other hand, favours the retention of the group Psilophyta subdivided into a number of families as used by Høeg (1967), arguing that largely due to heterobathmy there are too many plants showing a mixture of simple and advanced characters that are considered intermediates between the major subdivisions. Of the plants included here he uses *Renalia* and *Taeniocrada* to demonstrate the inadequacies of Banks' classification. Thus he illustrates a reduction series beginning with *T. decheniana* with terminal sporangia through *T. longisporangiata* to *T. langii* with lateral sporangia. Apart from our doubts on the validity of the taxa used by Schweitzer, we should like to raise two more general issues that stem from his conclusions. First in extending our concept of the Rhyniophytina, we neither find it necessary nor desirable to fit every species into a lineage. Secondly that such is the nature of the evolutionary process that there will always be plants showing a mosaic of characters, some being intergrading forms: these should be welcomed. It is the task of the taxonomist to accommodate them, and perhaps in meticulously accumulating data, workers on Silurian and Lower Devonian plants have paid too little attention to taxonomic theory and to processes involved in handling their data. In this case we have enlarged our concept of the Rhyniophytina to include the more complex forms and feel confident that given more information it will eventually be possible to delimit families within it. We do not feel that their existence provides good grounds for dismantling what to us are 'good' taxa—the Zosterophyllophytina and Trimerophytina; taxa whose concepts have been strengthened by recent advances in knowledge.

Acknowledgements

We thank Drs M. Fairon-Demaret and J. Tims for access to unpublished data and Professor W.S. Lacey for stimulating discussion. D.S.E. is grateful to the British Council for financial support.

References

Ananiev, A.R. and Stepanov, S.A. (1969) The first finding of the *Psilophyton* flora in the Lower Devonian Salairskij Ridge (western Siberia). In *Treatises of the Tomsk Order of the Worker's Red Banner*, State University, Tomsk: 203, 13–28. (in Russian).

Anderson, J. and Anderson, H. (1985) *Prodomus of South African megafloras, Devonian to Lower Cretaceous*. A.A. Balkema, Rotterdam.

Andrews, H.N. (1964) *Studies in paleobotany*. John Wiley, New York.

Banks, H.P. (1968a) The early history of land plants. In *Evolution and environment* (ed. E.T. Drake), pp. 73–107, Yale University Press, New Haven and London.

—— (1968b) Anatomy and affinities of a Devonian *Hostinella*. *Phytomorphology* **17**, 321–30.

—— (1975) Reclassification of Psilophyta. *Taxon* **24**, 401–13.

—— (1980) Floral assemblage zones in the Siluro-Devonian. In *Biostratigraphy of fossil plants: successional and palaoecological analysis* (eds. D. Dilcher and T.N. Taylor), pp. 1–24, Stroudsburg: Dowden, Hutchinson and Ross.

Bhutta, A.A. (1973) Observations on the sporangia of *Horneophyton lignieri* (Kidston and Lang) Barghoorn and Darrah 1938. *Pak. J. Bot.* **4**, 27–34.

Cookson, I.C. (1935) On plant-remains from the Silurian of Victoria, Australia, that extend and connect floras hitherto described. *Phil. Trans. R. Soc.* **225B**, 127–48.

Croft, W.N. and Lang, W.H. (1942) The Lower Devonian flora of the Senni Beds of Monmouthshire and Breconshire. *Phil. Trans. R. Soc.* **B231**, 131–63.

Daber, R. (1960) *Eogaspesiea gracilis* n.g. n.sp. *Geologie* **4**, 418–25.

Edwards, D. (1969) Further observations on *Zosterophyllum llanoveranum* from the Lower Devonian of South Wales. *Am. J. Bot.* **56**, 201–10.

—— (1970) Fertile Rhyniophytina from the Lower Devonian of Britain. *Palaeontology* **13**, 451–61.

—— (1976) The systematic position of *Hicklingia edwardii* Kidston and Lang. *New Phytol.* **76**, 178–81.

—— (1979) A late Silurian flora from the Lower Old Red Sandstone of south-west Dyfed. *Palaeontology* **22**, 23–52.

—— (1980a) Early land floras. In *The terrestrial environment and the origin of land vertebrates*. (ed. A.L. Panchen), Systematics Association Special Vol. 15, pp. 55–85, Academic Press, London.

—— (1980b) Studies on Lower Devonian petrifactions from Britain. 1. Pyritised axes of *Hostinella* from the Brecon Beacons Quarry, Powys, South Wales. *Rev. Palaeobot. Palynol.* **29**, 189–200.

—— Bassett, M.G. and Rogerson, E.C.W. (1979) The earliest vascular land plants: continuing the search for proof. *Lethaia* **12**, 313–24.

—— and Fanning, U. (1985) Evolution and environment in the late Silurian-early Devonian: the rise of the pteridophytes. *Phil. Trans. R. Soc. Lond.* **B309**, 147–65.

——, Feehan, J. and Smith, D.G. (1983) A late Wenlock flora from Co. Tipperary, Ireland. *Bot. J. Linn. Soc.* **86**, 19–36.

—— and Richardson, J.B. (1974) Lower Devonian (Dittonian) plants from the Welsh Borderland. *Palaeontology* **17**, 311–24.

Edwards, D.S. (1980) Evidence for the sporophytic status of the Lower Devonian plant *Rhynia gwynne-vaughanii* Kidston and Lang. *Rev. Palaeobot. Palynol.* **29**, 177–88.

Eggert, D.A.(1974) The sporangium of *Horneophyton lignieri* (Rhyniophytina). *Am. J. Bot.* **61**, 405–13.

El-Saadawy, W. El-S. and Lacey, W.S. (1979a) Observations on *Nothia aphylla* Lyon ex Høeg. *Rev. Palaeobot. Palynol.* **27**, 119–47.

—— and —— (1979b) The sporangia of *Horneophyton lignieri* (Kidston and Lang) Barghoorn and Darrah. *Rev. Palaeobot. Palynol.* **28**, 137–44.

Fairon-Demaret, M. (1985) Les plantes fossiles de l'Emsien du Sart Tilman, Belgique. 1. *Stockmansia langii* (Stockmans) comb nov. *Rev. Palaeobot. Palynol.* **44**, 243–60.

Gensel, P.G. (1976) *Renalia hueberi*, a new plant from the Lower Devonian of Gaspe. *Rev. Palaeobot. Palynol.* **22**, 19–37.

—— and Andrews, H.N. (1984) *Plant life in the Devonian*. Praeger, New York.

Gray, J. and Boucot, A.J. (1977) Early vascular land plants: proof and conjecture. *Lethaia* **10**, 145–74.

—— and —— (1978) The advent of land plant life. *Geology* **6**, 489–92.

Halle, T.G. (1916) Lower Devonian plants from Roragen in Norway. *K. svenska Vetensk Akad.* **57**, 1–46.

Heard, A. (1939) Further notes on Lower Devonian plants from South Wales. *Q. Jl. geol. Soc. Lond.* **95**, 223–9.

Hébant, C. (1977) *The conducting tissue of bryophytes*. J. Cramer, Vaduz.

Høeg, O.A. (1930) A psilophyte in South Africa. *Kgl. Norske Vid. Selsk. Forh.* **3**, 92–4.

—— (1967) Psilophyta. In *Traité de Paléobotanique* (ed. E. Boureau) Vol. 2, pp. 191–352, Masson et Cie, France.

Hueber, F.M. (1982) *Taeniocrada dubia* Kr. and W.: its conducting strand and helically strengthened tubes. Abstract. *Am. J. Bot.* **69**, 58–9.

—— (1983) A new species of *Baragwanathia* from the Sextant Formation (Emsian), northern Ontaria. *Bot. J. Linn. Soc.* **86**, 57–79.

Hughes, N.F. (1981) 'cf' symbol of an outmoded data-handling system. *I.O.P. Newsletter* **14**, 5–6.

Ishchenko, T.A. (1969) The *Cooksonia* palaeoflora in the Skal'skii horizon of Podolia and its stratigraphical significance. *Geological Journal.* **29**, 101–9 (in Russian).

—— (1975) *The Late Silurian flora of Podolia.* 'Dymka' Scientific Publishing House, Kiev (in Russian).

Kidston, R. and Lang, W.H. (1917) On Old Red Sandstone plants showing structure from the Rhynie Chert Bed, Aberdeenshire. Part 1. *Rhynia Gwynne-Vaughanii. Trans. R. Soc. Edinb.* **51**, 761–84.

—— and —— (1920a) On Old Red Sandstone plants showing structure from the Rhynie Chert Bed, Aberdeenshire. II. Additional notes on *Rhynia Gwynne-Vaughani*, Kidston & Lang, with description of *Rhynia major*, n. sp. and *Hornea Lignieri* n.g. n.sp. *Trans R. Soc. Edinb.* **52**, 603–27.

—— and —— (1920b) On Old Red Sandstone plants showing structure, from the Rhynie Chert Bed, Aberdeenshire. III. *Asteroxylon mackiei*, Kidston and Lang. *Trans. R. Soc. Edinb.* **52**, 643–80.

—— and —— (1921) On Old Red Sandstone plants showing structure from the Rhynie Chert Bed, Aberdeenshire. Part IV. Restorations of the vascular cryptogams, and discussion of their bearing on the general morphology of the Pteridophyta and the origin of the organisation of land plants. *Trans. R. Soc. Edinb.* **52**, 831–54.

Kräusel, R. and Weyland, H. (1930) Die Flora des deutschen Unterdevons. *Abh. Preuss. Geol. Landes. Neue Folge* **131**, 1–92.

Lang, W.H. (1937) On the plant remains from the Downtonian of England and Wales. *Phil. Trans. R. Soc. Lond.* **227B**, 245–91.

—— and Cookson, I.C. (1935) On a flora, including vascular land plants, associated with *Monograptus*, in rocks of Silurian age, from Victoria, Australia. *Phil. Trans. R. Soc.* **224B**, 421–49.

Lemoigne, Y. and Zdebska, D. (1980) Structures problematiques observées dans des axes provenant du Chert Dévonien de Rhynie. *Acta Palaeobotanica*, **21**, 3–8.

Li, Cheng-Sen. (1982) *Hsüa robusta*, a new land plant from the Lower Devonian of Yunnan, China. *Acta phytotax. sin.* **20**, 331–42.

Lyon, A.G. (1964) Probable fertile region of *Asteroxylon mackiei* K. and L. *Nature* **203**, 1082–3.

Niklas, K.J. (1976). The role of morphological biochemical reciprocity in early land plant evolution. *Ann. Bot.* **40**, 1239–54.

—— and Smocovitis, V. (1983) Evidence for a conducting strand in early Silurian (Llandoverian) plants: implications for the evolution of land plants. *Paleobiology* **9**, 126–37.

——, Tiffney, B.H. and Knoll, A.H. (1980) Apparent changes in the diversity of fossil plants. A preliminary assessment. In *Evolutionary Botany*, (eds. M.K. Hecht, W.C. Steere and B. Wallace), Vol. 12, pp. 1–89. Plenum press, New York and London.

——, —— and —— (1983) Patterns in vascular land plant diversification. *Nature, Lond.* **303**, 614–6.

Obrhel, J. (1962) Die Flora der Pridoli-Schichten (Budnany-Stufe) des mittelböhmischen Silurs. *Geologie* **11**, 83–97.

Pratt, L.M., Phillips, T.L. and Dennison, J.M. (1978) Evidence of non-vascular land plants from the early Silurian (Llandoverian) of Virginia, U.S.A. *Rev. Palaeobot. Palynol.* **25**, 121–49.

Raven, J.A. (1977) The evolution of vascular land plants in relation to supra-cellular transport processes. In *Advances in Botanical Research* (ed. H.W. Woodhouse), Vol. 5, pp. 154–219. Academic Press, London.

Satterthwait, D.F. and Schopf, J.W. (1972) Structurally preserved phloem zone tissue in *Rhynia*. *Am. J. Bot.* **59**, 373–6.

Schweitzer, H.J. (1980a) Die Gattungen *Taeniocrada* White und *Sciadophyton* Steinmann im Unterdevon des Rheinlandes. *Bonner Paläobotanische Mitteilungen* **5**, 1–38.

—— (1980b) Die Gattungen *Renalia* Gensel und *Psilophyton* Dawson im Unterdevon des Rheinlandes. *Bonner Pal*äobotanische Mitteilungen **6**, 1–34.

—— (1983) Die unterdevon Flora des Rheinlandes. *Palaeontographica* **189B**, 1–138.

Stebbins, G.L. and Hill, C.J.C. (1980) Did multicellular plants invade the land? *American Naturalist*, **115**. 342–53.

Stewart, W.N. (1983) *Paleobotany and the evolution of plants*. Cambridge University Press, Cambridge.

Taylor, T.N. (1982) The origin of land plants: a paleobotanical perspective. *Taxon* **31**, 155–77.

Tims, J.D. and Chambers, T.C. (1984) Rhyniophytina and Trimerophytina from the early land flora of Victoria, Australia. *Palaeontology* **27**, 265–79.

White, D. (1903) Description of fossil algae from the Chemung of New York with remarks on the genus *Haliserites* Sternberg. *New York State Mus. Ann. Rep.* 1901, **55**, 593–605.

Yurina, A.L. (1969) The Devonian floral of Central Kazachstan. *Materialy po Geologii Sentralnogo Kazakhstana* **8**, 1–208 (in Russian).

14. Comparative leaf architectural analysis of Cretaceous radiating angiosperms

R.A. SPICER

Abstract

Individuals of any group of organisms undergoing an initial radiation exhibit a plexus of morphological characters, which later in the evolution of the group are more discretely partitioned between taxa. During the initial radiation populations of whole organisms, or isolated organs, exhibit a continuum of form and the identification of natural taxa is often confounded. The palaeobotanical use of Linnean Systematics and the strictures of the Code compound these difficulties by imposing a hierarchy when none may be justified, and by perpetuating erroneous implied relationships.

In the course of studying mid Cretaceous leaves an alternative scheme has been devised that is independent of, but parallel to, Linnean Systematics. Individual specimens, described in detail, are grouped into morphological forms with alpha-numeric identifiers. Because of sampling limitations stratigraphic ranges are poorly known for macrofossils. Therefore these are not an integral part of the form descriptions. Transfer of specimens between forms is facilitated or forms abandoned or created should new information become available. This scheme allows the flexible categorizing of specimens until sufficient data are accumulated which would allow more reliable conventional taxonomic assignment.

Introduction

For biostratigraphic purposes the most useful material is that which is abundant, widespread, with limited stratigraphic range, and is not prone to reworking. Superficially, angiosperm leaf fossils meet all these criteria and have an additional advantage over palynomorphs in that they display inherent characteristics that appear to be directly correlated with environment, and climate in particular. Provided that facies bias due to depositional and preservational sorting and community structure within the source vegetation are taken into account, the climate signal itself can have biostratigraphic relevance. The large numbers of specimens utilized in biostratigraphic studies often reveal complex patterns of morphological change that can be often interpreted in evolutionary terms. However, it is precisely under these circumstances, when evolutionary changes are most clearly demonstrated, and when systematic schemes are most needed, that our present nomenclatural and taxonomic practices prove inadequate. In particular there are a number of nomenclatural and taxonomic problems that hinder the biostratigraphic use of mid and Late Cretaceous angiosperm leaf material.

The Late Cretaceous was a time when great vegetational changes were taking place on a global scale as the angiosperms underwent a major geographic and evolutionary radiation. In spite of climatic equability the Late Cretaceous saw considerable dynamic floristic provincialism as angiosperms evolved, spread, and often outcompeted non-angiospermous plants. It is now well established that mid and early Late Cretaceous angiosperm leaf forms exhibit increasing levels of vein organization and morphological diversity as they are recovered from sequentially younger rock units (Doyle and Hickey 1976, Hickey and Doyle 1977).

Characteristically, early Late Cretaceous angiosperm leaves possess a mélange of features which today are seen to be discretely partitioned between families or sometimes even orders. Populations of leaves may be assigned to artificial groupings based on architectural characters, but within these there is often a continuum of form.

At a very broad level complexes can be identified that appear to have some biological coherence in that they show consistent facies associations over broad geographic areas. Thus palmately veined forms with platanoid teeth (together with other characters) comprise the riparian (platanoid-like) communities, and the entire-margined, pinnately organized camptodrome forms comprise the interfluvial understorey (magnoliid-like) associations. Within these and other broad groupings, however, no clearly defined natural discontinuities exist that can be used as the basis for a first approximation to a natural classification.

Hybridization

The apparent homogeneity of the broad mid Cretaceous complexes could have been the result of morphological overlap between smaller scale breeding units. Such overlap and depositional mixing of organs derived from more than one breeding unit render it unlikely that the units could be recognized as discrete from the fossil record and the resulting morphological continuum would be similar to that observed. In some instances these smaller breeding complexes may have been equivalent to modern species, but this need not have been the case (Hughes 1976).

Alternatively, the fossil record may be reflecting true broad-scale intergradation of form resulting from the widespread hybridization expected of groups undergoing rapid evolution. Early in the evolution of a group genetic differences between individuals or populations are likely to be less than they would be after some divergence has taken place. This may be translated into a degree of morphological similarity that will confound the recognition of morphological or typological (*sensu* Mayr 1957) species. However, in order to understand this phenomenon more fully we have to consider the likely nature of evolutionary processes during a major radiation.

As well as the typological species Mayr (1957) recognized biological species which according to Dobzhansky (1970) are systems of populations between which gene exchange is limited or prevented by one or more reproductive isolating mechanisms. Although sympatric speciation processes are known to occur (for example by autopolyploidy), geographic of ecological isolation has been considered of major importance in restricting gene flow between populations. During periods of isolation inherent fertility barriers between populations become established by mutation, selection, and stochastic processes that alter the genotypes such that when the populations become sympatric once again, either by removal of a physical barrier or by range extension, successful interbreeding is prevented.

In modern populations fertility barriers need not be absolute to bring about eventual speciation. Selection against hybrid individuals is often strong because the combination of characters possessed by the hybrid is likely to render it less well adapted to its environment than either of its parents. This is the so-called Wallace Effect (Grant 1967) and serves to heighten weak fertility barriers and maintain the coherence of established species. However, phenomena observed in today's specialized and adapted plants may not reflect what was happening during the major angiosperm radiation. During the geographic spread of the angiosperms populations would have been initiated, and to some extent maintained, by the competitively successful combinations of features by which they are now characterized. To some extent their characteristics enabled them to escape from selection. Early angiosperm populations would have become fragmented not by the formation of physical geographic barriers but by chance long distance dispersal (Cretaceous seeds were mostly small; Tiffney 1984) initiating new colonies and by the geographic spread of other already genetically different angiosperm populations. Allopatric speciation would undoubtedly occur, but early in the radiation there would have been a greater degree of genetic homogeneity than seen today which, in conjunction with low levels of specialization in reproductive morphologies and strategies (Dilcher 1978), is likely to have resulted in high frequencies of successful hybridization.

Widespread frequent hybridization might be expected to slow down rates of evolution in that it blends character combinations. However, during early phases of major radiations the opposite may be true. By means of hybridization novel character combinations may arise that are better suited to their particular environment than those of either parent. The hybrid population may then outcompete the parental populations and eventually displace them. Nearer the origin of a novel group the parents may not be so well adapted as their descendants are later on: there is 'more room for improvement' and higher frequencies of successful hybrids are likely to occur. The plants are, as a whole, adaptable rather than adapted. The improved adaption of the hybrid populations to their specific environments eventually may lead to population segregation, isolation, and more well defined boundaries to both typological and biological species.

Naturally arising polyploids may or may not be morphologically similar to their ancestral diploids, but they are usually reproductively isolated from them. Populations of polyploids may be maintained by apomixis but may hybridize with other populations. This combination of polyploidy, apomixis, and hybridization is the cause of many problems in taxonomy of living angiosperms (as well as other groups) and are usually cited as noteworthy examples of evolution in action. For example, studies of European members of the genus *Sorbus* show dramatic variations in leaf organization between sexually reproducing species and intermediate apomictic forms, putatively the result of hybridization and polyploidy (Fig. 14.1) (Briggs and Walters 1984).

Under these circumstances even the biological species concept (which cannot by definition apply to apomictic populations) breaks down. Our concept of what constitutes the species in an evolutionary context has to be flexible to accommodate the complex, often reticulate evolutionary process itself. Neobotanists like Levin (1979) and Heywood (1980) are of the opinion that no single species definition can, nor should, be proposed because conceptual categories tend to obscure the actual pattern of evolutionary process. It has often been found in practice that groups of plants undergoing evolutionary change cannot be adequately accommodated within conventional taxonomy and have to be treated as special cases. Levin (1979) agrees with Cronquist (1978) that the purpose of taxonomy is to impose a mentally pleasing organization upon natural diversity rather than a framework that reflects phylogeny. This rather

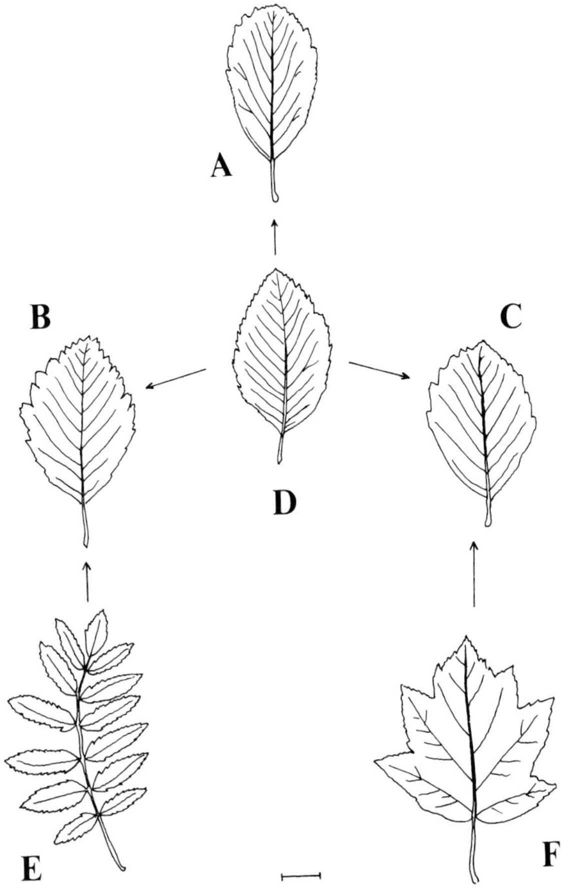

Fig. 14.1. Leaf variation in relation to sexual reproduction, hybridization and apoximis in *Sorbus*.

A. *S. rupicola*	apomictic
B. *S. anglica*	apomictic
C. *S. bristoliensis*	apomictic
D. *S. aria*	sexual
E. *S. aucuparia*	sexual
F. *S. torminalis*	sexual

Scale bar represents 2 cm. Redrawn from Briggs and Walters (1984).

pragmatic view among neobotanists ought to be somewhat disturbing to palaeobotanists, but fortunately or otherwise the inadequacy of our data often shields us from uncomfortable realities.

Phenotypic plasticity

Morphological plasticity in leaves is particularly pronounced in many angiosperms. It is an advantageous characteristic that improves photosynthetic efficiency. Measuring the extent of this plasticity from fossil material and separating it from possible variability due to hybridization and polyploidy is extremely difficult, if not impossible. Tracing the evolution of plasticity itself is therefore a significant problem.

Cretaceous leaves

Whether due to overlapping population samples, hybridization, polyploidy, or plasticity (or varying combinations of these phenomena) Cretaceous leaf fossils exhibit considerable morphological variation and intergradation (Fig. 14.2). This was noted by pioneers in the field of angiosperm palaeobotany, such as Lesquereux (1982) and Hollick (1930). For practical reasons they attempted to assign binomials to specimens, even though they were aware of the inadequcy of this approach, even suspecting that several 'genera' of leaves may have come from the same individual plant (Lesquereux, 1892). Lesquereux (1892) was driven to abandon the strict application of the binomial system, and adopted names incorporating 'double affinity'. For example, some leaves are referred to as *Sassafras (Araliopsis) platanoides*. These early workers were so impressed with the superficial resemblance of the fossils to modern taxa that they often assigned Cretaceous specimens to living genera as in the example just given. This eventually led to serious misconceptions regarding the angiosperm evolution (see Doyle and Hickey 1976) and has left a legacy of inappropriate names and groupings of forms that have obscured genuine evolutionary pattern, stratigraphic utility, and palaeoenvironmentally useful information.

The existence of a formal nomenclatural system to 'legitimize' specimens encourages premature taxonomic assignments which increase the likelihoold that subsequent revisions would be necessary. That same nomenclatural system subsequently tends to restrict amendments and to perpetuate inappropriate names through the principle of priority. In biostratigraphic work large numbers of specimens need to be examined and related one to another. To repeatedly revise taxa on a formal basis is both time-wasting and expensive.

For theoretical and practical reasons formal nomenclatural and conventional taxonomic procedures appear to be inappropriate for Cretaceous angiosperm leaf studies. However, some kind of framework is required by means of which specimens may be referred to unambiguously and categorized, however loosely, while data are being

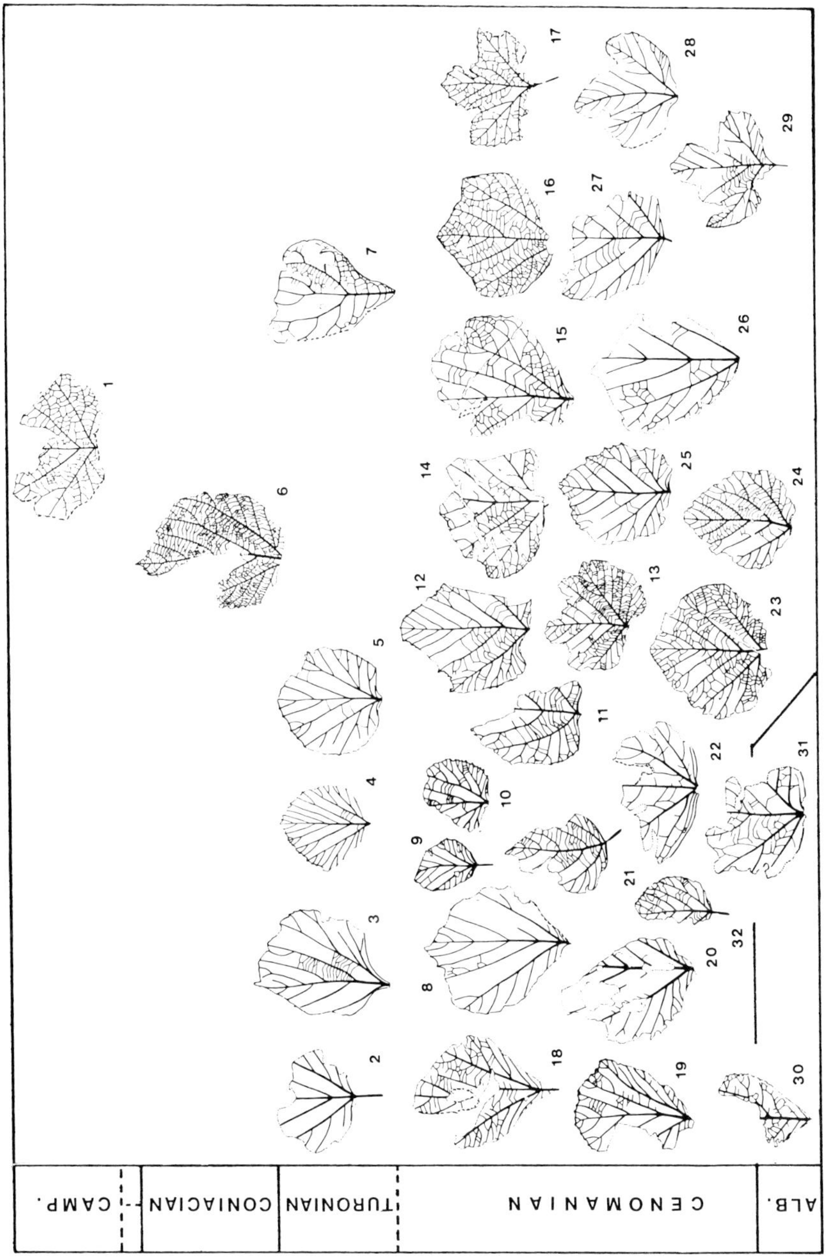

Fig. 14.2. Illustration of the morphological variation and intergradation of form seen in Late Cretaceous 'platanoid-like' leaves from Alaska.

accumulated for both biostratigraphic and evolutionary studies and from which specimens or groups of specimens may be transferred to the conventionl taxonomic system when the risk of wholesale and frequent taxonomic revision is deemed to be acceptably low.

The Form System

One system that has proved useful in practice for Cretaceous leaves has been to group specimens not into species but into forms (Spicer, work in progress). Individual specimens are assigned unique identifiers consisting of a locality identifier (usually a number) separated from a specimen number by a decimal point (Fig. 14.3). The specimen number is unique within a locality. The system of identifying localities differs from institution to institution. For this reason the locality identifier is prefixed by an alphabetic institution code. Thus the complete specimen identifier could indicate where the specimen came from (locality) and where it is kept. This system also allows specimens to be uniquely identified without the need for allocation of specimen numbers of an international basis.

Specimens are then grouped into forms on the basis of shared suites of architectural characters. Forms are allocated numbers which in themselves have no significance other than to uniquely lable the forms. A form consists of a single specimen or several specimens. Specimens can be freely moved from one form to another as patterns of morphological relationships (which may or may not reflect phylogenetic relationships) develop with the accumulating data. A form has no formal written definition but is defined by the specimens it contains. A form ceases to exist only when all specimens have been transferred out of it.

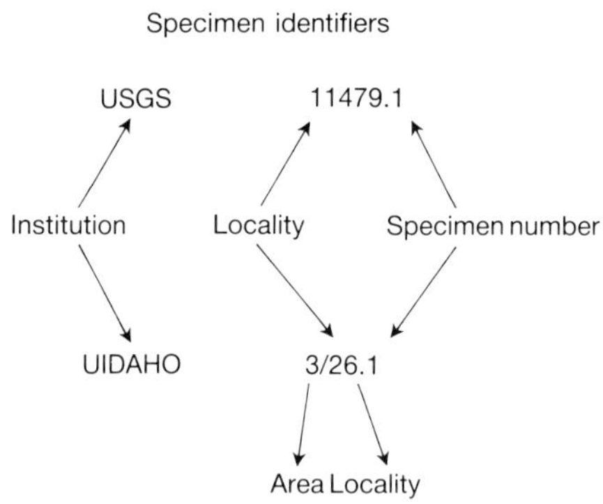

Fig. 14.3. An example of specimen identifiers composed of information as to where the specimen is kept, its origin (locality), and a number unique within the locality. Such identifiers do not prejudice subsequent taxonomic assignments.

Groups of forms that are morphologically similar are allocated alphabetic codes that reflect their major architectural characteristics. In a recent study of Alaskan leaves (Spicer, in press) the codes used to identify groups of forms were, in many cases, derived from modern subclass names (broadly based on Takhtajan 1969) not because of any supposed phylogenetic relationship, but merely because the general character combinations were readily comprehended from familiarity with the extant leaf forms. In cases where no suitable modern subclass existed the form identifier was constructed solely from the description of the architectural features themselves. In some ways this is the more preferable system. Thus leaves with RAdiating PEctinal veins (lateral primaries of Hickey, 1973) were grouped into a form designated as RAPE. There are no rules for constructing the codes for labelling the groups. 'Proper' names could be assigned if desired, although these names should be clearly distinguishable from more conventional binominals. The important thing is that the groups consist of forms that share a declared similarity. In the Alaskan study these declarations were kept simple and broad to accomodate the observed morphological variation as in the following examples:

Angiospermous Leaves:

Code	Description	Equivalence
MA	Magnoliidae-like: camptodromous entire margined with intersecondary veins but lacking any pectinal veins.	Broadly equivalent to Laurophylls of Krassilov (1977) (MA and MAPE)
MAPE	Magnoliidae-like: as above but with pectinal veins, usually acrodromous.	
RAPE	Nymphaealaen-like: peltate or cordate with numerous radiating pectinal veins; similar modern leaves are included in the Magnoliidae by Hickey and Wolfe (1975).	Equivalent to Nymphaeaphylls of Krassilov (1977)
HATR	Trochodendroidean hamamelididaean-like: cordate to ovate leaves usually with crenate (but may be finely serrate to entire) margins, acrodromous pectinal veins.	Equivalent to Trochodendroides of Krasilov (1977)
HAPLP	Hamamelididaean-like leaves with Platanoid teeth and peltate bases.	Equivalent to Platanophylls of Krassilov (1977) (HAPLP, HAPLTL and HAPLD)
HAPLTL	Hamamelididaean-like leaves with Platanoid teeth and truncate to lobate bases.	
HAPLD	Hamamelididaean-like leaves with Spinose teeth and decurrent bases.	
HASPD	Hamamelididaean-like leaves with Spinose teeth and decurrent bases.	

RO	Rosidaean-like leaves.	Equivalent to Rosiphylls of Krassilov (1977)
DI	Dilleniidaean-like leaves.	
DIPE	Dilleniidaean-like leaves with pectinal veins	

The use of forms allows specimens to be grouped together in a temporary 'holding pattern' while the relationships between specimens are being elucidated. These relationships might be purely artificial, based on morphological or architectural similarity, or there may be reason to believe the groupings may reflect biological taxa with greater or lesser degrees of phylogenetic significance. The use of the word 'species' instead of 'form' would be counterproductive and confusing in that it would have too many unwarranted connotations. A form may or may not be regarded as having a similar taxonomic or biological standing to species, but this would depend largely on what was known about the specimens presently included in the particular form. Ultimately, possibly when there was a consensus to do so among several palaeobotanists, specimens within a form could be described under the umbrella of conventional systematics and the form designation would give way to a more formal taxonomic rank.

Conclusions

We have little direct knowledge of the evolutionary processes that take place during a major radiation. Our understanding of these processes is based on extrapolation of short-term studies of living taxa carried out against an evolutionary environment that may be very different to that of a major radiation. Our only direct evidence of what has happened is the fossil record. Apart from the inadequacies of preservational sampling and specimen recovery, our interpretation of fossil information is hindered, distorted, and often obscured by an inappropriate systematic, taxonomic and nomenclatural framework.

The pressure to name a specimen first and understand it second is very strong. There is a considerable temptation to squeeze fossil material into extant taxa which inevitably increases the antiquity of modern groups and obscures evidence of evolutionary change. This is particularly true for angiosperms. The conventional taxonomic system as currently applied is inappropriate for handling specimens which represent isolated organs from individuals which are a part of a plexus of individuals in an evolutionary mélange.

There is therefore, in my opinion, a strong case for alternative, parallel, non-Linnean taxonomic systems that allow for more flexible specimen groupings. Individual specimens are the ultimate source of reference, and should be uniquely identified in a neutral fashion that will not prejudice any taxonomic assignments. These alternative schemes, and the form system herein described is but one example, must allow for specimens to be informally grouped and transferred between groups as relevant data are accumulated and evaluated. They should provide a temporary 'holding pattern' for specimens and an experimental 'scratch pad' for testing artificial or natural relationships. Only after considerable investigation, analysis, and even experimentation when groupings have been demonstrated to be robust in a variety of circumstances, should transfer be made to the Linnean system. The adoption of this approach would reduce the necessity for frequent, time-consuming, and nomenclaturally confusing taxonomic revisions, the perpetuation of misleading and inappropriate names, and even periodic revisions of the *International Code of Botanical Nomenclature.*

References

Briggs, D. and Walters, S.M. (1984) *Plant Variation and Evolution* (Second Edition). Cambridge University Press, Cambridge.

Cronquist, A.(1978) Once again, what is a species? In *Biosystematics in Agriculture* (ed. J.A. Romberger), pp. 3-20, Allanheld, Osmun and Co., Montclair, N.J.

Dilcher, D.L. (1978) Early Angiosperm reproduction, an introductory report: *Rev. Paleobot. Palynol.* **27**, 291–328.

Dobzhansky, T. (1970) Genetics of the Evolutionary Process. Columbia University Press, New York. 505 pp.

Doyle, J.A. and Hickey, L.J. (1976) Pollen and leaves from the mid Cretaceous Potomac Group and their bearing on early angiosperm evolution. In *Origin and Early Evolution of Angiosperms* (ed. C.B. Beck), pp. 139-206, Columbia University Press, New York.

—— Jardiné, S. and Doerenkamp, A. (1982) *Afropollis*, a new genus of early angiosperm pollen, with notes on the Cretaceous palynostratigraphy and palaeoenvironments of northern Gondwana: *Bulletin Centres Recherche Exploration Prod. Elf-Aquitaine.* **6**, 39–117.

Grant, V. (1967) The selective origin of incompatibility barriers in the plant genus *Gilia*: *Am. Nat.* **100**, 99–118.

Heywood, V.H. (1980) The Impact of Linnaeus on botanical taxonomy—past, present and future: *Veroffentlichungen der Joachim Jungius-Gesellschaft der Wissenschaften*, Hamburg **43**, 97–115.

Hickey, L.J. (1973) Classification of the architecture of dicotyledonous leaves. *Am. J. Bot.* **80**, 17–33.

—— and Doyle, J.A. (1977) Early Cretaceous fossil evidence for angiosperm evolution: *Bot. Rev.* **43**, 3–104.

—— and Wolfe, J.A. (1975) The bases of angiosperm phylogeny, Vegetative Morphology. *Ann. Mo. bot Gdn*. **62**, 538–89.

Hollick, A. (1930) The Upper Cretaceous floras of Alaska. *U.S. Geological Survey Professional Paper* **159**, 1–123.

Hughes, N. (1976) *Palaeobiology of angiosperm origins*. Cambridge University Press, Cambridge.

Kovach, W.L., and Dilcher, D.L. (1984) Dispersed cuticles from the Eocene of N. America. *Botanical Journal of the Linnean Society (Lond)* **88**, 63–104.

Krassilov, V.A. (1977) The Origin of angiosperms. *Bot. Rev*. **43**, 143–176.

Lesquereux, L. (1982) The flora of the Dakota Group. *U.S. Geological Survey Monograph* **17**, 1–400.

Levin, D.A. (1979) The nature of plant species: *Science* **204**, 381–4.

Mayr, E. (1957) Species concepts and definitions. In *The species Problem* (ed. E. Mayr), pp. 1–22, American Association for the Advancement of Science Publication 50.

Takhtajan, A. (1969) *Flowering Plants, origin and dispersal*. Oliver and Boyd, Edinburgh.

Tiffney, B.H. (1984) Seed size, dispersal syndromes and the rise of the angiosperms. Evidence and hypothesis 1. *Ann. Mo. bot Gdn*. **71**, 551–76.

15. The problems of data-handling for early angiosperm–like pollen

N.F. Hughes

Abstract

Evidence for evolutionary succession of the earliest angiosperms must depend almost entirely on fossil pollen distribution data because many more specimens are available. Pollen, like leaves, seeds or even flowers, represents only one organ of a plant otherwise unknown. Only rarely can two organs be recorded from the same bed and thus be inferred as belonging to the same plant.

Current methods of handling dispersed pollen data are unsatisfactory both in lack of precision and unnecessary loss of usable data. An Alternative Proposal is made in three steps for data-handling to avoid all the recognized difficulties; these relate to (1) Records of occurrence, (2) Definitions of taxa, and (3) Classification for fossils. All of the proposals apply equally to megafossils and pollen.

Using these proposals, it is suggested that the angiosperm ancestor problem may now be approached and solved by a systematic and sustained search for mid-Cretaceous dispersed pollen.

Introduction

The purpose of detailed study of the dispersed very small grains of angiosperm–like pollen of Cretaceous age, is the elucidation of the origin and early evolution of angiospermous plants. Regrettably it is most unlikely that megafossils of these plants will be found in the future in a significantly greater number of localities than are already known, although the quantity of fossils from each locality and the quality of interpretation of them will clearly improve with advancing knowledge. The chance distribution of these megafossil localities in time and space will remain a severe constraint on interpretation of evolution. The dispersed pollen, however, is obtainable from very many of the numerous strata occurring between the megafossil localities both in rock successions and in palaeogeography, and consequently it will provide the predominant and usually the critical evidence.

The problems relate to making satisfactory 'alternative' arrangements for the most efficient use of this available and abundant palynologic data at all stages of data-handling and interpretation. The methods of data-handling in current use were devised through history and through other special considerations and have now reached a development barrier which appears to be insurmountable without major changes of procedure. The suggestions made below have been devised primarily for the purpose of solving the so-called 'mystery' of early angiosperm evolution, although any such solution will also include improvement of techniques for stratigraphic correlation in general as part of the necessary action. All of the suggestions made are compatible with continued use of past data, and there is no reason why application of these methods should not be used for the solution of many other palaeobiologic problems. The suggestions amount to an 'Alternative Approach' to taxonomy and plant classification as called for in the Chapter titles within this volume.

Search for ancestors: causes of failure

This is admittedly a view with the benefit of hindsight, but the two main causes of failure for over 100 years appear now to have been the lack of proper attention to stratigraphic succession, and the many uninspired 'uniformitarian' reconstructions of Mesozoic fossils based on very inadequate evidence. Sometimes no reconstruction was actually drawn but as in the case of Mesozoic fan-shaped leaves resembling those of living *Ginkgo biloba* L., the name *Ginkgo* was deliberately used for fossils implying the presence of a whole plant for which no evidence whatever existed beyond the leaf. This was particularly unfortunate because the living genus *Ginkgo* is monospecific, and the Mesozoic has now proved to contain a large diverse group of such plants with entirely different and unexpected characters. Consequently it is necessary to begin again with genuine neutral descriptions and naming of fossils; the previous rejection of this group as possible angiosperm ancestors is invalid, and the topic must remain wide open. Similar discussion can be applied to almost all Mesozoic cycadophytes and coniferophytes, and the position now is that virtually any or all of them may be angiosperm ancestors. A comparable confusion has dominated treatment of plant fossils in the Tertiary (Boulter 1979).

In the years up to about 1970, which may be known as the 'megafossil investigation' era, botanists worked through and then steadily abandoned all the megafossils as being too rare, which was correct: and too incomplete, which was premature. Despairingly they grasped at almost any account of morphology of parts of a fossil plant from Carboniferous to Cretaceous time, without even analyzing the data and conspicuously without result. Eventually and understandably they turned with more hope to comparative morphology of living plants, which was being expanded rapidly into microscopic, genetic and chemical fields, and to distributions in Recent plant geography. Additionally the discomfort of having 350 families of angiosperms without any adequate agreed classification, was clearly due to a lack of phylogeny which could only come from the fossils. Of late, the problem has been belaboured by cladists and other evolution-theorists using numerical analyses, but significant progress is unlikely; it is comparable with study of the surface of the Moon before the Apollo programme.

A particularly unfortunate extension of the comparative morphology of living plants has been the still popular dictum that the angiosperms were monophyletic in orgin. When so very few of the potentially relevant microfossils have yet been examined, there is no excuse for use of such a damagingly narrow hypothesis; a synthesis of that kind should follow rather than guide and precede the main investigation. The principal damage is caused by the concentration of speculation about all the known attributes of angiosperms on a single gymnosperm group at a time, a procedure which is inevitably unsuccessful, especially when the information about each fossil group is so incomplete. Possibly the palaeobotanists who still favour an assumed monophyletic origin imagine that the majority of fossils have already been surveyed, but to do so ignores the main mass of dispersed palynomorph information, some studied but much yet to come.

Background to study of Mid-Cretaceous land-plant evolution

From Barremian to Turonian time inclusive, the total number of land-plant taxa living at any one instant of time appears to have increased very rapidly, perhaps by an order of magnitude. This increase can be seen in palynomorph assemblages, but perhaps less clearly in megafossils because of the scattered nature of their occurrences in world Cretaceous strata.

The base from which this expansion of angiospermous plants stemmed, appears to have been the surprisingly small number of mid-Mesozoic seed-plant taxa known as the fossil gymnosperms. All available groups of these plants have been worked on many times this century and have been assessed with great care; Caytoniales, peltapserms, corystosperms, *Glossopteris*, Bennettitales, Nilssoniales, Gnetopsida, *Czekanowskia*, *Classopollis*, *Eucommiidites* and several others less well known, have all failed to convince as ancestors because there was never enough evidence to support the monophyletic obsession.

It seems probable that the scope and variety of mid-Mesozoic floras built from these taxa have been persistently under-estimated, because for a century the description of fossils of these floras has been based conceptually on the even smaller number of living gymnosperm survivors and the very restricted variety of these extant relicts; they may be all we have, but the Mesozoic was much richer. Several prominent palaeobotanical authors since about 1950, have tried to extend the recognized spectrum of Mesozoic gymnosperm taxa and to express this by creating new families, but it is surprising how narrow scepticism and unimaginative classification can still blunt such initiative.

A much healthier approach to the ancestor problem would seem to be an assumption that each and every Mesozoic group of gymnosperms could have given rise to some plants with one or more angiosperm characters and that only a few were successful. Plants with two or more angiosperm characters in the style of living examples could have arisen later, whereupon it becomes necessary to attempt to define what is meant by a pre-Cenomanian or a pre-Campanian angiosperm.

The overall distribution of dispersed angiosperm pollen for the whole Cretaceous world, assuming that stratigraphic time-correlation to any distance from western Europe can be greatly improved, should eventually provide a framework for defining the evolutionary progress from 'gymnosperm' to 'angiosperm state' in post-Hauterivian time.

To make full use of the dispersed pollen it is necessary to treat taxonomy, nomenclature and classification without the waste of information involved in present methods; this is discussed here as a data-handling problem.

Alternative data-handling approach

This proposed alternative approach is divided for convenience and clarity into separate considerations appropriate at three successive levels of data-handling: (1) the records of individual occurrences of fossil pollen (2) the definitions of base-taxa (species level), and (3) the systematic use of higher taxa and their classification in time.

At each level a brief outline of current procedures and their disadvantages is offered first, and this is followed by the relevant alternative proposal for avoiding the difficulties. The three levels of discussion are interrelated, and are designed both to interlock as a compatible whole and to diverge as little as possible from current practice.

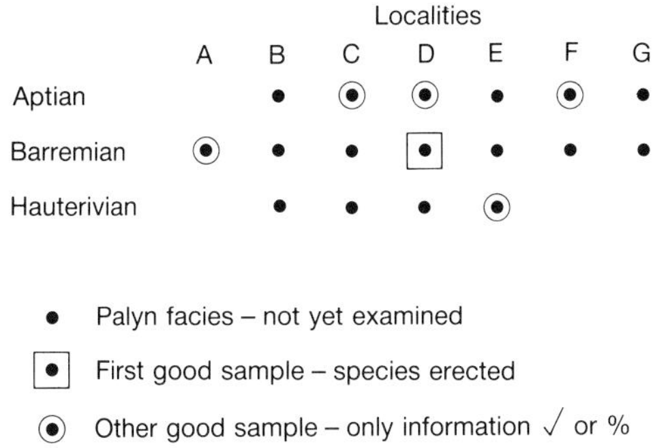

Fig. 15.1. Diagram to illustrate the customary method of recording the presence of a selected type of palynomorph, the total occurrence of which in seven rock successions (A–G) is shown by black dots. The taxon topotypes and other attributed specimens are indicated by surrounded dots. The only material fully described is from the Barremian at d; material from the five ringed levels is normally published in a locality species list, usually without quantification and always without description or comment on variation.

1. *Records of occurrence*

At present the usual pattern of investigation is for the first (or an early) discovery of pollen to be elaborately figured and described as a new taxon (species) based upon these first records; such data presents no problem and can be readily retrieved from publication or from data-storage using current methods. As suggested however in Fig. 15.1, all the other discoveries (often many) of comparable pollen are simply attributed to the published taxon, but are seldom figured and never described; details of all of these others are only retrievable through 'species lists' of unknowable quality that are usually even devoid of any facts about observation quantities. Virtually all this widespread information is dependent on 'identification' by a string of different observers of unknown competence; a new worker with new objectives is obliged to retrieve as best he can, and in effect much potentially useful earlier observation is unavailable to him in any useful form and is thus wasted.

Alternative proposal. The requirement is to build a system in which the precise details of each individual occurrence of fossils will be equally well recorded and equally retrievable in their entirety; the records of occurrence can then each be used independently, as necessary. Such a system has been developed (Fig. 15.2) through the use of the Palaeopalynologic Data-Handling Code (PDHC) (Hughes 1984; Appendix to this paper) which allows for this retrieval requirement by employing a separate complete record form description of each occurrence, including illustrations and character variation; for example each black dot on Fig. 15.1

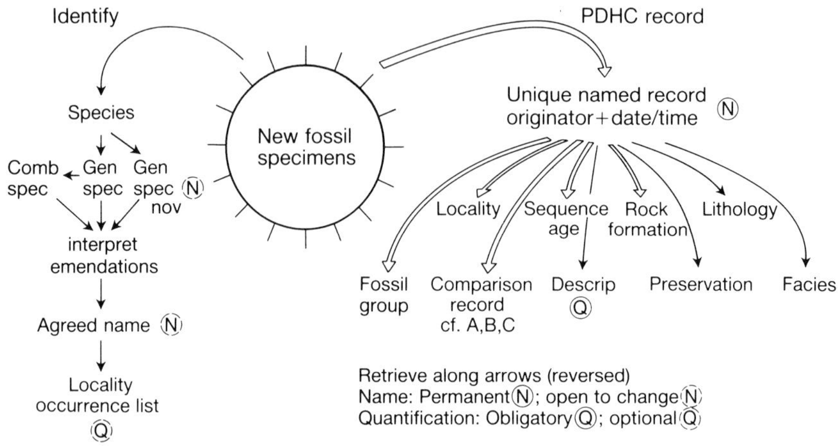

Fig. 15.2. Diagram to illustrate (right) the making from new material of a *PDHC* record with its unique author/date identifier and the usual information headings through which retrieval may be sought; also (left) for comparison is the existing procedure under present custom. N = degree of permanence of the name involved; Q = quantification statement.

would be so recorded. Automatically each record receives an identifier consisting of the observing author's name plus date/time of recording, which combination is unique for retrieval without any need for use of a taxon name for this purpose. It is not envisaged that all such material and data would be formally published in printed journals, but would merely be adequately advertized as incorporated with illustrations in designated data-banks.

2. *Definitions of taxa*

Taxa of fossil specimens do not parallel those of living specimens; they are entirely man-made and have no other significance (Hughes 1978); they should therefore be designed to be as convenient as possible for all users in the sense of arranging freedom from any confusion of definition. At present, emendations of (generic) name can cause difficulty but as these emendations are based on a form of precision in the *International Code of Botanical Nomenclature* (*ICBN*) their correct use is only a matter of labour and time, however unrewarding this may prove to be. There are, in addition, two other sources of confusion which are imprecise and from which currently there is no easy relief: (a) attributions of new specimens to a taxon (species) are customarily made without any qualification, with the result that the understood compass of the taxon enlarges constantly and changes its meaning without control; as it is not practicable to document this process fully and accurately the result is a 'balloon taxon', of ever decreasing usefulness for stratigraphic correlation (Fig. 15.3) or for any other precise purpose; and (b) emendation of the taxon meaning (definition) by official process (*ICBN, Articles 47, 51*) is certain to occur as better-preserved specimens are discovered and as techniques of observation evolve; only the latest change is normally signalled with the name, and there are no standards for recording changed statistics in such cases; the result is that users are seldom clear (Fig. 15.4) about the new meaning (around the holotype) and even less often do they state the position with clarity; in theory, unusually diligent adherence to the *ICBN* Rules could clarify the position, but it is not to be expected in practice.

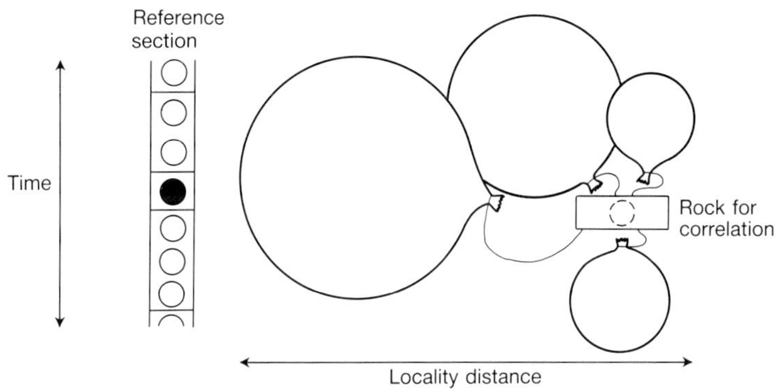

Fig. 15.3. Diagram to suggest the complete lack of usefulness for stratigraphic time-correlation of any taxon (species) which has been allowed to grow in meaning by attribution, in both time and in a geographic sense, without any clear separation of the original species record.

Your friends
1983 Southpole 1 or 2
1982 Confused 1 or 2 or 3

1972 Sceptic 1
1971 Speedy 1 or 2

State
1980 3 More topotypes
More characters
B —— S.E.M. Study
1970 2 Topotypes
A —— Extra specimens
1960 1 Original Holotype
Paratypes ?

Fig. 15.4. Diagram to show the uncertainties of definition of a taxon. Left column shows three successive states of a taxon originated in 1960, following normal finding of extra specimens, and ultimate use of Scanning Electron Microscope (SEM). Right columns suggest uses of the various states which may be employed by serious well-meaning palynologists; South-pole indicating lack of adequate access to the literature.

Alternative proposal. All these difficulties can be completely avoided by using as a taxon the biorecord (Hughes 1976, p. 26) which cannot be changed once erected, and which may only be 'improved upon' by describing and naming a new separate biorecord with the new definition. All specimen attributions have to be made in the form of comparison records (Hughes 1976, p. 27) prefaced by the special symbols cfA, cfB, cfC (Fig. 15.5), each with a clearly stated usage and purpose. Thus at all times the taxon definition and name remain clear and unchanged. Most occurrence records will bear a comparison record name through which retrieval may be effected if desired (Fig. 15.2), although the record identifier may more often be used alone.

3. *Systematic level*

Classification is necessary to provide a very broad retrieval framework for data on fossils for use in all geologic and palaeobiologic purposes; the requirement is only for a simple logical arrangement allowing for retrieval by locality, rock formation and approximate geologic age. However, it has long been customary to employ a full biologic classification based on the details of the extant flora and fauna, with fossils of all ages added; the result is that all retrieval is subject to the vagaries of

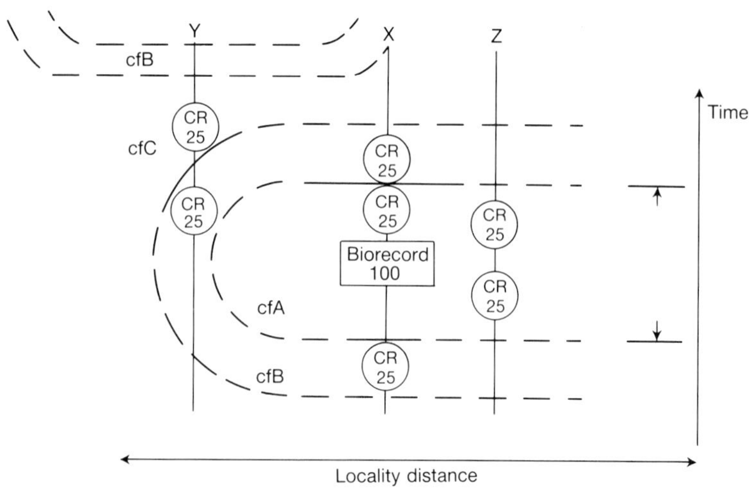

Fig. 15.5. Diagram to illustrate the employment, in connection with an immutable biorecord of say 100 topotype specimens, of graded comparison records (each of 25 specimens) indicating different degrees of time and space correlation potential. The time-slice (right) with arrowed bracket represents the degree of useful correlation between sections X and Z (but not Y) using presence and absence of cfA records of this particular biorecord. cfA = similarity to biorecord in all respects except for one minor quantitative character; cfB involves a single qualitative character difference.

biologic developments. The single biologic classification further has the disadvantage of narrowing investigations into past occurrences of organisms by distorting what must have been the position at the time with great quantities of irrelevant and misleading information. Mid-Cretaceous gymnosperms have to be divided into allies of cycads and conifers although the variety present stretched far beyond these living flora concepts. Such angiosperms as existed in Albian times have to be viewed in a framework of dicotyledons, monocotyledons, Magnoliatae and so on, despite minimal fossil critical evidence on which to base any such decisions.

The use of Recent taxa from time plane '0' has no relevance to Albian life at − 100 Ma. Comparable would be the use of mid-Jurassic taxa from the Yorkshire flora (−180 Ma) for discussion of a Permian flora at − 280 Ma. Neither is acceptable or useful practice, and both should be ruled out.

Alternative proposal. It is suggested for a beginning that there should be a fresh independent palaeobiologic classification in each geologic period or perhaps more conveniently every 50 million years (Fig. 15.6). In each intermediate period classification the levels of higher taxa employed should be related generally to the approximate numbers of base taxa

Ma	Age	Est. no. of species	gen. 10	fam. 100	order 1000	class 10K	div. 100K	regn. M
150	Tithon (JR)	1K–2K	Brachyph. Eucomm	Gymno	Embryo.			
100	Albian (CR)	<10K	Clavatipoll. 'Magnolia' leaf	Dicot.	Angio. (Gymno.)	Embryo.		
50	Eocene	<100K	'Chloranth.'	Laurac	Dicot.	Angio.	Embryo.	
0	Recent	>400K	Ascarina Pinus	Chloranthac. Abietac.	Magnolial.	Dicot.	Angio (Gymno.)	Embryobionta

Fig. 15.6. Diagram to illustrate in outline the new proposal for a separate classification of plants every 50 million years (Ma) to match the scope of plants existing at each selected time, to dispose of meaningless hierarchies and to express better the probable development of radiation at the time; the diagram is only intended to convey the purpose of the proposal and is not intended at this stage to be a useable classification. Estimated average numbers of species usefully included are suggested to rise by an order of magnitude per hierarchical rank.

(species) believed to have existed at any one relevant instant of time. Fig. 15.6 illustrates these points and includes the suggestion that at say Albian time there should be no false continuation of hierarchy above Class Embryobionta. The standard pattern of hierarchy of principal taxon ranks would be used and two classifications adjacent in time could be related through changing ranks. Additionally further classifications could be inserted, should they be needed.

4. Feasibility

The three parts of the 'alternative proposal' are compatible with each other and all can be introduced without disruption of palynologic work or significant loss. Because the new system uses the same approach to taxa, it can easily be developed in parallel with the old for any introductory period. Assembly of new or old records into old-style (*ICBN*) taxa can take place optionally at any time, but is independent of the record base (Fig. 15.7).

Old records would in no way be lost but would be re-assessed for new style records. Their value in detail would then be available to all potential users and any lack of critical data would be visible in the record making discard for any particular purpose an easy matter to arrange.

More exacting work would certainly be necessary to complete all the individual records, but presumably the actual observations were already made for fossil lists on the old system, and writing the details would simply be a matter for good organization.

Those studying evolution would have to relate the various classifications but almost certainly this would ultimately provide a more accurate account than present procedures, of the actual history.

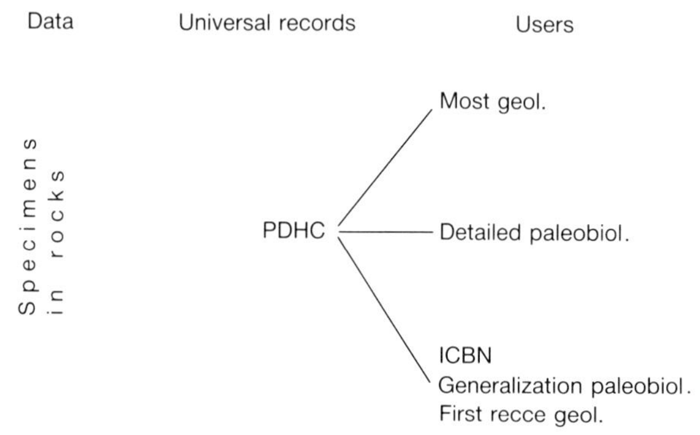

Fig. 15.7. Diagram to show under the new proposals the intention to place all data on the fossils first into universal immutable records. The various subsequent uses include traditional placement in biologic taxa for very generalized biologic and geologic purposes, but entirely independently of the record making the therefore no longer prejudicing retrieval.

Cretaceous angiosperm-like fossils

Significance of pollen grain fossils

It is likely that many botanists may consider dispersed pollen grains to be too small, too independent of the plant and too little understood, to act as a satisfactory basis for interpreting early angiosperm evolution and thus ultimately for building a truly phylogenetic classification of angiosperms from fossil successions.

Use of SEM and TEM observations has recently transformed understanding of pollen functions in connection with harmomegathic changes and electrostatics and much more progress is likely. The question of the independence of the pollen organ from the plant becomes much less of a difficulty when the function of pollen is more fully interpretable. In addition the total quantity of available dispersed pollen is so large, that the main problem is how to make significant use of the abundance.

Ultimately the fossil 'supply' situation is such that the only option is to make use of what nature provides and to eschew previous patterns of human thought as impractical.

Plant megafossils

Flowers, seeds, leaves will remain as individually important fossils and as check points in all interpretations but because their time and space provenance cannot be significantly improved each megafossil discovery should demand a massive palynologic background matrix of fossil occurrences to support and exploit it.

Concluding remarks

The alternative data-handling scheme outlined above is in use but the author will welcome constructive suggestions for further development. Solution of the angiosperm ancestor problem can now be envisaged as a product of a comprehensive search for dispersed pollen of mid-Cretaceous age, but it will require continuous organized work on this widespread material. Intermediate classifications of Mesozoic seed plants between Jurassic gymnosperms and late Cretaceous angiosperms will almost certainly require new categories for seed-plants with only one angiospermid character and perhaps even for those with two as steps in a general polyphyletic evolution.

Appendix

Paleopalynologic Data-handling Code (PDHC)

The *Palaeopalynology Data-handling Code* (1984) is presented as a brief set of Principles, followed by a *Multi-Purpose Record Form* and by a *Taxon Description Form*, both accompanied by explanatory notes. No other wording is strictly necessary; further explanations will be found in 'author's commentary' below.

Maintenance and modification of this Code are proposed as functions of the International Federation of Palynological Societies (IFPS).

Principles of PDHC

(1) *Occurrence of data.* Each and every item of recorded data on the occurrence of palynomorphs in rock has equal importance as a record of a past existence.
(2) *Numbering of records.* Use of the name of the originator plus the date/time as indicated on the record forms provides a unique identifier.
(3) *Retrieval of data.* Each record is retrievable by the name of the originator and the date/time of filing, by locality, by rock formation, by stratigraphic date, by general morphographic description, or equally by any one of these descriptors alone.
(4) *Availability of a record.* This is effected in the first instance by publication or more probably by its transmission, to any palynologist who is not or has not been a member of the institution or company to which the originator belongs, or to a nationally recognized scientific library. Brief notice of the existence of the record should be published in a regular scientific journal as early as possible.
(5) *Stability.* A once-stored record is unalterable.
(6) *Updating.* New information of any kind is stored in a new complete record, including a reference to any relevant old record.
(7) *A rock sample* is a collected unit of rock, designated and described.
(8) *A sample locality* is described adequately so that a further and contiguous sample may be located and collected.
(9) *Applicability of PDHC.* All Pre-Quaternary palynomorph occurrences are comparable for data-handling purposes, and are included.

Notes on preparation of Multi–Purpose record (Record Form A)

- Entries A,B,C,R,S,T are essential for all records.
- Entry consists of line–letter and all words to right of double vertical line, except for standard terms in parentheses (—).
- All line–letters to be transmitted, even when no entry on that line.

A Originator: one only, per record; family name, followed by initials. Address of institution maintaining relevant records.

B Origination date: date of preparation of record for use by others (not date of publication). Use date/time in alternate figures and words e.g. 1984 MAR 30 SEVENTEEN 45, in sufficient detail to provide unique entry which (with A) acts as a unique identifier.

C Locality. Global Latitude and Longitude or (specified) National Grid reference; with height in metres (or otherwise specified) with reference to appropriate Mean Sea Level (above or below).

D Rock formation: full name (may be unknown subsurface).

E Sample position: quantified relation of sample position to formation boundary or to other indicated level. Alternatively add the term FLOAT, if such is the origin of the sample.

F Sample lithology: description.

G Sequence age: formal stratigraphic time division in the order Era Period Age Chron carried as far as appropriate in each case, without terms or punctuation e.g. Mesozoic Cretaceous Aptian.

H Radiometric age (if available): in million years (Ma), with error range also in Ma. Paleomagnetic signature (if known): normal reversed or transition.

J Group name for fossils of organisms: explicit rather than formal e.g. Miospores trilete, Pollen polysaccate, Acritarchs polygonomorph.

K Graded comparison with taxon: cfA or cfB or cfC (no other), followed by taxon binomial (biorecord or species). Note: separate 'Taxon description' form is used for erection of any taxon.

L Number of specimens included in the variation described from the single sample quoted.

M Description of set of specimens including size mean and range of size; give quantified variation of one or more characters to indicate basis of record comparison. Illustrations. Photographs and other figures labelled M_1 etc and followed by slide/stub numbers and stage/stub grid co-ordinates.

N Variation. Give details of character variation pattern among specimens.

P Preservation comment. Include any restriction of information e.g. possible suspicion of re-working.

Q Facies information: list of principal associated palynomorphs.

R Repository for specimen material.

Palaeopalynologic Data-handling Code (PDHC)

RECORD FORM
A

MAIN IDENTIFIER	Guide	MULTI-PURPOSE RECORD OF PALYNOMORPHS
	Originator	A (Name) (Address)
	Date/Time	B (Year)--------- (month) (day)----(hour)
ROCK SAMPLE	Locality	C (Grid) (Co-ords) ±
	Rock formation	D
	Sample position	E
	Sample lithology	F
	Sequence age	G (Era) (Period) (Age) (Chron)
	Numerical age	H Ma; (Error range) Ma. Paleomag:
FOSSILS	Group of organisms	J
	Graded comparison	K cf
	Number of specimens	L
	Description	M
	Variation	N
	Preservation	P
	Facies	Q
REFERENCE	Specimen repository	R
	Earlier records	S
	Record ends	T ENDS

Palaeopalynologic Data-handling Code (PDHC)

RECORD FORM
B

	Guide	REFERENCE TAXON DESCRIPTION
IDENTIFIERS	Group of organisms	J
	Sequence age	G (Era) (Period) (Age) (Chron)
	Originator	A (Name) (Address)
	Origination date	B (Year)--------- (month) (day)----(hour)
	Taxon name	K
FOSSIL DESCRIPTION	Description	M
	Variation records	N
	Number of specimens	L
ROCK DATA	Locality	C (Grid) (Co-ords) ®
	Rock formation	D
	Sample position	E
	Sample lithology	F
	Preservation	P
REFERENCE	Repository	R
	Earlier records	S
	Conclusion	T ENDS

S Earlier records: quote here originators and dates of relevant old records when any specimen or topotype (from same original sample) has been used previously in a record now updated by this new record.
T Terminates record.

Notes on preparation of taxon description (Record Form B)

- Entries are required on all lines.
- Entry consists of line letter and all words to right of double vertical line, except standard terms in parentheses.

K Enter binominal taxon name (species or biorecord)

All other entries as for Multi-Purpose Record.

Author's commentary on proposed PDHC

1. Past attempts to ameliorate these problems

Palaeopalynology became an important study area at much the same time as the *International Code of Botanical Nomenclature* became prominent with its then (pre-1959) provisions of a separate Appendix for fossil plants which was essentially for megafossils only. Palynologists were immediately confronted with the standard microfossil problem of selecting from abundance rather than the usual megafossil search for enough adequately preserved specimens; they were so busy that for good order they acquiesced in the treatment of palynomorphs simply as small specimens of fossil plants, without realizing that from 1959 onwards *ICBN* was also moving steadily towards treating fossil plants in every possible way as similar to recent plants. This movement was excused with the assertion that *ICBN* was only concerned with nomenclature in the pure sense (Schopf 1978) and by implying that taxonomy thus remained a field of free enterprise; a glance, however, at *Articles 2–5* on taxa and their ranks in any recent edition of *ICBN* shows how misleading this assertion had become.

The first printed reference to my unease (Hughes 1964) over this situation in palaeopalynology amounted to a modest attempt to strengthen the fossil plant section of *ICBN*; this was rejected at Edinburgh in 1964. After various attempts to work both inside the articles of *ICBN* (Hughes and Moody-Stuart 1966) or outside (Hughes and Moody-Stuart 1969), the logic of the latter eventually predominated (see Hughes 1976). One last but totally unsuccessful attempt (Hughes 1978) to persuade the Fossil Plant Committee of *ICBN* to move towards achieving the necessary flexibility without disturbing the main rules was made at the meetings for *ICBN* in Sydney in 1981.

Despite great advances in the subject it is probably true that more *published* palaeopalynology work is still concerned with general exploration and to a lesser extent with biologic synthesis than with geology-based exploitation of the correlation potential of such universally occurring fossils; consequently until recently there has been inadequate support pressure from palynologists to improve data-handling by avoiding the continuance of some obsolete practices connected with *ICBN*.

This present proposal therefore is designed to pursue the more general and stratigraphic needs separately from the more classificatory needs while preserving a strong interface of compatibility.

2 *Timeliness*

Such an approach has become more timely because of three developments: (1) the economics of palaeopalynologic work have been progressively changed by advances in Automated Data-processing (ADP) although admittedly the bottleneck of tube or screen/keyboard hours still remains and retrieval of data across national frontiers is not yet ideally organised (2) enough work has been done to provide general understanding of the main groups of palynomorphs so that absolute surprises are rare and a generalized filing system of informal major groups has become feasible, and (3) palynologists have become more receptive to the custom of presenting data in ways which others may subsequently evaluate, rather than of providing only unquantified statements of their own opinions.

3. *Distinction in meaning of taxa of fossils from those of recent organisms*

Whereas living organisms are in general both complete and real, can be visited in their environment, and can be tested genetically, fossils have none of these attributes. A fossil can be taken to represent a past taxon which had a time distribution whether this be known in detail or not. An author erecting such a taxon for past organisms, provides, by the act of issuing a description, geologic time-limits for the taxon regardless of his intention or belief in the matter. Such limits inevitably cut across the genetic continuity of descent from parent to progeny although this may not immediately be revealed by collecting. Taxonomic arrangement of fossils is thus to be seen as an activity of convenience, without in the general case any further significance.

4. *Priority of discovery*

Although by custom handling records of living and fossil organisms has been kept similar, the true requirements are very different. With living organisms the biologist visits a new living species, investigates it and records it for posterity by erecting a taxon which can be checked by others, and for which he is rightly credited with priority of discovery. In the case of fossils the exploring geologist/paleontologist discovers, investigates and records new data about fossils of past organisms and should be credited with these discoveries. He should *not* however by credited with any priority for the placing of fossils into taxa which is a purely subjective exercise on which opinion will continue to evolve, particularly in the light of new techniques and of further discoveries.

5. *Graded comparison records*

The object of these records is to oblige the author to express a valuable opinion on resemblance at the time of observation: cfA = resemblance to base-taxon, but may include small quantitative difference in one character; cfB = less resemblance which may include one qualitative character difference; cfC = some resemblance but differences would justify erection of a new base-taxon if that was needed.

6. *Supposed disadvantages of PDHC*

(a) More routine observation and recording work will be necessary. This is true, but the investment is expected to be profitable.
(b) Numbers of names will be excessive. This is unlikely to be true because production will be slow on account of the higher standards required. Parsimony in the use of names tends to produce confusion of meaning, and appears to represent false economy in an age of constantly improving data-handling.
(c) Individual failure to comply with *PDHC* requirements, through ignorance or disagreement. Omissions from a record would weaken it, but at the same time the precise deficiencies would be clearly labelled and sub-standard records could be avoided in retrieval.

7. *Clear advantages of the use of PDHC*

(a) Equality of treatment of all records.
(b) Complete stability of meaning of base-taxa.
(c) Endurance of author's name on records, related to their quality.
(d) Records automatically acquire a unique number, without the use of any serial register.
(e) Retrieval of data through several identifiers, rather than through one elusive generic name.

8. The Future

As palaeopalynologists will be aware, the logic of the provisions described is equally applicable to all other sectors of palaeontology.

References

Boulter, M.C. (1979) Taxonomy and nomenclature of fossil pollen from the Tertiary. *Taxon* **28**, 337–44.

Hughes, N.F. (1976) *Palaeobiology of angiosperm origins*. Cambridge University Press, Cambridge xii, 242pp.

—— (1978) Proposals for a new palaeobotanical appendix for the International Code of Botanical Nomenclature. *Taxon* **27**, 497–504.

—— (1984) Presentation of a Data-Handling Code for Palaeopalynology. *Abstr. 6 Int. Palyn. Conf. (Calgary)*, **68**. Revised version in Appendix to this paper (above).

Schopf, J.W. (1978) Commentary on the Hughes proposals. *Taxon* **27**, 505–12.

16. Systematics of pre-Quaternary palynomorphs

D.J. BATTEN

Abstract

During the past 25 years both our knowledge of morphological variation in plant microfossils and our understanding of their stratigraphic and geographic distribution have greatly increased. However, unsatisfactory generic and species descriptions coupled with inadequate comparisons between, and distinctions from previously published forms continue to appear in the literature. As a result, additional synonyms are produced regularly and mis-identifications are common. Sole use of the scanning electron microscope in taxonomic studies is creating new problems. Important locality and stratigraphic data are often missing. Even careful analyses of numerous specimens recovered from large numbers of samples can lead to taxonomic difficulties. These usually centre around the characters on which to base identifications of genera and species where there is almost continuous morphological gradation between one form and another. Such 'complexes' often do not figure prominently in biostratigraphic zonations because these tend to rely, with good reason, on the occurrence of marker species and associations of key taxa that are comparatively easy to identify. Although this may be adequate for dating and correlation at Age/Stage level, it can be limiting if more detailed subdivision is desired. Discrimination of several types within 'populations' usually referred to a single species can provide a basis for refinement. It has to be accepted that there will be at least a few palynomorphs in most assemblages which are unidentifiable for various reasons and that a two-tiered separation of morphotypes is generally necessary to accommodate the majority satisfactorily. This can be done using traditional taxonomic methods and the Linnean system of nomenclature.

Introduction

Published diagnoses and descriptions of genera and species of dispersed plant microfossils vary considerably in quality and hence in practical value. Considering the rapid development of pre-Quaternary palynology in many parts of the world since the 1950s, this is not surprizing. The majority of papers that appeared before 1970 and many in the more recent literature encompass large stratigraphic intervals, and the species considered have tended to be either broadly circumscribed or based mainly on the holotype and closely similar forms.

Despite these limitations, our knowledge of morphological variation within and between dispersed spore, pollen, dinoflagellate cyst, acritarch and other palynomorph species, and of their stratigraphic and geographic distribution has greatly increased during the past 25 years. A vast amount of literature now exists to which palynologists can refer. It is unfortunate that much is of only marginal value to the conscientious taxonomist, particularly in the field of spores and pollen. The reasons for this are considered herein, and a few suggestions for improving the quality of descriptions and illustrations of genera and species are made. These are aimed at increasing their usefulness for stratigraphic, palaeoenvironmental and evolutionary studies. Some emphasis is placed on ways of dealing with morphological complexes at both generic and species levels.

My article is not liberally sprinkled with references to the work of others. The difficulties to which I refer must be only too evident to those palynologists who seriously attempt to use the taxonomic literature and it would be an invidious act to single out individual papers that I consider inadequate. The discussion is aimed mainly at spores and pollen grains but some is equally relevant to other acid (HCl and HF) resistant microfossils.

Problems with the literature

In many parts of the world increasing demands are being made on palynologists to produce refined biostratigraphic breakdowns within, and correlations between depositional basins, particularly those prospective for hydrocarbons. Initially the published literature is usually of value in providing the basis of a broad framework from which to develop detailed palynological subdivisions. It is during this latter phase of analysis that many papers have little direct application, usually because they lack sufficient biostratigraphic and taxonomic precision. Much early work is now of only limited use simply because it has become outdated. It is a sad fact, however, that despite all that has been learned about dispersed plant microfossils since the pioneering papers were written, far too many articles appearing today are still unsatisfactory for a variety of reasons. There is no excuse for this because for all geological periods exemplary taxonomic and biostratigraphic papers have been produced which set a standard for others to follow.

Some of the inadequacies rest with their author's ignorance of the literature; new forms are described that are clearly synonymous with pre-existing taxa about which no mention is made. Others demonstrate an inability to identify accurately what is seen under the microscope. Too many palynologists insist on describing new species that are not properly distinguished either from each other or from previously published taxa. A common fault is the inadequate provision of background information. There may not be any clear indication of sampling localities and horizons, indeed sometimes geological data are totally lacking. Although details may have to be sparse for reasons of Company confidentiality or because stratigraphic problems have yet to be resolved, there can often be no such reasonable excuse.

Papers that merely provide descriptions and illustrations of new forms of palynomorphs and those whose main purpose is to revise, invalidate, or change the names or status of taxa without also aiming to solve stratigraphic or other problems have little value. The finding of new genera and species in palynology is often a relatively trivial discovery. When the diagnoses are incomplete and the illustrations substandard, purely taxonomic papers are a positive nuisance. This is because, provided their authors have adhered to the rules of the *International Code of Botanical Nomenclature (ICBN)*, an attempt has to be made, during the course of later investigations, to take into account the observations presented. Almost inevitably problems of partial synonymy are generated which have to be dealt with eventually by more competent palynologists. There is no doubt that better quality original work is hampered by the appearance of these inadequate papers. Literally hundreds of dispersed miospore species have little or no taxonomic or biostratigraphic value. They merely serve to clutter and confuse the literature. As a result, pre-Quaternary palynologists have to spend more time engaged in taxonomic pursuits than they should.

A recent development has been the illustration of plant microfossils using only scanning electron micrographs. This is acceptable when the figures are solely for illustrative purposes in a paper that does not contain descriptions of new taxa. It is not, however, where these are included because it is commonly difficult, if not impossible, to relate them to specimens as seen in transmitted light. Although it is certainly true that speciation of miospores, particularly angiosperm pollen, may require the use of characters only visible under the scanning electron microscope (SEM) the appearance of the grains in ordinary light should not be ignored. Indeed, since the basic instrument of the palynologist is the light microscope, and for most practical purposes is likely to remain so for the foreseeable future, new taxa based only on observations made under the SEM should be considered inadequately described. It is usually possible to recognize in transmitted light the group of species to which a particular form is referable even though precise identification requires recourse to the SEM. Analysis under the latter can even lead to the identification of characters in normal illumination that were not previously noticed or considered significant. The fact that megaspores, seeds and other relatively large palynomorphs are less photogenic in transmitted and reflected light than they are under the SEM should also not be used as an excuse to avoid illustrating them in this way. A binocular microscope is always needed, at least initially, when working on these 'mesofossils'.

Approaches to nomenclature

Whether palynomorphs are to be used for biostratigraphic analyses, determinations of depositional environments, climatic assessment or other purposes, it is obviously important that they are not only adequately described but also properly named. Nomenclature and taxonomy are separate issues although some authors have confused the two, and it can be difficult to maintain a clear distinction in some areas, especially in the context of names used in hierarchical classification.

Palynologists usually abide by the rules and recommendations of the *ICBN*. Despite its inadequacies, it does provide international stability in the naming of fossil plant remains. In a series of papers Hughes has proposed a taxonomic and nomenclatural procedure in which the use of Linnean binomials is bypassed (summarized in Hughes 1976). Although a few palynologists have followed his lead, in the interests of maintaining a widely used convention most have not seen fit to abandon the 'traditional' well-tried system. This has the additional advantage over the alternatives that have been proposed to date in providing an eminently satisfactory means of communication.

Use of genus and species

Hughes (1976 and elsewhere) has asserted that conventional methods of data-handling have failed to improve the general standard of correlation and evolutionary studies using macro- and micro-plant remains. One of the aims of his alternative procedure has been to remove the problem of having to deal with genera and species whose morphological and stratigraphic ranges are continually being extended (giving rise to 'balloon taxa'; Hughes 1970) or emended. His approach is, in general, directed towards making palynologists do more careful, and therefore more useful systematic work. A respectable number probably do, in fact, put several of his recommendations into practice. His proposals have not met, however, with widespread approval. This is probably because they are not seen to generate a wholly satisfactory alternative to, and thus a significant improvement on traditional approaches to systematics.

In 'conventional' palynology, as in palaeontology generally, the primary taxonomic units are the genus and species. The grouping of various bits of plants, which have been separately named, in an attempt to reconstruct an entire organism, and the relating of genera to families are secondary acts on the part of a palaeobotanist/palynologist. A purely morphological diagnosis of a genus or species of dispersed palynomorphs is usually more satisfactory than one which takes into account possible affinities with either modern or fossil taxa. As far as is reasonable, consideration of botanical relationships should be kept separate from both descriptions and the basis of naming because these depend on subjective assessments which may have to be altered later when new facts are discovered. In my opinion, this should be the approach of all who deal with pre-Quaternary plant remains, in the interests of taxonomic and nomenclatural stability. Unfortunately many taxa have been described, especially from Cretaceous and Tertiary strata, that demand or imply acceptance of affinity via either their diagnoses or their names, or both.

Genus- and species-complexes

Another problem that taxonomists concerned with dispersed palynomorphs frequently encounter centres around the almost continuous gradation in morphology not only between one species and the next but also between genera. This situation is exacerbated by the fact that published taxa are not consistently defined; in some instances it is unclear whether apparent differences are merely a function of state of preservation. As a result it is reasonable to describe the taxonomy of several large groups of dispersed palynomorphs, such as Late Cretaceous and Tertiary angiosperm pollen grains, as being in a chaotic state.

This has led to an increasing tendency to abandon the use of species for dispersed pollen of flowering plants (Boulter and Wilkinson 1977, Wilkinson and Boulter 1980) or at least to rely more on genera (Zaklinskaya 1978) and forms recorded in open nomenclature (Robaszynski *et al.* 1985) for stratigraphic correlation. Unfortunately, even if this is done, many decisions on what genera to use and what forms to include in them must be based on subjective opinions and compromises. An alternative approach, adopted by some palynologists (e.g. Médus *et al.* 1980, Médus 1981), has been to cite species *sensu* particular authors rather than attempt to use, or rely on the original diagnoses and accompanying illustrations to guide identification.

Precision in speciation

It is often necessary to work at more than one level of precision in palynological taxonomy for various reasons. The need to obtain an adequate scientific reward for the time expended means, for example, that it is usually not worth attempting to separate more than a few types from the bulk of bisaccate pollen grains and smooth-walled triradiate spores. Diagnostic characters may be obliterated in poorly preserved material with the result that certain species, genera or even larger groupings cannot be identified with certainty. It is sometimes impossible to be confident in naming forms that occur only as isolated specimens; this is especially true when dealing with numerous species whose characters overlap, such as those referable to the late Mesozoic genera *Cicatricosisporites* and *Trilobosporites*. A few genera, such as the angiospermous *Tricolpopollenites* and *Tricolporopollenites*, were so broadly defined originally that it no longer makes sense to place in them all species exhibiting their few diagnostic characters. *Quercoidites* can be usefully distinguished from the former, and *Ilexpollenites* is worth removing from *Tricolporopollenites*.

A similar two-tiered separation is often necessary in practice to enable the majority of palynomorphs in an assemblage to be identified. Some of the broadly circumscribed species from the older literature can be used as 'portmanteaus' to accommodate specimens that are not readily referable to any of the newer, more precisely defined taxa. This situation arises when (1) recourse to the SEM, which is not always feasible, is essential for identification (2) specimens are poorly preserved and thus have lost some of their diagnostic features and (3) the characters displayed are morphologically intermediate between those of two or more species. If broadly based species are used for such palynomorphs in a paper it is advisable to make this clear to the reader in the text, and perhaps to add the word 'group' to the name, as in the *Classopollis hammenii*-group of gymnospermous (cheirolepidiacean) affinity.

An alternative approach to dealing with broadly circumscribed species in the older literature is to emend their diagnoses. Forms originally considered to be representative of the species which differ significantly from the holotype and/or from a more restricted concept of the taxon may be removed from it and placed in, or used as a basis for other species. This has commonly been done but should be avoided if possible because, in the process, new taxonomic problems are often created.

Many published palynostratigraphic breakdowns rely on the occurrence of 'marker' species and associations of known 'key' taxa that are relatively easy to identify. Members of large groups within which there is much morphological overlap between the species that have been described, tend to be under-used not only because of problems in identification but also owing to difficulties in determining stratigraphic significance in these circumstances.

When dealing with such 'complexes' it is necessary to accept that there must be considerable morphological overlap between species and always some specimens that are not precisely identifiable. This is hardly surprising in view of what has been learned from actuopalynological studies. Many plant species cannot be determined from mere bits of tissues or organs. It is well known that certain structures and sculptural forms of spore and pollen exines may occur in more than one genus, let alone species, and even be in common at family or order level (as in the Ginkgoales, Bennettitales and Cycadales/Nilssoniales). On the other hand, a single plant, or even sporangium or anther, may yield spores or pollen that in the dispersed state can be referable to more than one palynomorph species or genus.

In the interest of stable taxonomy, sampling must be adequate if new species are to be satisfactorily based. This applies particularly to those whose characters must inevitably overlap; unfortunately, discrete features are often unavailable. Anderson and Anderson (1983) emphasize the value of 'palaeodemes' ('fossil populations') of plant megafossils in a similar context, although their use of species and *formae* is not generally appropriate for palynomorphs. It is also helpful if the stratigraphic ranges of the genera employed are relatively narrow. Many are, in fact, restricted to only one or two geological periods or parts thereof, but there are some which, because of their simple morphology (e.g. the verrucate *Converrucosisporites* and the smooth-walled *Leiotriletes*) or owing to difficulties in using their characters satisfactorily for taxonomic purposes (e.g. several of the non-striate bisaccates such as *Alisporites*) are inevitably long-ranging. The ages of the strata from which the species attributable to them were described must be taken into account when attempting to identify specimens in a new assemblage in order to avoid extending unrealistically their stratigraphic ranges.

It does not necessarily follow that the accumulation of a large amount of data on a particular group of palynomorphs will make it much easier to determine the bases on which to erect new species and hence to identify members in other assemblages. For example, the range of variation in size, shape and distribution of verrucae and in the development of exine thickenings in equatorial radial regions of *Trilobosporites* is so great that as many as half of the specimens encountered in a sedimentary

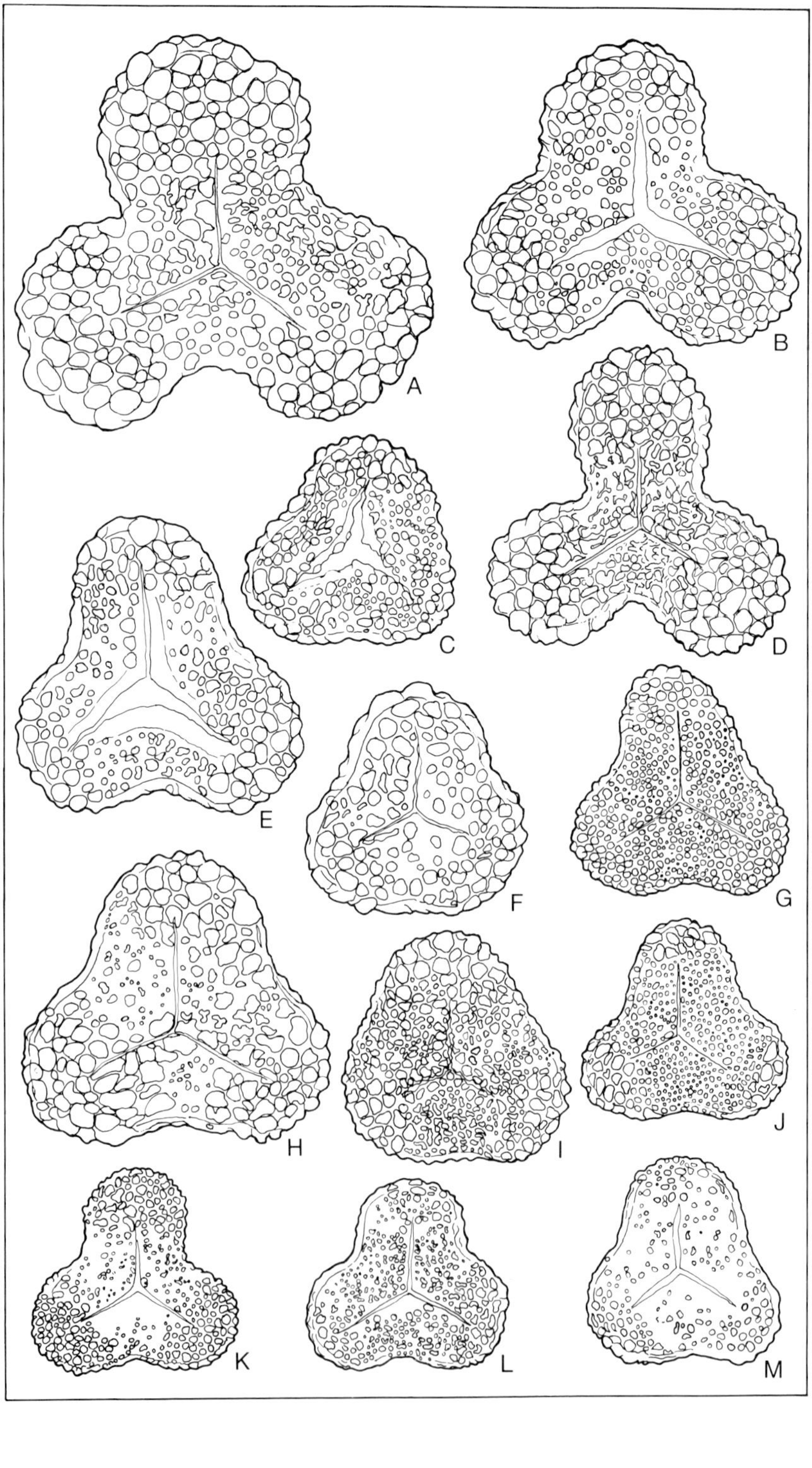
A
B
C
D
E
F
G
H
I
J
K
L
M

sequence can reasonably be placed in more than one species. This does not mean, however, that speciation should not be attempted; indeed, careful analysis of forms that would normally be lumped under one name, such as *T. apiverrucatus* (Fig. 16.1), may be found to consist of more than one species which have hitherto unsuspected stratigraphic significance. Even in relatively small groups of taxa comprising some well characterized forms there may be a few specimens in any given assemblage that are not positively identifiable, as recently demonstrated (Batten, in press) for the Normapolles genus *Vancampopollenites* from the Upper Cretaceous of Portugal.

Conclusion

A number of palynologists, including Staplin (1981) and Caratini (1983), have complained in the past about the state of palynological systematics and stressed the need for more research of lasting value. I do so again here because there are still too many papers being published in which the standard of documentation and illustration of plant microfossils is lower than it should be. This state of affairs will only improve when, in particular, it is recognized that new genera and species must be (1) based on sufficiently large 'populations' of specimens to allow descriptions to fully encompass their morphological range, (2) illustrated with informative photographs taken in transmitted light as well as with scanning electron micrographs and line drawings where these clarify sculptural and structural details (3) properly compared with and distinguished from other taxa which resemble them and (4) described from samples for which adequate locality and stratigraphic data are made available.

These recommendations may seem to state the obvious, yet all too often they have clearly not been borne in mind. This frustrates the more competent palynologist who endeavours to make full use of the literature. Unlike some of my predecessors, I do not seek an improvement by advocating a new method because, in my opinion, there is no such need. The binomial system of nomenclature should not be abandoned; it provides a very satisfactory means of communication and no useful purpose

---

Fig. 16.1. Diagrammatic illustrations of specimens of *Trilobosporites* (*sensu lato*) from English Wealden strata of Early Cretaceous age. All magnifications × 536. Considering both the range of morphological variation that is encountered within this genus of granulate to verrucate spores and the circumscription of *T. apiverrucatus* Couper 1958, the majority can be referred to that long-ranging species. However, A, B and D; C, E, F and H; G, I and J; and K–M may, in fact, each represent distinct species which have stratigraphic potential.

is served by ignoring taxa that have been well described in the past. The SEM should not be used to the exclusion of the light microscope which is the basic tool of the palynologist and likely to remain so for many years yet. It is not only more logical to begin any palynological study using transmitted light rather than scanning microscopy but also easier. The SEM can be employed subsequently, as well as a transmission EM if necessary, to enhance or clarify morphological interpretation. There is no doubt that, when carried out with care, traditional taxonomic methods can be successfully applied to palynomorphs.

References

Anderson, J.M. and Anderson, H.M. (1983) *Palaeoflora of southern Africa. Molteno Formation (Triassic).* Vol. 1 *Part 1. Introduction, Part 2. Dicroidium.* A.A. Balkema, Rotterdam.

Batten, D.J. (in press) The Cretaceous Normapolles pollen genus *Vancampopollenites*: occurrence, form and function. *Spec. Pap. Palaeont.* **35**.

Boulter, M.C. and Wilkinson, G.C (1977) A system of group names for some Tertiary pollen. *Palaeontology* **20**, 559–79.

Caratini, C. (1983) Nomenclature and taxonomy: supports to palynology or rhetorical pursuit. Comments on a recent paper on polycolpate fossils. *Rev. Palaeobot. Palynol.* **40**, 227–31.

Couper, R.A. (1958) British Mesozoic microspores and pollen grains. *Palaeontographica B* **103**, 75–179.

Hughes, N.F. (1970) The need for agreed standards of recording in palaeopalynology and palaeobotany. *Paläont. Abh. B* **3/4**, 357–64.

—— (1976) *Palaeobiology of angiosperm origins.* Cambridge University Press, Cambridge.

Médus, J. (1981) Pollens Normapolles de coupes stratotypiques du Crétacé supérieur des Charentes et du Sénonien du Portugal. *Comun. Serv. Geol. Portugal* **67**, 19–28.

—— Boch, A. Parron, C. Lauverjat, J. and Triat, J.M. 1980. Turonian Normapolles from Portugal and southern France: correlations. *Rev. Palaeobot. Palynol.* **31**, 105–53.

Robaszynski, F. Bless, M.J.M. Felder, P.J. Foucher, J-C Legoux, O. Manivit, H. Meesen, J.P.M.Th. and Van der Tuuk, L.A. (1985). The Campanian-Maastrichtian boundary in the chalky facies close to the type Maastrichtian area. *Bull. Centres Rech. Explor.-Prod. Elf-Aquitaine* **9**, 1–113.

Staplin, F.L.(1981) Twilight of palynology? *Rev. Palaeobot. Palynol.* **32**, 445.

Wilkinson, G.C. and Boulter, M.C. (1980) Oligocene pollen and spores from the western part of the British Isles. *Palaeontographica B* **175**, 27–83.

Zaklinskaya, E.D. (1978) Die Floren der Oberkreide und des Frühen Paläogen (nach palynologischen Angaben). In *Paläozoische und mesozoische Floren Eurasiens und die Phytogeographie dieser Zeit* (eds. V.A. Vakhrameev, I.A. Dobruskina, S.V. Meyen and E.D. Zaklinskaya), 195–229, VEB. Gustav Fischer, Berlin.

17. Time and evolution

P.R. GRANT

Abstract

A systematic change in morphology of otherwise identical fossils obtained from a stratigraphically sequential series of sampling positions may be construed as the manifestation of evolutionary change.

A problem arises should we wish to compare one evolutionary change to any other. Say, for the purposes of investigating relative response to some putative palaeoecological pressure.

The significance, implications and monitoring of morphological features in fossils should fall to the more biologically able palaeontologist, whereas the concepts of time should be more natural to the geologically inclined palaeontologist. Here is where I fly my flag and pursue aspects of time as they may affect interpretations of evolutionary rates.

A true rate requires that a measurement of the change in some phenomenon (or its manifestation) is made with respect to a time framework.

Time is a nonspatial continuum in which events occur in an apparently irreversible succession from past through to future.

Physical measures of Geological Time

Several physical phenomena are used to monitor geological time, the most direct of which is the radioactive decay of certain unstable isotopes. Those that are particularly useful are potassium, rubidium, uranium, lead, samarium, and carbon. By measurement of proportions of daughter products and remnant original radioactive source elements the period of time from the genesis of the radiogenic source to its present state can be computed (Odin 1982a).

Other physical techniques are indirect. These include:

(a) the determination of the orientation of the magnetic field induced in rocks at the time of their formation; and

(b) the determination of the stable isotope composition of minerals and mineral tests of organisms. The most commonly used stable isotopes are of oxygen, sulphur, and carbon. Both indirect techniques rely on matching new data to a pattern established by previous compilation. As more data are accumulated then higher precision will become available for comparing 'unknowns' against master curves of isotope (or magnetism) values matched to a radiometric time scale. The geochemical methods are reviewed in Odin, Renard and Vergnaud Grazzini (1982), while magnetostratigraphy is reviewed in Channell (1982).

Techniques appropriate to sediments

Radio-isotope dating of igneous rocks associated with sedimentary as well as dating of the sediments themselves has lead to the temporal calibration of the Geological Record (Odin, Curry, Gale and Kennedy 1982). However, this does not mean that within this framework we can specify an actual age for any but a very few exposures of rock. A very small percentage of these would be practical direct use to the evolutionary palaeontologist.

The one isotope that can be used to determine the age of organic matter is radio-carbon; but it may be used only in relatively recent sedimentary regimes (younger than about 5 Ma). Webb (1982) has shown the potential of combined pollen analysis and radiocarbon dating with interpolated time resolution of five to fifty years. Even the best techniques with high analytical precision yield errors of one to two per cent at best. This produces an error of 0.65–1.35My in the best cases for a 65Ma old sample. This does not compare favourably with the 'average' Tertiary biozonal duration of about 1My (Hardenbol and Berggren 1978).

As older sediments are considered then, although the analytical error may remain constant the absolute error in terms of time inevitably increases. Also it is unfortunate that sediments are particularly problematical to date radiometrically. The most common technique, that of 'glaucony', is still scorned by some but does seem to be regaining its credibility (Keppens and Pasteels 1982). The technique relies on diagenetic glauconites becoming closed chemical systems within the sediment. On this assumption it is then reasonable to analyse rubidium, strontium, potassium and argon isotopes in order to determine a radiometric date. Severe problems arise if the glauconites were not totally closed systems or if the sedimentary basin has suffered significant heating. Also suitable sedimentary facies must be developed to obtain glauconite.

Magnetostratigraphy of Tertiary, and Late Jurassic–Early Cretaceous can yield good temporal resolution independent of sedimentary facies or of local radiometric determinations. For application at the K–T boundary see Dingus (1984). Assessing sediments in this technique requires that there are sufficient magnetic clasts that have been deposited in an alignment controlled by the contemporaneous global magnetic field.

The sedimentary record and time

Dingus (1984) reviewed 'stratigraphic completeness' and pointed to sufficient evidence that not only are there gaps in the sedimentary record but that they are of significant dimension that any serious investigation of evolutionary rates must attempt to take these into account. Such an exercise is essential to establish a sound temporal scale against which evolutionary (morphological) change can be monitored.

Dingus states that '. . . there is no way to determine the position and duration of all the gaps in a particular transitional interval'. I concur with this statement in general but with reservation. The correlation technique described below has a potential for highlighting the most subtle hiatuses in sequences.

Biochronology

The division of the Geological Record into time slices by consideration of vertical and lateral distribution of fossils is underpinned by the assumption that an agency has caused a unidirectional change in fossil morphology and assemblages. Hancock (1977) gives a review of the development of the concepts and applications of biostratigraphy. The tranches of the Geological Record that can be established palaeontologically are much finer average time units than can be achieved by any other technique. For example note the Tertiary example given above, along with the 21 fold palynozonation of the Carboniferous of NW Europe (Clayton *et al.* 1977) compared with the time duration of the Carboniferous of 70my and the concomitant errors of +/− ~7.5 my that may be expected analytically.

Figure 17.1 illustrates how a section of geological time may be divided into time slices by use of fossil assemblages. Now there is nothing wrong with the suggested interpretation provided that the rock sequence from which it was obtained was complete, and as we have already seen, this is most unlikely.

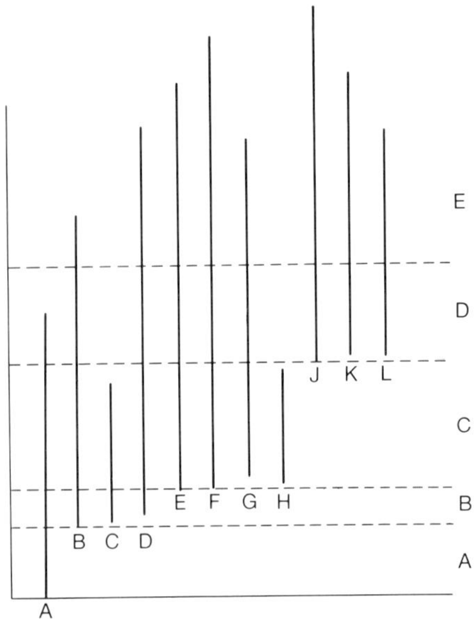

Fig. 17.1. Most workers would agree that this time period could be divided into four sub units.

A. . . pre BCD
B. . . after incoming of BCD and pre FGH
C. . . after incoming of FGH and pre JKL
(H would almost certainly be considered an 'index fossil' for this zone)
D. . . after the incoming of JKL and after the extinction of C and H

Any time/fossil combination more complex than this would invariably be treated differently by separate individuals, and Kauffman (1977) doubts the consistency that any individual worker would have in the interpretation of such diagrams.

Figure 17.2 demonstrates the effect of an hiatus on the apparent temporal distribution of fossils. Since d'Orbigny's time the pragmatic approach has been to search for breaks in the continuity of the palaeontological record and ascribe to that break the status of a 'zone' boundary. To the rigorous, zones may have a key or index form but a true zone comprises numerous forms which endow the zone with an integrity and robustness that allow it to be used for regional correlation notwithstanding locally induced phenotypic variations in the assemblage. The construction of zonal schemes is however rather subjective and a number of criteria are utilized. For example occurrences of high proportions of particular taxa are used to define acme zones. These may be satisfactory in certain circumstances but due regard for ecological control of such 'zones' must be borne in mind—the correlation of densospores with the durainous parts of coal seams in the British Westphalian (Smith and Butterworth 1967, p. 87) for example.

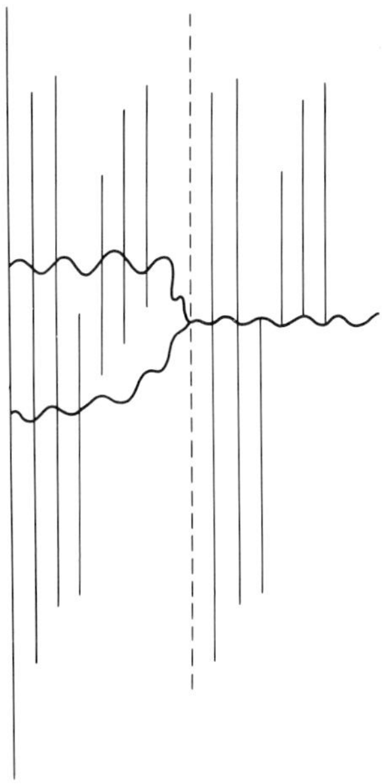

Fig. 17.2. (modified after Haq and Worsley, 1982). This shows spurious zonal limits imposed on the recovered fossil record from an incomplete lithological sequence.

I contend that the plethora of zonal epithets (Hedberg 1971) is a result of individuals realizing that there is stratigraphic resolution finer that the sturdy zone, but some form of homeliness with the zone concept has constrained more radical approaches.

A Biostratigraphic Method with computerization potential

Background

Firstly the technique relies on two assumptions:
(a) that time is continuous and non reversible—a scientific axiom.
(b) that evolution is non reversible (Eldredge and Gould 1977).

Secondly, the technique relies on the fact that at any one instant in time there is a global assemblage of organisms. Should one of these die or the assemblage be augmented by a birth then a new distinct assemblage emerges.

Clearly, in the fossil record births and deaths of individuals are extremely rarely recorded. However the incoming and extinction of taxa can be identified. Each taxon divides time into three segments:
(i) pre-taxon time
(ii) taxon time, and
(iii) post taxon time.

If we plot the time of existence of taxa one against another then we produce the familiar range chart. This range chart indicates the fossil record of evolutionary quanta. These evolutionary quanta are the periods of time between the incomings or extinctions of taxa and do not occupy equal time slices but are phenomenologically identical. No preference is given here to either incomings or extinctions, although in practice, for example, if bore-hole data is being used then extinctions are likely to be more reliable than records of incomings.

Figure 17.3 shows the difficulty in observing ranges only locally. It can be seen from the diagram that a new taxon emerges at a particular point in time and space. It may migrate into other geographic localities—but will take time to do so. There then may be a period of evolutionary stasis in terms of geographic distribution though 'minor' genetic evolution may be occurring which has no significant effect on the preservable morphology of the fossil. It is possible that some geological event may disturb the geographic distribution of the taxon. In this particular case the taxon does not recover from such a dramatic change and goes into decline, becoming extinct at point D in time and space.

It can clearly be seen that the total (global) time range of this taxon would not be recorded at a single location in space. In order to overcome this facies control of the temporal distribution of taxa Hazel (1977) introduced the concept of 'range-through'. In practice what this means is that if an assemblage is being analyzed for its stratigraphical potential and a taxon is not present at one level but is known to exist stratigraphically above and below the sample under investigation then the absent taxon is considered to be present to expedite the analysis.

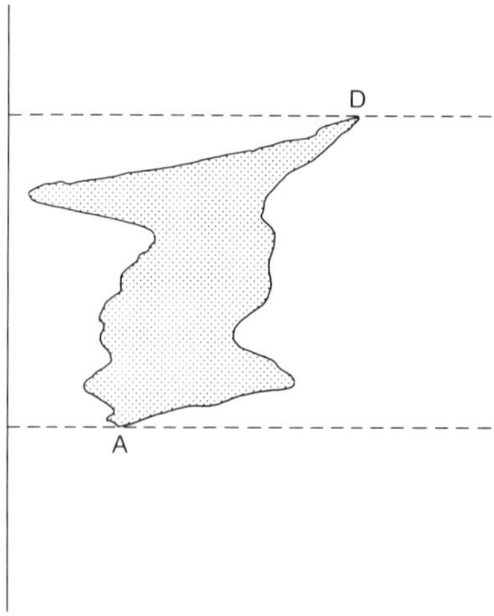

Fig. 17.3. (After Bignot 1985) See text for discussion.

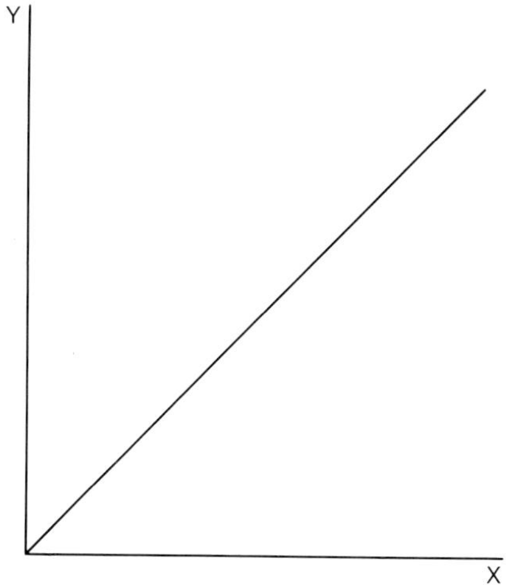

Fig. 17.4. A plot of two identical sections against each other.

Methodology

If we have a stratigraphic section and we compare it to itself we will obtain a perfect correlation. Regardless of any paucity of data or fluctuations in sedimentation/evolution rates, hiatuses etc. They will all match exactly. Thus if we plot one sequence against itself we would produce a correlation graph with a unit slope: see Fig. 17.4.

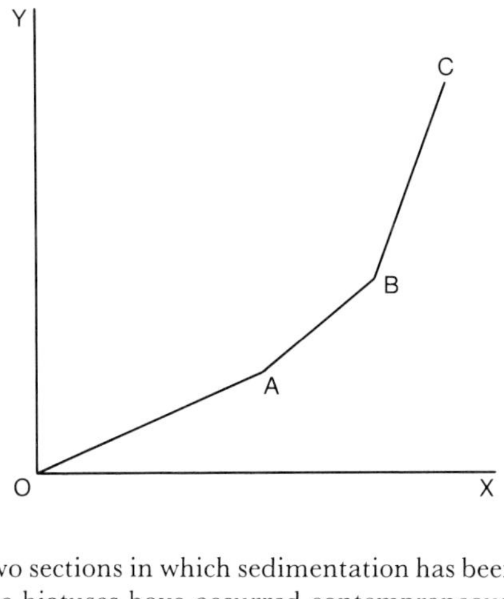

Fig. 17.5. A plot of two sections in which sedimentation has been continuous but not at equal rates, and where hiatuses have occurred contempraneously in both sections. In section X the sedimentation rate: is greater than Y for the period OA; is greater than Y, but less than OA, for the period AB; and, is less than Y during BC. See text for discussion of the implication of evolution rates.

If we have continuous sections that have not accumulated at the same rate then differential slopes will occur, Fig. 17.5. The effect on the slope of the correlation line of sedimentation rate is inverse to that of evolutionary rate of the assemblage. That is an increase in slope may be due to an decrease in evolutionary rate in the 'Y' section compared to the 'X' section or it may be due to a reduction in the sedimentation rate in the 'X' section compared to that in the 'Y' section.

These graphs are similar in their interpretation to those presented by Shaw (1964). However they are obtained by plotting statistical measures of comparability of assemblage composition rather than defining a correlation function based directly on the stratigraphic positions of bottoms and tops of ranges (Shaw 1964) and (Miller 1977).

Various measures of similarity are discussed by Simpson (1960) and a

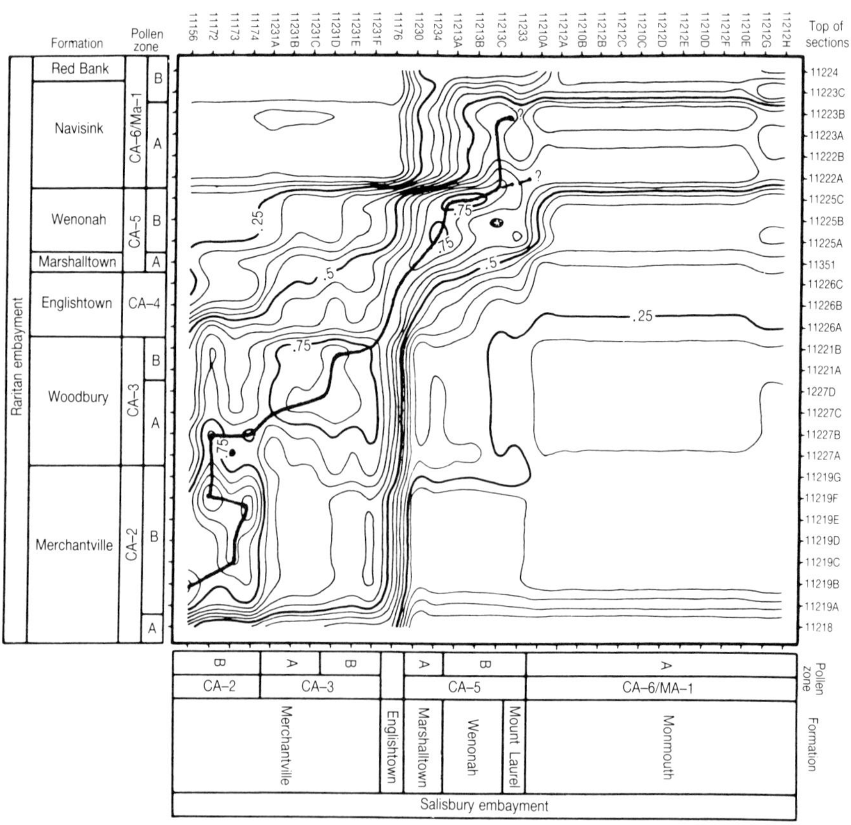

Fig. 17.6. A contoured similarity map of the relationships between the palynological assemblages of the upper Cretaceous of the Raritan and Salisbury Embayments of the Atlantic Coastal Plain with the trajectory of 'maximum correlation' superimposed (after Christopher 1978).

Similarity technique was presented by Christopher (1978) for the palynostratigraphy of the uppermost Cretaceous of the Atlantic Coastal Plain.

By tracing the maxima on Christopher's Similarity maps it is possible to derive a maximum 'correlation line' between two sections, Fig. 17.6. The characteristics of such lines are:

(i) that there is a measure of certaintity of correlation as indicated by the value of the similarity coefficient.

(ii) that variations in positive slope indicate variations in sedimentation/evolution rates between the sections.

(iii) that correlation lines parallel to an axis indicate a hiatus in the other section.

(iv) that a laterally displaced correlation line indicates a fault, and.

(v) that a correlation line with a dog-leg including a negative slope is indicative of reworking.

Comparison of numerous sections would lead to the construction of a composite section devoid of breaks and similar in concept to that of Shaw (1964) and Miller (1977), and in the spirit of Oppel's 'ideal profile' independent of any local palaeontologic or lithic succession (Hancock, 1977).

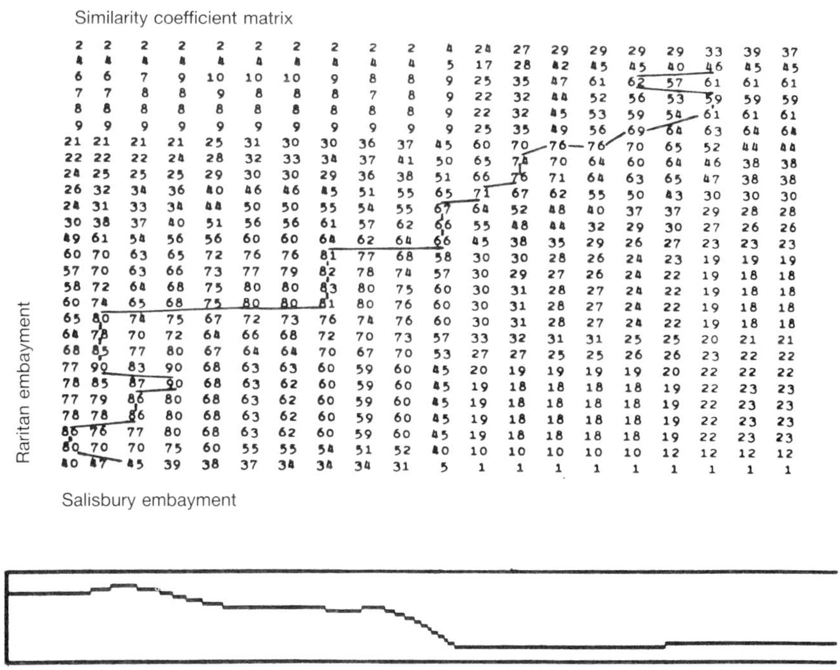

Fig. 17.7. A numerical version of Fig. 17.6 with a line of section marked 'AA', and the Similarity profile along this section at the bottom of the matrix.

Because of the presence/absence data that is required as initial data, the numerical nature of the intermediate data required for map making or calculation of correlation trajectories this technique is suitable for computerization.

Examples of output from a rudimentary computer program to analyze such data is presented as Figs. 17.7 and 17.8.

Discussion

For the purposes of biostratigraphy it is not necessary to understand the mechanisms of evolution, just that the process is immutably unidirectional. However, the evolutionary palaeontologist does require information on the relative and absolute time scales over which morphologically manifest genetic changes occurred. Both branches of palaeontology require such analyses of lithological sequences, mainly to establish the degree of continuity but also as an aid to correlation and as a primary stage in dating.

Although the newly described technique does have the potential for a 'continuous' biochronology the limitations will be set by the closeness of sampling and the efficiency of taxonomic identification. With the advent of personal computers and the ease with which they can be attached to large mainframe computers, along with the burgeoning of voice recognition units there is no reason why, say, palynologists should not be as

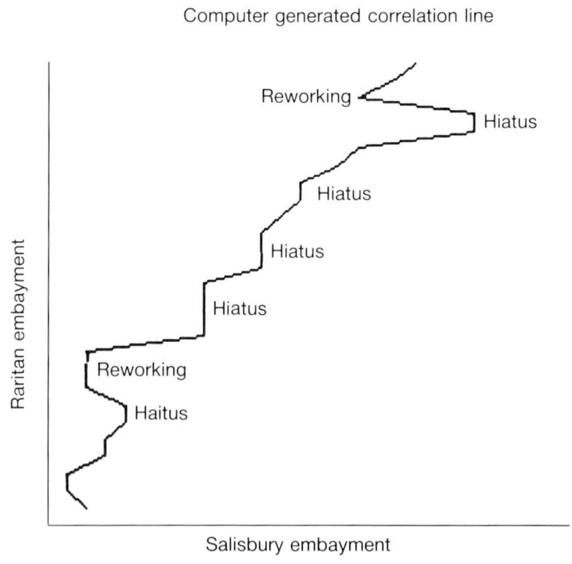

Fig. 17.8. A crude computer trace of correlation through the Similarity Matrix—compare with the correlation trace through the numerical matrix of Fig. 17.7.

rigorous in their taxonomic differentiation as advocated by Hughes and Moody-Stuart (1967, 1969).

The technique described above illustrates the superiority of biological division of Geological Time compared to that feasible using other methods. Unfortunately, whatever the approximation to continuous monitoring of biological change, the crucial missing factor is time. It is important to avoid circularity of argument with respect to discussions of evolutionary rates by keeping a clear distinction between temporal and biological divisions of the Geological Record.

References

Bignot, G. (1985) *Elements of Micropalaeontology* Graham and Trotman, London.

Channcll, J.E.T. (1982) Palaeomagnetic stratigraphy as a correlation technique In *Numerical Dating in Stratigraphy* Odin G.S. (ed) Wiley, New York.

Christopher, R.A. (1978) Quantitative Palynological Correlation of three Campanian and Maestrictian sections (Upper Cretaceous) from the Atlantic Coastal Plain *Palynology* **2**, 1–27.

Clayton, G. Coquel, R. Doubinger, J. Gueinn, K.J. Loboziak, S. Owens, B. and Streel, M. (1977) Carboniferous Miospores of Western Europe: illustration and zonation *Meded. Rijks. Geol. Dienst.* **29**, 1–71.

Dingus, L. (1984) Effects of stratigraphic completeness on interpretation of extinction rates across the Cretaceous–Tertiary boundary *Paleobiology* **10**, 420–38.

Eldredge, N. and Gould, S.J. (1977) Evolutionary Models and Biostratigraphic Strategies In *Concepts and Methods of Biostratigraphy* (eds. Kauffman E.G. and Hazel J.E.), Dowden, Hutchinson and Ross, Stroudsburg, Pa.

Hancock, J.M. (1977) The Historic Development of Biostratigraphic Correlation In *Concepts and Methods of Biostratigraphy* (eds. Kauffman E.G. and Hazel J.E.), Dowden, Hutchinson and Ross, Stroudsburg, Pa.

Hazel, J.E. (1977) Use of Certain Multivariate and other Techniques in Assemblage Zonal Biostratigraphy: Examples Utilizing Cambrian, Cretaceous, and Tertiary Benthic Invertebrates In *Concepts and Methods of Biostratigraphy* (eds. Kauffman E.G. and Hazel J.E.), Dowden, Hutchinson and Ross, Stroudsburg, Pa.

Haq, B.U. and Worsley, T.R. (1982) Biochronology-Biological Events in Time Resolution, their Potential and Limitations In *Numerical Dating in Stratigraphy* (ed Odin, G.S.), Wiley, New York.

Hohn, M.E. (1982) 'Properties of Composite Sections Constructed by Least Squares' In *Quantitative Stratigraphic Correlation* (eds Cubitt J.M. and Reyment R.A.), Wiley, New York.

Hedberg, H.D. (ed) (1971) Preliminary Report on biostratigraphic Units *Int Geol Conq* **24** ISSC Rep 5.

Hughes, N.F. and Moody-Stuart, J.C. (1967) Proposed Method of Recording

Pre-Quaternary Palynological Data *Rev. Palaeobotan. Palynol* **3**, 347–58.

—— and Moody-Stuart, J.C. (1969) A Method of Stratigraphic Correlation Using early Cretaceous Spores *Palaeontology* **12**, 84–111.

Kauffman, E.G. (1977) Evolutionary Rates and Biostratigraphy In *Concepts and Methods of Biostratigraphy* (eds. Kauffman E.G. and Hazel J.E.) Dowden, Hutchinson and Ross, Stroudsburg Pa.

Keppens, E. and Pasteels, P. (1982) A Comparison of rubidium-strontium and potassium-argon aparent ages on glauconies In *Numerical Dating in Stratigraphy* (ed. Odin GS) Wiley, New York.

Miller, F.X. (1977) The Graphic Correlation Method in Biostratigraphy. In *Concepts and Methods of Biostratigraphy* (eds. Kauffman E.G. and Hazel J.E.) Dowden, Hutchinson and Ross, Stroudsburg, Pa.

Odin, G.S. (ed) (1982) *Numerical Dating in Stratigraphy* Wiley, New York.

—— (1982a) Uncertainties in evaluating the numerical time scale In *Numerical Dating in Stratigraphy* (ed. Odin G.S.) Wiley, New York.

—— Curry, D. Gale, N.H. and Kennedy, W.J. (1982b) The Phanerozoic time scale in 1981 In *Numerical Dating in Stratigraphy* (ed. Odin, G.S.) Wiley, New York.

—— Renard, M. and Vernaud Grazzini, C. (1982) Geochemical events as a means of correlation In *Numerical Dating in Stratigraphy* (ed. Odin G.S.) Wiley, New York.

Shaw, A.B. (1964) *Time in Stratigraphy* McGraw-Hill, New York.

Simpson, G.G. (1960) Notes on the Measurement of Faunal Resemblance *Am. J. Sci.* **258-A**, 300–11.

Smith, A.H.V. and Butterworth, M.A. (1967) Miospores in the Coal Seams of the Carboniferous of Great Britain *Spec Pap in Pal* No. 1 Palaeontological Association, London.

Webb, T. (1982) Temporal Resolution in Holocene pollen data *Third N. Am. Paleontol. Conv. Proc.* **2**, 569–72.

18. Suggestions on the designation and terminology of palaeobotanical type material

C.H. SHUTE

Abstract

Two aspects of palaeobotanical type material are discussed in relation to its designation and terminology. One concerns the use by palaeobotanists of the term *specimen* as applied to rocks containing more than one plant fossil, as opposed to the term applying to individual specimens of plant fossils. The other considers a more precise terminology for holotype, lectotype and neotype material.

Introduction

In palaeobotanical literature it is not unusual to find that a holotype, lectotype or neotype consists of more than one specimen of plant fossil; the term 'specimen' having been applied either deliberately or inadvertently to a piece of rock which contains more than one plant remain conforming to the type description. At present, according to the *International Code of Botanical Nomenclature* a lectotype should be chosen from one of these plant remains. A suggestion is made here that such rock specimens could in appropriate instances usefully serve as the type.

A palaeobotanical 'primary' type (holotype lectotype, neotype) can be divided into many pieces, of different status, but all of it type material. At present there is no formal terminology for these pieces, apart from simply 'type'. Here a terminology is suggested which although a little more complex would give a more precise designation to them. Also a name is suggested for the totality of such material. These suggestions are meant as a practical terminology for use in research and curation.

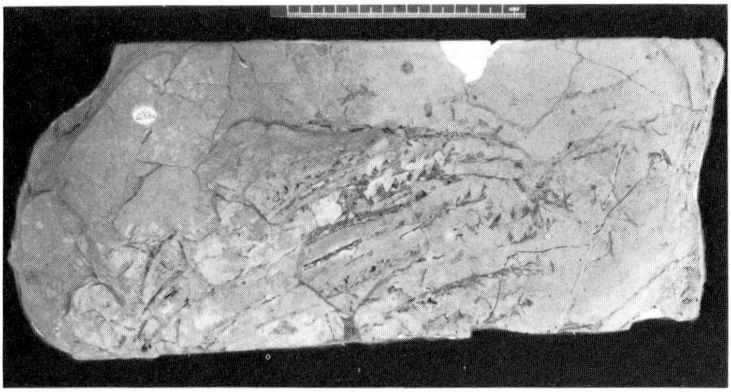

Fig. 18.1. *Zosterophyllum llanoveranum* Croft and Lang [Holotype], × 0.3.

Use of the term Specimen

The term specimen has been applied by palaeobotanists both to rocks bearing several plant remains and to individual plant fossil remains, sometimes rather loosely. When typification is involved, variability of meaning by authors has in some cases led to confusion in interpretation of the literature. For instance, in their description of the holotype of *Zosterophyllum llanoveranum* Croft and Lang (1942, pl. 10, Figs. 32–34, t. Fig. 88) described plant specimens on a piece of fractured rock, the major part of which is shown here in Fig. 18.1. On this rock are numerous *Zosterophyllum* spikes. These are not demonstratably from one individual plant since the distal end of the axes terminate either discretely on the surface of the rock, or are discrete when terminated by the edge, as was recognized by the authors. *Article 7.3* of the 1983 *International Code of Botanical Nomenclature* (Voss *et al.* 1983) states that 'A holotype is the one specimen or other element used by the author or designated by him as the nomenclatorial type'. The only way that the specimen selected by Croft and Lang could be considered the holotype is when the whole block of rock containing the plant specimens is considered to be the 'one specimen or other element used by the author'. A similar example is the specimen chosen as the holotype of *Lerichea krystofovichii* Stockmans (1968, pl. 9, Figs. 2, 2a) here illustrated in Fig. 18.2. Again, either there is no holotype, since there is more than one plant fossil, or the holotype is to be considered as the whole rock specimen. Figure 18.3 shows a rock specimen regarded in the literature as the holotype of *Neuropteris tenuifolia* (Schloth.) Sternbg. (Schlotheim 1820, Sternberg 1833, Crookall 1965). But on the rock specimen are numerous pinna fragments and isolated pinnules conforming to the type description, no single one of which has been designated as type.

In neobotany there is provision in *Article 9.1* of the *ICBN* for 'small herbaceous plants' to be represented by more than one individual in typification, these specimens to be conserved on one herbarium sheet. In a similar way palaeobotany could, with regard to holotypes, lectotypes and neotypes, treat the rock specimen which contains more than one plant fossil of the same species as the type, when in other respects the description is valid and when no single plant specimen has been selected as type. This seems consistent with a broad interpretation of the 'other element' of the ICBN (clearly, plant fossils on the rock not conforming to the diagnosis, should not be regarded as a portion of the type). The major difference between a type herbarium sheet and a type rock slab would be that the former is put together by design and the latter by chance.

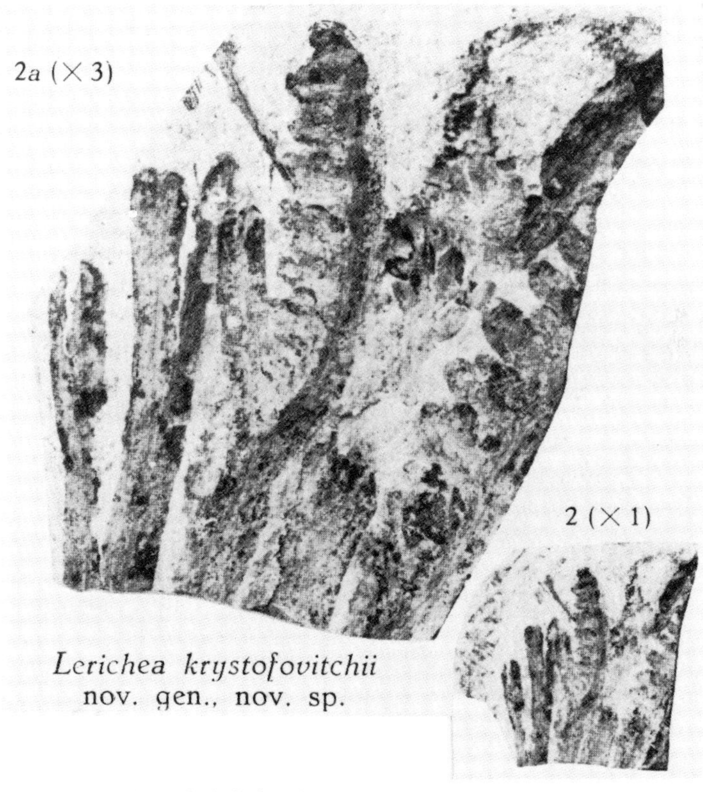

Fig. 18.2. *Lerichea krystofovichii* Stockmans [Holotype] (from Stockmans, 1968, with kind permission of the Institut Royal des Sciences naturelles de Belgique).

Fig. 18.3. *Neuropteris tenuifolia* (Schloth.) Sternbg. [Holotype] (from Schlotheim 1820).

This interpretation of holotype, lectotype and neotype is not intended to preclude the designation of one particular plant fossil specimen as the type. For example in their description of a new species of Charophyte, *Palaeonitella tarafiyensis*, Hill and El Khayal (1983, Fig. 12) illustrated a piece of rock here reproduced in Fig. 18.4. It displays many charophyte stems conforming to the description of the new species, but one of them is arrowed on the figure; the figure caption and the text state clearly that this particular plant specimen is the selected holotype. It is also common practice to arrange the photograph of a holotype in the protologue so as to display a limited area of the total rock, and thus define one selected plant fossil amongst many. In the earlier literature a similar purpose was achieved by line-drawings. A combination of photography and line-drawings can be used. For example Harris (1964) used this combination to illustrate *Pteroma thomasii*. His photographs of the hototype (1964, pl. 7, Figs. 10, 11) show more than one plant fossil conforming to the type description, but his line-drawings (Figs. 66A, G) show only one, which he designates as the holotype.

To summarize, I have suggested that where a rock specimen has been selected as the type (whether holotype, lectotype or neotype) on the basis of containing identical, but organically unconnected plant specimens, the presently required selection of a single plant specimen from them as a lectotype may be of use, but is perhaps not essential. Either the whole block could be designated as the type, or a particular specimen could be

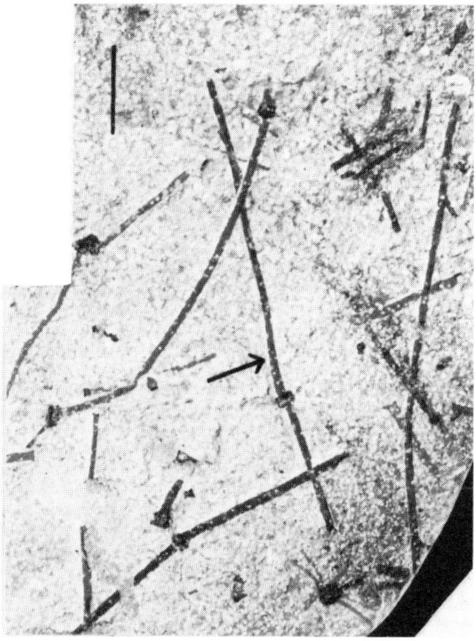

Fig. 18.4. *Palaeonitella tarafiyensis* Hill and El Khayal [Holotype] (from Hill and El Khayal 1983, with kind permission of the Trustees of the British Museum — Natural History).

designated as the type. Above all it would be important, in cases of possible ambiguity, that authors made clear which option was being chosen.

Terminology of Type Material

The palaeobotanical type is often subject to intense investigation because sometimes the type is the only specimen, or one of a few, and also because each specimen of the same species can, because of preservational differences, provide complementary information. This intense investigation can mean that the type specimen is divided into many portions and preparations. *Article 9.4* of the ICBN is devoted to this aspect of palaeobotany. It states: 'One whole specimen used in establishing a taxon of fossil plants is to be considered the nomenclatorial type. If this specimen is cut into pieces (sections of fossil wood, pieces of coal-ball plants, etc.), all parts originally used in establishing the diagnosis ought to be clearly marked'.

In the Palaeobotany Section of the British Museum (Natural History) we use an informal system of designation and terminology related to the bits and pieces that often constitute a type. Examples of the terminology are terms such as 'from the holotype', for example, for figured or unfigured cuticle preparations, and 'portion of holotype' for an unfigured piece of wood. But this terminology is for our use in the Museum and terminologies used elsewhere vary.

In practice it is helpful if portions of the type *not* used in establishing the diagnosis are also clearly marked. Palaeobotanical type material is increasingly subject to re-investigation as techniques improve and new questions are asked. This trend means that more characters are likely to be examined than were available to the author of the protologue and that more of the type material becomes used.

Suggested modifications to the term type, as applied to holotype, lectotype and neotype which would enable more formal attribution to be given to type material and provide a terminology of more general use, are as follows:

(1) Only material figured or designated in the protologue to be called Type.
(2) All material prepared from the same specimen as the Type (including the Type itself) to be regarded as the Integral Type.
(3) Material from the Integral Type, contemporary with, but not figured or designated in the protologue to be called pars-Type.
(4) Material from the Integral Type prepared subsequent to the protologue to be called ex Type or ex pars-Type.

So (1) and (3) would define 'parts used in establishing the diagnosis', and distinguish those portions figured or designated in the protologue—the 'Type', and portions not figured or designated in the protologue—the 'pars-Type'. Historically, palaeobotanical type material has been distributed between individuals and institutions on the principle of disseminating type material. Although all of it is type material, the distributed pieces can have different type status. This can lead to complications. For example, as a result of such distribution the British Museum (Natural History) has material from Cotta's (1832) petrified axes of *Psaronius* which are, in fact, mirror images of the holotype material of his protologue. At present we have no formal designation for the Museum material which would distinguish it from the material figured in the protologue, but with my new suggestions it would be referred to as pars-holotype, and the originally figured specimen as the holotype.

Item (4) above would identify portions of the type *not* originally used in establishing the diagnosis, that is, portions of the type not available to the author of the protologue. Also, item (4) distinguishes whether this different material is prepared from pieces of the type figured or designated in the original description—the 'ex Type'—or prepared from pieces of the type not figured nor designated in the original description—

the 'ex pars-Type'. To return to the Cotta example: if slides were prepared from Cotta's original collections they would be called ex holotype; if prepared from the British Museum (Natural History) specimens they would be ex pars-holotype.

Item (2) would serve as a reminder that the type, as such, is *all* the available material of the specimen designated as type. For the Cotta example the total number of pieces and preparations would collectively represent the Integral Holotype. In the case of coal-ball and similar fossils only pieces and preparations of the plant specimen under discussion would form the 'Integral Type', since unlike the Cotta example of petrified axes, there would normally also be other pieces and preparations unrelated to the plant under discussion. Replicas, photographs and line-drawings would not, however, be regarded as part of the 'Integral Type'.

Conclusions

These suggested additions to the terms holotype, lectotype and neotype are more elaborate than present formal requirements, but have the merits of practicability both for purposes of research and curation. In this paper my objective is hopefully to promote discussion amongst palaeobotanists and neobotanist who deal with type material, and to promote more care over how people designate their type material within the context of the existing rules of nomenclature.

Acknowledgements

I should like to thank Dr C.J. Cleal, Dr C.R. Hill and Dr C.J. Humphries for critically reading the manuscript.

References

Cotta, C. Bernhard (1832) *Die Dendrolithen in Beziehung auf ihren inneren Bau.* Dresden and Leipzig.

Croft, B.A. and Lang, W.H. (1942) The lower Devonian flora of the Senni Beds of Monmouthshire and Breconshire. *Phil. Trans. R. Soc. Series B,* **231**, 85–216.

Crookall, R. (1965) Fossil plants of the Carboniferous rocks of Great Britain. (Second section). *Mem. geol. Surv. Gr. Br. Palaeont.*, **4**, 85–216.

Harris, T.M. (1964) *The Yorkshire Jurassic Flora. 2. Caytoniales, Cycadales and Pteridosperms.* Brit. Mus. (Nat. Hist.), London.

Hill, C.R. and El Khayal, A.A. (1983) Late Permian plants including Charophytes from the Khuff Formation of Saudi Arabia. *Bull. Br. Mus. nat. Hist.* (Geol.), **37**, 105–12.

Schlotheim, E.F. (1820) *Die Petrefactenkunde auf ihrem jetzigen Standpunkte durch die Beschreibung seiner Sammlung Versteinerungen. Ein Beitrag zur Flora der Vorwelt.* Gotha.

Sternberg, K. von (1833) *Versuch einer Geognostisch-botanischen Darstellung der Flora der Vorwelt.* Vol. 2, pts. 5, 6, Leipzig.

Stockmans, F. (1968) Vegetaux Mesodévoniens récoltés aux confins du Massif de Brabant. *Mém. Inst. r. Sci. nat. Belg.,* **159**, 1–49.

Voss, E.G. Burdet, H.M. Chaloner, W.G. Demovlin, V. Hiepko, P. McNeill, J. Meikle, R.D. Nicholson, D.H. Rollins, R.C. and Silva, P.C. (1983) *International Code of Botanical Nomenclature*; Thirteenth International Botanical Congress, Sydney, 1981. *Regnum veg. III*

19. Computerized palaeobotanical data bases: the way forward?

R.A. SPICER

Abstract

The growing volume of collected plant fossil material, the development of sophisticated comparative morphological studies, and the acquisition of more detailed geological data relevant to specimens are placing increased demands on palaeobotanists and exposing weaknesses in conventional nomenclatural and taxonomic practices. Computerized data bases (CDBs) offer a logical solution to problems associated with the curation and comparison of large numbers of specimens. The structure of an operational data base designed for palaeobotanical use is discussed.

The introduction of CDBs, however, may impose potentially unacceptable constraints upon innovation. They demand a degree of formalization of input, and if they are to be used for taxonomic or identification purposes standard character sets are required. The problems of devising detailed character sets are discussed. To maximize their efficiency CDBs need to be large and internationally accessible. This imposes practical problems of consistency between those responsible for loading information. Automated character scoring in order to achieve objectivity and remove loader dependency is discussed. Future possibilities may include character assessment using laser technology.

Introduction

The ultimate aim of palaeobotanical systematics is to arrange specimens in a way that reflects phylogeny. We strive towards a natural classification. However, our organizational framework for achieving that goal rests on a classification approach that was devised in the eighteenth century when species were considered immutable and the purpose of taxonomy was to reveal the pattern of Creation. Furthermore, the classification system was designed to categorize whole organisms not isolated organs. Our nomenclatural practices continue to encourage the allocation of names to isolated specimens at an early stage in evaluating their true relationships. Once these names are formally allocated, altering them in the light of new evidence is often a tedious if not impossible task. Quite clearly our methods of nomenclature and classification may not be the most appropriate for elucidating or even complementing evolutionary pattern under all circumstances.

Traditionally the main thrust of palaeobotanical work has centred on anatomical and morphological studies. Under these circumstances individual specimens are intrinsically of prime importance and require names. More recently attempts have been made to broaden palaeobotanical investigation by introducing objective or quantitative sampling methods and detailed studies of assemblages in a sedimentological context have yielded a variety of new material and information (for example Scott 1977, Spicer and Hill 1979). In some respects the intrinsic value of individual specimens is lessened by this approach in that, for practical reasons, all specimens are not retained in perpetuity and the emphasis moves towards the relationship between specimens and sedimentological phenomena. In some respects more traditional nomenclatural practices are inappropriate for this type of work, involving as it does the handling of populations of organs exhibiting wide morphological variation. Many specimens, although morphologically similar to each other and to type specimens, may fall outside already proscribed taxonomic limits. The examination of larger numbers of specimens demanded by the quantitative approach has already been shown to lead to an increased yield of forms previously considered rare (Spicer and Hill 1979). If quantitative studies proliferate they could conceivably give rise to numerous taxonomic revisions.

Increasingly we see the need to evaluate global biotic patterns in time. Plate tectonic and palaeoclimatological research require palaeobotanical input as a line of constraint, and yet our established practices are hardly appropriate for integrating and evaluating plant fossil data on a global scale, even at relatively narrow time slices, or maintaining up-to-date revisions of taxa based on specimens from different geographical areas.

Established taxonomic practices in palaeobotany are clearly often inadequate for the demands now placed upon them. They are centred around individual specimens or suites of specimens that often do not have appropriate curatorial or locality data associated with them. Nomenclature often obscures this inadequate situation and hinders its rectification. For a more detailed critique of established practices see Hughes (1970, 1978 and this volume). Computerized data bases (CDBs) and a coherent research effort into possible alternative systematic approaches have the potential to solve these problems and are clearly overdue, but computerization of palaeobotanical data is not straightforward and in some respects may be disadvantageous.

Existing palaeobotanical data bases

Palaeobotanical data bases have been in existence for a very long time. Scientific collections of specimens in museums are physical data bases. Their essential feature is that the material is arranged in an orderly way that facilitates the location of any desired specimen or specimens. Each specimen, apart from its intrinsic qualities, is associated with a range of information that increases the value of the specimen. This information usually consists of some kind of nomenclatural label (a name), an authority, and locality information. The process of retrieving a specimen within the collection involves an understanding of how the collection is arranged.

Museum collections are, of course, fundamental to taxonomic studies and will always remain the ultimate source of reference. However, they do suffer from serious drawbacks. They are physically bulky, which puts a limit on the speed with which specimens can be located and compared with each other. Individual collections are geographically separate, making direct comparison of specimens a somewhat tedious process. As more specimens are described and the volume of collected material increases these problems become more acute.

Another limiting factor is the human brain. In order to effectively utilized these collections, large numbers of specimen attributes have to be memorized and evaluated in an objective manner. An intimate knowledge of several hundred specimens may be achieved by some workers but for many detailed familiarity of this kind is impossible to achieve. Direct comparison of two specimens on a one to one basis is relatively easy, but the problem increases rapidly as more specimens are involved.

Curatorial applications of computerized data bases

To a large extent computerized data bases can overcome these problems. The purpose of a CDB can be divided into two parts: curatorial and taxonomic. The most obvious application is purely curatorial. All information relating to the identity and source of individual specimens can be stored and accessed with great speed and accuracy. The type of information concerned here is shown in Fig. 19.1. Each specimen has a unique accession number that can be related to any amount of attribute information such as genus and species names, the author who named and described the specimen, locality information, including age and

SPECIMENS	FAMILY	GENUS	SPECIES	AUTHOR	LOCALITY
0001		LEPIDODENDRON	DICHOTOMUM	STERNBERG	NR BARNSLEY
0002		LEPIDODENDRON	ACULEATUM	STERNBERG	BOHEMIA
0003					
0004					

Fig. 19.1. Curatorial information stored in an inefficient 'Ledger' format where there is considerable repetition of record elements.

geographic origin. A potential user of the data base might want to list all the specimens belonging to a given genus or species or that were recovered from a given locality. To be universally useful, however, the program responsible for operating and maintaining the data base (the data base manager) must be capable of allowing a search on any or all attributes and returning any or all attributes to the user. If the data base were arranged in the same way as a specimen ledger or card index (Fig 19.1), it would be extremely wasteful of computer storage space. For example, if there were twenty specimens of the genus *Lepidostrobus* in the data base the name *Lepidostrobus* would have to be stored twenty times. Any attribute would have multiple records and a large part of the data base would be repetitive. Multiple attribute entries of this type would also slow down searches because the attribute columns or files would be the same length as the total number of specimens in the collection. Should data need to be appended or deleted from any attribute file, the entire data base would need to be accessed perhaps even reorganized.

Most modern CDBs are constructed somewhat differently to this simple and inefficient tabular form. They may take the form of multi-dimensional tree structures or even more complicated plexes where diagrammatically they appear as highly complex webs. While these may be suitable for certain applications, the flexibility demanded of a palaeobotanical data base dictates that the most efficient organization is one based on relational algebra (Codd 1970, 1972). Known as relational data bases, the information they contain is conceptually organized in a

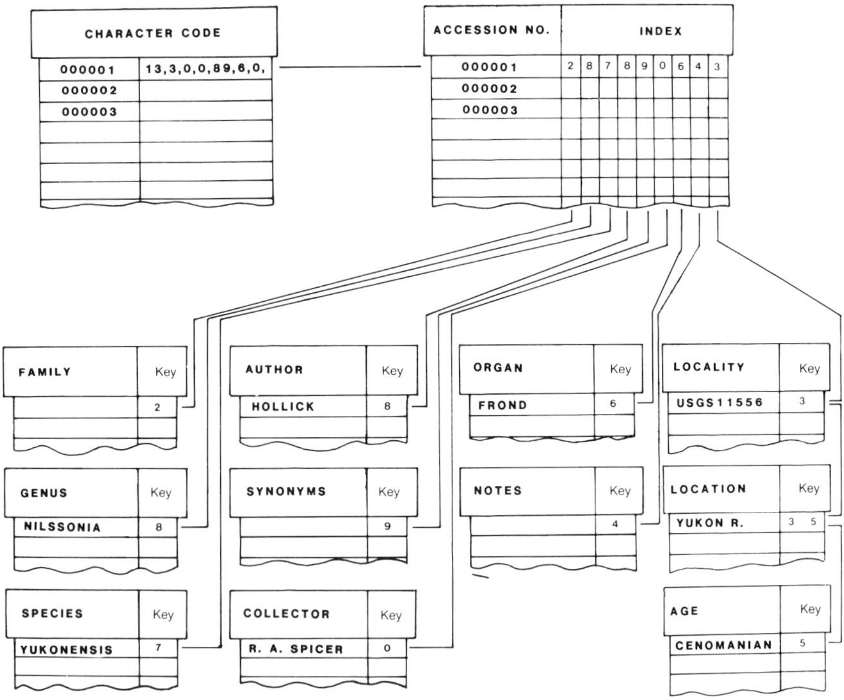

Fig. 19.2. Diagram of one possible way in which a Palaeobotanical Data Base might be organized to allow maximum flexibility of search and retrieval. See text for explanation.

two-dimensional form that maximizes flexibility both in searching and subsequent addition of data, and provides the optimum solution to the requirements of high speed searching and efficient storage. The following outline of a CDB (Fig. 19.2) is based upon an actual system that I designed and programmed to handle the curatorial aspects of the angiosperm Cleared Leaf Collection assembled by J.A. Wolfe of the US Geological Survey. The CDB was designed and made operational in 1978 and probably now could be improved upon, but it serves as an example.

Every specimen must be uniquely identifiable. This is achieved by allocating to each specimen an accession number which may be an actual number or a mixture of alphabetical and numerical characters. These are stored in an accession number file which is the longest file in the data base in terms of numbers of items. Linked to each accession number is an index file. This file contains a set of numeric keys which relate to attributes of the specimens stored in other files. The sequence of the keys in the index is fixed so that the first key relates to, for example, the file containing the generic names, the second key to the species file and so on.

Very few of the attributes are determined by another. Species names are not determined by the genus name because several genera may have the same specific epithet associated with them, for example '*thomasii*' or '*pustulosa*'. Similarly, an authority is not uniquely associated with a genus name because whereas a given genus was only ever defined by one authority, other authors use that genus to describe their own species. Because a prime requirement for the data base is flexibility and we want total freedom of search pattern each attribute file is independent of the others and accessed only through the index keys. In relational data bases ultimate flexibility is achieved by having all the information stored as binary files (files with two attributes in each record) (Martin 1977). Inevitably this leads to some inefficiency as regards storage in that there is a proliferation of keys.

Locality information is somewhat different. Each locality uniquely defines a geographical location and a geological age. This information is therefore related and for storage and search purposes forms an identifiable group of attribute files. The unique identifier in this set of files is the locality number which also relates to the specimens (and all the other information) via the index file.

In an ideal world the organ type would be uniquely related to the genus name, for example *Lepidostrobus* would be a cone by definition. However, there are instances where generic names have been used to refer to more than one type of organ. An example here might be *Pterospermites* which originally was a name of a seed but was subsequently applied to leaves (see Wolfe 1966). It is not the purpose of a CDB to correct erroneous nomenclature that already exists but merely to store it. This means that the CDB has to be capable of storing information that does not necessarily conform to the current rules of nomenclature.

Search procedure

Suppose we wish to search our data base for all occurrences of the genus *Lepidodendron* and find out which species have been recovered from which locality. The first step in the procedure for the data base manager would be to call the search program (Fig. 19.3) and examine the genus file for the word *Lepidodendron* to find out its unique key. Having found the key, the second column of the index file is searched to find all the occurrences of that key. When a hit is scored, a copy of the accession number and the complete index entry for that specimen is placed in a temporary store, called a search buffer (Fig. 19.3). Now we would like to know which species have been found at which locality. In order that the output be meaningful at a later date we may as well request that the genus name be returned as well. The search program scans the index

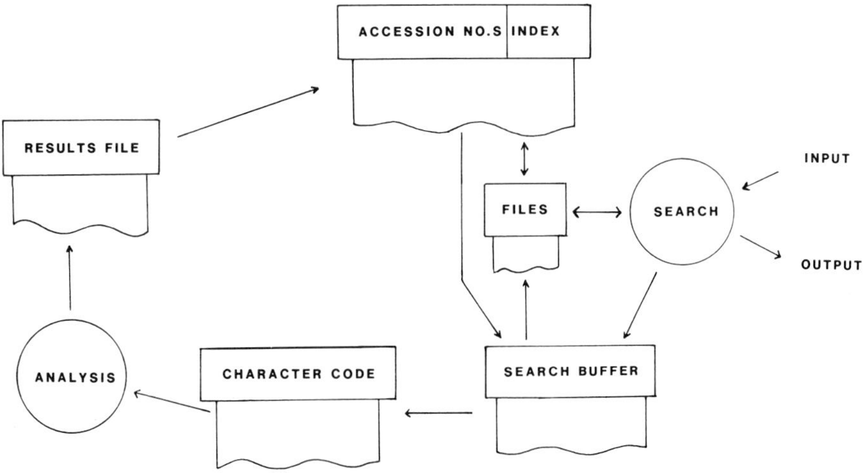

Fig. 19.3. Diagram of search procedure. See text for explanation.

files in the buffer for the keys stored in columns 2, 3 and 9 and looks up the names associated with each set of keys in the respective attribute files and returns this information to the output device which is usually the terminal. Having seen this information we may decide that we would like to know which museum currently holds these specimens, so a further search on the indices stored in the search buffer would provide the museum keys from column 8 and the museum names would be returned from the museum attribute file.

The actual process of searching is far more complicated than is apparent from this simple outline. It involves instructions as to how to scan the index file, comparing each value stored with those requested, locating the appropriate attribute files, ensuring search is started at the beginning of a file and so on. A good data base manager not only should allow for total flexibility of search and data retrieval in the shortest time, but it must also be easy to use. The process of searching and even the structure of the data base itself should not be apparent to the user and need not be known before efficient successful searches can be carried out.

If a palaeobotanical data base were to be set up and operated on an international scale, carrying out a search should be as straightforward as possible. In fact, the nature of palaeobotanical information is such that it would be possible for the data base manager to enquire of the user what his or her wishes are at every step in the process so that no prior knowledge of the system and no programming experience are required.

In the data base manager designed for the Cleared Leaf Collection the search program itself interacted with the user by asking a series of simple questions to guide the search. If erroneous or unintelligible answers were given, the user was politely requested to try again. An example of the search using the Cleared Leaf Data Base is given as follows with computer generated statements indicated by an asterisk:

*How many files do you wish to search on?
1.
*File 1?
Genus
*Element?
Magnolia
*Information to be returned
*Enter number of files
2.
*File 1?
Genus
*File 2?
Species

**Magnolia grandifolia*
**Magnolia stellata*

*2 individuals
*Do you wish to perform statistics?
No
*End of search

Data was loaded into the system via an input program. This program not only allowed the construction of new files, and the addition of information to existing files, but incorporated an editor that facilitated correcting and updating of files within the data base. In an open access data base use of this program would be strictly limited to those responsible for data base construction and maintenance.

The organization and formalization of data into the file structure outlined here (which are characteristic of most modern data bases) impose some limitations. Anyone who has used a bibliographic data base knows how critical it is to specify unambiguous terms. For example, if one wanted information on fossil floras in polar regions and requested references on 'high latitude plants', one should not be too surprised to find among the useful references articles on iron smelters in northern Norway!

The effectiveness of searches is clearly reduced if there is a large proportion of data missing. This problem is not unique to CDBs however. Perhaps more important is that CDBs cannot exercise judgements as we know them. Searches can only be carried out from the information present in the attribute file. This may sound ridiculously obvious, but it could be very frustrating searching for all material from, for example, Cayton Bay, when the locality file contains some specimens recorded as coming from the Gristhorpe Beds near Scarborough. Such specimens would not be retrieved. Not only must the information be as complete as possible it should be in the same format. International agreement as to the format used is essential and yet may pose the single most important obstacle to data base construction.

Experimental systematics

One part of the palaeobotanical CDB that has not yet been discussed is the file (or files) containing the 'character code'. This is the description, in digital form, of the particular fossil under consideration. It allows the data base to become more than just a catalogue of curatorial information. The properly constructed character code could allow not only the refinement of our perception of morphological relationships for conventional taxonomic purposes (identifying unknown specimens, moving specimens between existing taxa, or erecting new taxa) but could provide the basis for establishing more natural classifications for refining phylogenies and the development of parallel innovative taxonomic systems more suited for handling dispersed isolated organs for biostratigraphic purposes.

Whether one adopts a phenetic approach to classification or attempts a cladistic analysis, the characteristics of the specimens need to be translated into a form which can be readily stored and manipulated by the computer. Translating descriptive terminologies (e.g. Hickey 1979) into digital form (Dolf 1976) is fundamentally inappropriate because descriptive terminologies are usually designed for the specific purpose of maximizing the efficient use of text while at the same time minimizing ambiguity. Characters recognized for these purposes are unlikely to be useful for phenetic classification because they may well be interlinked or redundant in mathematical terms when assessing relative similarity or dissimilarity in specimen populations of fluctuating size and composition. Furthermore, 'terminological' characters may be composites of more fundamental character units, and whether fundamental or composite terminological characters may not be homologous and therefore inappropriate for cladistic analysis. A better approach is a character scoring scheme for numerical taxonomic purposes (e.g. Hill 1980).

Ideally there should be a fundamental set of characters to define a specimen, but even if such characters were to be identified they could not be universally utilized because the underlying rationale for various classificatory techniques is different. Supposing it were possible to agree on a specific classificatory technique, would it be a realistic proposition to construct a character code file for data base use? Theoretically the answer is probably yes, but there are immense practical problems if the data base is to be large enough to be worthwhile. The primary constraint is the time taken to score or numerically describe each specimen. Scoring has traditionally been carried out by inspection and requires the expertise of a skilled taxonomist. In spite of claims to the contrary this approach involves a degree of subjectivity. Decisions have to be made based on judgement and to a certain degree on prior knowledge of character form and distribution within the population under study.

The number of specimens involved in any reasonably sized data base dictate that it would be impossible for any one person to construct the entire character code file. The consequence of this is that more than one person would be required to score the specimens which in turn means that personal bias in scoring becomes a factor. This is significant when attempting to achieve consistent taxonomic partitioning. To overcome this it would be necessary to delegate responsibility for character scoring to the computer itself. This could be achieved by having relatively unskilled (cheaper) personnel digitizing (either by tracing on as digitizing pad or by optical means) predetermined aspects of the specimens in a straightforward manner and using a programmed library of diagnostic patterns within the computer to assign the character states. The intense research effort directed at pattern recognition and artificial intelligence makes it likely that suitable cost-effective technology will soon be available for palaeobotanical research efforts in this area, although it would require a period of experimentation in a data base environment before systems were acceptable for routine palaeobotanical work. One of the main causes of problems in direct recognition of pattern inherent in specimens is the variation due to biological or preservational factors. Any system which depends upon critical measurements of distances, thicknessses, or angles is likely to yield spurious data due to factors other than those of direct taxonomic significance (for example preservation state). Although that is not to say that variation and lack of consistent pattern are not in themselves taxonomically useful.

A possible future solution to these problems, and one which also retains the possibility of using variation in a constructive manner, could well be based upon laser technology. Cairns *et al.* (1977, 1979) described a method of identifying diatoms using auto and cross-correlations of patterns generated by passing coherent light through frustules (Fig. 19.4). This technique involved comparing variations in signal strength when photographs of diatoms on 35 mm film were illuminated by laser light and the resultant image passed through various holographic images of other diatoms. Pattern repetition is a feature of many plant fossils and it may prove possible to utilize diffraction and interference effects to characterize specimens. The advantages here are twofold. Firstly, the complexity of pattern in the specimens is simplified to an array of more or less discrete points or lines which can be described in

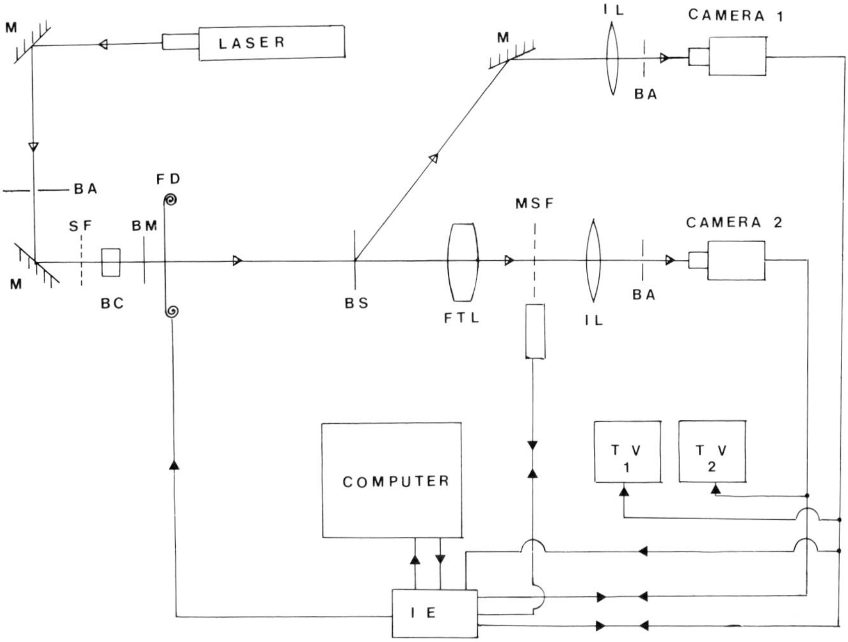

Fig. 19.4 Diagram of a coherent optical processor. M-Mirror, BA-Beam attenuator, SF-spatial filter, BC-Beam collimator, BM-Beam modulator, FD-Computer controlled 35 mm film drive, BS-Beam splitter, FTL-Fourier transform lens, MSF-Computer controlled array of matched spatial filters, IL-Imaging lens, IE-Interface electronics. For a description of the processor see Cairns *et al.* 1977 and 1979 (redrawn from Cairns *et al.* 1979).

mathematical terms far more easily than can the pattern possessed by the specimen itself. Secondly, variation or the departure from regular pattern repetition is eliminated in terms of diffraction pattern organization (specimens can therefore be compared more readily) but is preserved in a degraded difinition of that pattern. The relative intensity of points or lines between diffraction patterns from different specimens could be used as a comparative estimate of variation or levels of pattern organization exhibited by the specimens themselves. Put in another way, variation is separated from inherent pattern and the degree of pattern organization (for example as in leaf rank; Hickey 1971) can be quantitatively assessed.

Conclusions

The concept of investigating alternative systematic approaches may seem radical, but palaeobotany could hardly be regarded as a science if an obviously inadequate code of pratice were not re-evaluated and scrutinized for possible improvement when appropriate technology became available to do so. Computational techniques, and in particular computer data bases, provide a means for expanding and improving our research capabilities, as well as providing a vehicle for experimentally testing the appropriateness of the established systematic framework. In their curatorial role CDBs offer high speed retrieval of information and access to large amounts of data which must improve and speed up the nomenclatural revisions, even within the existing code of practice. This information can be made globally available (apart from political constraints). Biostratigraphy, palaeoecology, and palaeobiogeography are areas of research that could be significantly enhanced depending upon the quality of locality and associated geological data. The inherent constraints of data base organization search procedures would force more critical and detailed accummulation of associated information. Perhaps the most severe problems connected with the establishment of a workable palaeobotanical data base would be the need for international cooperation and standardization of data. Even if agreement could be reached as to a workable format for data entry, the formalization of data in this way could be perceived as a limitation on the flexible application of the data base system itself.

With respect to innovation in systematic approaches to palaeobotany there would seem to be practical constraints of various types at the present time. Existing approaches to character scoring and manipulation need to be rigorously tested across the entire range of organ types and evaluated in a data base environment. Nevertheless, even if technical problems could be overcome the practical limitations of manual character assignment are formidable, being both expensive and dependent to some extent on the individual coding the characters. These problems could, however, be overcome by automated character assignments. Variation due to biological or preservational factors cannot be ignored and may be an integral part of defining taxa, but these could present significant problems to machine-based character recognition and scoring. Techniques which separate variation from fundamental pattern are required and advances in this field may come from applied laser technology.

References

Cairns, J. Jr., Dickson, K.L. and Slocomb J. (1977) The ABC's of Diatom Identification Using Laser Holographs. *Hydrobiologia*, **54**, 7–16.

——, Dickson, K.L., Pryfogle, Peter, Almeida, S.P., Case, S.K., Fournier, J.M. and Fujii, H. (1979) Determining the accuracy of coherent optical identification of diatoms. *Water Resources Bulletin. American Water Resources Association* **15**, 1770–5.

Codd, E.F. (1970) A Relational Model of Data for Large Shared Data Banks. *Comm. ACM 13, No. 6., ACM (Association for Computing Machinery).* New York, London and Amsterdam.

—— (1972) Further Normalization of the Data Base Relational Model. In *Courant Computer Science Symposia V6: Data Base Systems* (ed. R. Rustin). Prentice-Hall Inc., Englewood Cliffs, N.Y.

Dolf, G.E. (1976) Taxometric partitioning of Leaf Collections. *Palaeontographica*, Abt. B **156**, 65–86.

Hickey, L.J. (1971) Evolutionary significance of leaf architectural features in the woody dicots (Abstract). *American Journal of Botany* **58**, 469.

—— (1979) A revised classification of the architecture of dicotyledonous leaves. In *Anatomy of the Dycotyledons* (C.R. Metcalfe and L. Chalk), Clarendon Press, Oxford.

Hill, R.S. (1980) A numerical taxonomic approach to the study of angiosperm leaves. *Botanical Gazette* **141**, 213–29.

Hughes, N.F. (1970) Remedy for the general data handling failure of palaeontology. In *Data Processing in Biology and Geology* (ed. J.L. Cutbill), Systematics Association Special Vol. 3, pp. 321–30.

—— (1978) Proposal for a new Palaeobotanical appendix for the International Code of Botanical Nomenclature. *Taxon* **27**, 497–504.

Martin, James (1977) *Computer Data-Base Organization*. Second Edition. Prentice-Hall, Englewood-Cliffs, N.J.

Scott, A.C. (1977) A review of the ecology of upper Carboniferous plant assemblages with new data from Strathclyde. *Palaeontology* **20**, 447–73.

Spicer, R.A., and Hill, C.R. (1979) Principal Components and Correspondence Analyses of quantitative data from a Jurassic plant bed. *Rev. Palaeobotany and Palynology* **28**, 273–99.

Wolfe, J.A. (1966) Tertiary Plants from the Cook Inlet Region, Alaska. *U.S. Geological Survey Professional Paper.* **398–B**.

20. The evolutionary state of palaeobotany

K.L. ALVIN

Abstract

The adoption of a biological and increasingly experimental approach is having a profound effect upon palaeobotanical studies. These approaches, together with more critical and disciplined methods of assessment of evolutionary relationships, greatly increases the quantity and quality of information obtainable from plant fossils. The papers presented in this volume show how modern techniques of chemical, microscopical and numerical analysis offer exciting potential for progress in plant fossil systematics and taxonomy.

An overview

A symposium on systematic and taxonomic approaches provides a timely opportunity for those intimately involved in the pursuit of knowledge within their particular area of the subject to stand back from the intimacies and to view the broader aims within the discipline and to assess how their involvement contributes to some of the basic and fundamental problems in palaeobotany.

In palaeobotany, as with any intellectual discipline, attitudes and approaches change with time. Problems may persist but the ways in which they are viewed and tackled change. Although it may seem facile to recognize labelled phases in the development or evolution of a subject, historians have found it a useful means of concentrating the mind on those features which best characterize given periods. I would label the present phase in the evolution of palaeobotany the biological phase. Palaeobotanists, most of whom nowadays have probably been trained as biologists, are more deeply aware than they have generally been in the past of the fact that they are dealing with living organisms. They are therefore less interested in the static morphology and tidy utilitarian classification of their specimens and much more so in them as representing the remains of whole plants with physiological, biochemical, and genetic characteristics and interacting with their environment as members of communities of living organisms. Classification and taxonomy, of course, remain of basic importance and the long term aim of achieving understanding of plant evolution is central, but today's palaeobotanist increasingly questions traditional conventions in the approaches to these and seeks progress on the basis of understanding the plants in biological terms. After all, it is populations of whole living plants that evolve, not morphologically defined organs.

At every phase in the evolution of his subject the palaeobotanist has been conscious of the special difficulties presented by his material compared with living plants. He has tried to devise a generally acceptable taxonomic system by adapting and modifying the current system used for living plants. In this volume all the main aspects of this problem are reappraised.

One of the main aims of the palaeobotanist and one of crucial importance in the context of the biological approach is to reconstruct whole plants from the scrappy, imperfectly preserved, and usually distorted remains. Since most fossil plants are preserved as compressions or impressions, an understanding of the ways in which these are formed is fundamental to an accurate assessment of the original morphology and habit. Experimental modelling techniques, such as those described by Rex, are a valuable, essentially objective means of reconstructing the life form of individual organs from compressions. Far too little has been done in this direction and the interpretation of the original form of organs, particularly those preserved as impressions where there is a lack of any original organic substance, has been little better than sheer guess work. The importance of considering not only the effects of pressure and different kinds of sediment but also internal structural features in this exercise cannot be overstressed. Material of more or less uniform composition will not behave in compression in the same way as a plant organ of highly heterogeneous anatomy and we particularly need more results from experiments in order to establish some principles concerning such effects.

Comparison of different preservational forms presents a major and familiar challange of special importance perhaps in the context of the late Palaeozoic where both permineralized and compression forms of equivalent structures occur commonly. Potentially, and indeed in practice, the availability of more than one preservational form represents an advantage in terms of the biological information that is available in the material and it is noteworthy that it is particularly where both good compression and permineralized forms of the same biological entities have been linked that our knowledge of plants as whole organisms is most complete. Perhaps for this reason the problem of devising an acceptable taxonomic practice to cope with this situation is seen as one of special importance. Galtier, discussing this problem, categorizes correlations and the taxonomic level at which it is possible to recognize them, taking examples from various plant groups. Reproductive organs, such as fern sporangia and pteridosperm pollen-bearing parts, offer especially valuable means of correlation at both specific and generic levels. The conservatism of the vegetative anatomical organization in some of the more important Carboniferous groups, especially the calamophytes and lepidophytes, emerges strongly from such comparisons. Relative conservatism in one or more aspect of organization within a group or even between major groups such as, for example, wood anatomy in gymnosperms, presents a major problem in plant reconstruction.

The taxonomic difficulties consequent upon plant dismemberment and different preservational states with their differing levels of information are discussed by several other contributors. At the conference, Chaloner, Collinson, Cleal, Hughes, Leistokow*, Shute, Spicer, and Stockey* all focused on different aspects and pointed to a number of shortcomings in the existing International Code. Even the concept of the 'Type Specimen' as pointed out by Shute, needs clarification and tightening. What do we do if the designated 'type' is a block with perhaps numerous fragments which may have come from as many individual plants?

The importance of establishing a code of good practice on the basis of examples was stressed by Chaloner who commented on the limited parallelism between the nomenclatural problems presented by fossil plants with those of living fungi with pleomorphic life-cycles*. It is surely important in the interests of precision of meaning that names established for different organs or different preservational forms continue in use within the confines of their definition. At the generic level for example, name pairs like *Archaeopteris/Callixylon* should never be thought of as synomyms, otherwise we make assumptions which may be unjustified. The state of our knowledge is that certain species of *Archaeopteris* foliage are known to have the wood anatomy characteristics of *Callixylon*. We cannot assume that all trees with *Callixylon* wood had *Archaeopteris* foliage or *vice versa*. The likelihood as to whether they did or did not must be continually reassessed as our knowledge of examples increases. At the specific level, the evidence on which attribution of one organ to another is based must always be borne in mind before formal taxonomic decisions about naming the combined taxon or its parts are taken. Although such evidence as repeated association and anatomical

*in oral presentation

resemblances can provide very high probability of identity, the only absolute proof is organic connection. The keynote of establishing a code of good practice for formal taxonomic purposes must be wariness.

The need to view the definition and bounds of taxa as essentially temporary is something that palaeobotanists have often found too hard to accept. The kind of progress that has been made recently on those early land-plants we now call Rhyniophytina, and summarized here by Edwards and Edwards, underlines the need to keep under continual review our concept of named taxa, especially where we are dealing with ones which appear to be in a state of relatively rapid adaptive radiation.

In the context of Cretaceous and Tertiary angiosperm leaves, fruits, seeds, and pollen, some of the basic taxonomic problems are highlighted by Collinson, Hughes, and Spicer. The problem of whether to employ the name of the equivalent living taxon for fossil material is discussed critically by Collinson. Not to do so right up to the species level in the case of archaeological or Quaternary material (always given that the ancient material provides enough information) would surely be ridiculous and similarly, not to do so with older material would be unjustified and would result in a reduction in the value of plant records in palaeoclimatic and palaeoenvironmental interpretations. A geological or other time boundary is not a satisfactory basis upon which to change taxomic procedure. What is of prime importance is that the information contained in the fossil material and revealed by palaeobotanical work should not be obscured by nomenclatural practice and that the system used should provide an accurate basis for reference, comparison, and assessment of the limitation of the given record.

The implications of the fact that we may often be dealing with remains of organisms that are phenotypically variable or genetically unstable (rapidly evolving?) with the result that taxonomic boundaries will be blurred is also discussed by both Spicer and Hughes. A workable and practical system of nomenclature is essential, and if the traditional Linnaean system imposes boundaries and hierarchies where none may exist, then clearly this must be modified, or perhaps replaced by one which more accurately conveys the true status of the material that it labels.

An under-explored potential tool to increase characterization of plant fossils, and one which may eventually help in problems of classification and the correlation of different organs, is biochemical analysis. Thomas stresses the paucity of literature in his area of palaeobiochemistry and points to the more promising groups of substances such as flavonoids, phenolics, and terpenoids. As the presence or absence of a particular substance or even a group of substances is seldom of diagnostic value, palaeobotanists must be more ready than they have generally been in the past to employ numerical analytical methods for classification and comparison. Niklas and his co-workers (for references, see Thomas) are among the few to have utilized chemotaxonomic characters in palaeobotany and have seen this need.

Numerical or morphometric approaches are beginning to be employed particularly with material such as Tertiary leaf fossils. Burnham demonstrated at the conference the potential powers of such methods not only in the critical comparison of fossil material with extant species, but also in giving a valuable assessment of the degree of homogeneity of populations of fossil plant organs. The objectivity of such methods is their chief strength and the reward of the considerable time and effort inevitably required in collecting the raw data for analysis. There are numerous potential applications of these methods in palaeobotany and it is surprising that there have yet been so few carried out.

Numerical and computer methods have wider potential. They offer means of comparison and classification of fossil taxa particularly within large and complex groups where individual entities (species or populations) may not be readily distinguishable on the basis of 'spot' characters. Further, an increasing need to collect and store data from both fossil and comparative living material in order that it may be readily available for future use and to be of continuing and cumulative value requires the construction of complex data banks. Spicer in his discussion of this need and his assessment of available methodologies pointed tantalizingly to the futuristic idea of automated methods of fossil identification.

The value of simple computer methods to produce correlation lines among complex and geographically dispersed sequences of palynological assemblages is stressed by Grant who also provides something of a reminder that we, as predominantly biologically orientated palaeobotanists, have responsibility for the basis of biochronology which remains for most purposes the preferred method of division of geological time.

Problems of data handling press increasingly in palaeobotany just as in other branches of science where there is great dependence on accumulated information. We must clearly pursue vigorously the development of methods of convenient data storage. Such methods offer the possibility of utilizing more finely tuned data about fossil plants, their variation and distribution in time and space. Perhaps given adequate data handling methods a more 'record'-orientated resolution of information (in Hughes' sense) will become more workable and therefore of greater practicality than the traditional 'species' system.

Whether palaeobotanists will ever be of one mind concerning the value of cladistics as a discipline for assessment of evolutionary relationships is doubtful but as both Crane and Doyle demonstrated at the symposium in somewhat different contexts, it is a method of data processing combined with an intellectual exercise which can produce impressive looking diagrams, even tantalizing, especially when pertaining to the paranoia-provoking topic of angiosperm relationships! The extent to which major shifts in the relative positions of angiosperms, Gnetales, Bennettitales, *Caytonia*, and coniferopsids may be produced, according to Doyle's analysis, in cladograms by only small changes in levels of parsimony is worrying to the sceptic. It seems, as far as the problem of angiosperm origin is concerned, that we are really thrown back in an all too familar way on the need for a greater amount of factual information, not only about key fossil groups, but indeed also about relevant living groups. It is to be hoped that in the future this information will include chemistry, fine structure, and biology as well as morphology and anatomy.

However, the discipline of cladism need not be preoccupied with the production of cladograms. Rothwell tackles with the tool of cladistic thought the evolution and classification of the earliest gymnosperms and identifies the synapomorphies that define these plants, shattering the long established thesis that ovule symmetry is a character of prime importance. He convincingly argues that it is the organization of the pollen-receiving mechanism, an aspect of reproductive biology essentially, that characterizes the primitive gymnosperm. Other later gymnosperm groups possessed more derived mechanisms.

In the assessment of biological attributes of fossil plants we come to the heart of the modern approach to fossils. Such assessment is far from straightforward and is heavily dependent upon technique both in the field, to obtain the best preserved specimens (and enough of them), and also in the laboratory. Taylor and Stubblefield, in their discussion of reproductive organs and classification, and Rothwell within the more specific area of Carboniferous gymnosperm reproductive organs, most notably the pollination biology of their ovules, all stress the importance of good preservation to maximize this kind of information. The application of electron microscopy, especially TEM, which has been adopted relatively late by palaeobotanists, had led to considerable understanding of pollen development. Similar work on the ultrastructure of cuticles and the gymnosperm megaspore membrane, provided it is backed up by comparable work on extant plants, promises to yield further information of potential biological significance.

As Hill points out, modern microscopical techniques enable us to produce an enormous amount of information from such objects as pieces of plant cuticle and pollen grains. How can we make best use of this information? Modern methods of analysis and comparison are essential for good systematic and phylogenetic studies but also structural and morphological information must be regarded as potentially of value in throwing light on the ways in which the plants function and interact with their environment. Unfortunately we know far too little yet about the functional significance of morphological (including micromorphological) and anatomical features. In recent years we have gained considerable insight into the function of certain patterns of pollen morphology and exine structure. We also know that there is a clear relationship between aspects of wood anatomy and environment and between some angiosperm leaf features and climate (leaf physiognomy), but we have much to learn. Of special significance in relation to fossil plants is the acquisition of some understanding of the function of leaf architecture, leaf surface features, including hairs and perhaps especially stomatal configuration and cuticle organization. To what extent may such features in fossil plants provide clues about the ways in which the plants interacted with their environment and therefore the evolutionary pressures to which they were subjected? We may often bemoan the paucity of information of the kind we want about living plants, but to be constructive we should perhaps be more prepared than we have generally been in the past to look at living plants ourselves and to study them in ways which are going to produce this information.

And what more of the future? Systematic and taxonomic approaches today are a good deal more versatile and questioning that at any previous time in the evolution of our subject. I believe that the reason for this has been the fuller realization among palaeobotanists that their subject is essentially a biological one. The implications of this attitude of mind will continue to have a powerful influence on the way fundamental problems are tackled. We may expect to see greater use made of more objective methods of typification, comparison, and classification, such as are offered by modern data-handling techniques, and we may equally confidently expect increasing applications to be made of new techniques which will produce more detailed information, not only about the plants we study, but also about the ways in which the material we have at hand for study was formed from the original living organisms.

Index

Systematics Association Publications

1. Bibliography of key works for the identification of the British fauna and flora *3rd edition* (1967)
   Edited by G.J. Kerrich, R.D. Meikle and N. Tebble
   Out of print
2. Function and taxonomic importance (1959)
   Edited by A.J. Cain
3. The species concept in palaeontology (1956)
   Edited by P.C. Sylvester-Bradley
4. Taxonomy and geography (1962)
   Edited by D. Nichols
5. Speciation in the sea (1963)
   Edited by J.P. Harding and N. Tebble
6. Phenetic and phylogenetic classification (1964)
   Edited by V.H. Heywood and J. McNeil
   Out of print
7. Aspects of Tethyan biogeography (1967)
   Edited by C.G. Adams and D.V. Ager
8. The soil ecosystem (1969)
   Edited by H. Sheals
9. Organisms and continents through time (1973)†
   Edited by N.F. Hughes

Published by the Association

Systematics Association Special Volumes

1. The new systematics (1940)
   Edited by Julian Huxley (Reprinted 1971)
2. Chemotaxonomy and serotaxonomy (1968)*
   Edited by J.G. Hawkes
3. Data processing in biology and geology (1971)*
   Edited by J.L. Cutbill

4. Scanning electron microscopy (1971)*
Edited by V.H. Heywood
5. Taxonomy and ecology (1973)*
Edited by V.H. Heywood
6. The changing flora and fauna of Britain (1974)*
Edited by D.L. Hawksworth
7. Biological identification with computers (1975)*
Edited by R.J. Pankhurst
8. Lichenology: progress and problems (1976)*
Edited by D.H. Brown, D.L. Hawksworth and R.H. Bailey
9. Key works to the fauna and flora of the British Isles and north-western Europe (1978)*
Edited by G.J. Kerrich, D.L. Hawksworth and R.W. Sims
10. Modern approaches to the taxonomy of red and brown algae (1978)*
Edited by D.E.G. Irvine and J.H. Price
11. Biology and systematics of colonial organisms (1979)*
Edited by G. Larwood and B.R. Rosen
12. The origin of major invertebrate groups (1979)*
Edited by M.R. House
13. Advances in Bryozoology (1979)*
Edited by G.P. Larwood and M.B. Abbot
14. Bryophyte systematics (1979)*
Edited by G.C.S. Clarke and J.G. Duckett
15. The terrestrial environment and the origin of land vertebrates (1980)*
Edited by A.L. Panchen
16. Chemosystematics: principles and practice (1980)*
Edited by F.A. Bisby, J.G. Vaughan and C.A. Wright
17. The shore environment: methods and ecosystems (2 Volumes) (1980)*
Edited by J.H. Price, D.E.G. Irvine and W.F. Farnham
18. The Ammonoidea (1981)*
Edited by M.R. House and J.R. Senior
19. Biosystematics of social insects (1981)*
Edited by P.E. House and J.-L. Clément
20. Genome evolution (1982)*
Edited by G.A. Dover and R.B. Flavell
21. Problems of phylogenetic reconstruction (1982)*
Edited by K.A. Joysey and A.E. Friday
22. Concepts in nematode systematics (1983)*
Edited by A.R. Stone, H.M. Platt and L.F. Khalil
23. Evolution, time and space: the emergence of the biosphere (1983)*
Edited by R.W. Sims, J.H. Price and P.E.S. Whalley

24. Protein polymorphism: adaptive and taxonomic significance (1983)*
 Edited by G.S. Oxford and D. Rollinson
25. Current concepts in plant taxonomy (1983)*
 Edited by V.H. Heywood and D.M. Moore
26. Databases in systematics (1984)*
 Edited by R. Allkin and F.A. Bisby
27. Systematics of the green algae (1984)*
 Edited by D.E.J. Irvine and D.M. John
28. The origins and relationships of lower invertebrates (1985)‡
 Edited by S. Conway Morris, J.D. George, R. Gibson, H.M. Platt
29. Infraspecific classification of wild and cultivated plants (1986)‡
 Edited by B.T. Styles
30. Biomineralization in lower plants and animals (1986)‡
 Edited by B.S.C. Leadbeater and R. Riding
31. Systematic and taxonomic approaches in palaeobotany (1986)‡
 Edited by R.A. Spicer and B.A. Thomas
32. Coevolution and systematics (1986)‡
 Edited by A.R. Stone and D.L. Hawksworth

*Published by Academic Press for the Systematics Association
†Published by the Palaeontological Association in conjunction with the Systematics Association
‡Published by the Oxford University Press for the Systematics Association